AF740705

Principles of Electromagnetism

Principles of Electromagnetism

Ophelia Burgess

Larsen & Keller
www.larsen-keller.com

Principles of Electromagnetism
Ophelia Burgess
ISBN: 978-1-64172-142-4 (Hardback)

© 2019 Larsen & Keller

Larsen & Keller
Published by Larsen and Keller Education,
5 Penn Plaza,
19th Floor,
New York, NY 10001, USA

Cataloging-in-Publication Data

Principles of electromagnetism / Ophelia Burgess.
p. cm.
Includes bibliographical references and index.
ISBN 978-1-64172-142-4
1. Electromagnetism. 2. Physics. I. Burgess, Ophelia.
QC760 .P75 2019
538.3--dc23

This book contains information obtained from authentic and highly regarded sources. All chapters are published with permission under the Creative Commons Attribution Share Alike License or equivalent. A wide variety of references are listed. Permissions and sources are indicated; for detailed attributions, please refer to the permissions page. Reasonable efforts have been made to publish reliable data and information, but the authors, editors and publisher cannot assume any responsibility for the validity of all materials or the consequences of their use.

Trademark Notice: All trademarks used herein are the property of their respective owners. The use of any trademark in this text does not vest in the author or publisher any trademark ownership rights in such trademarks, nor does the use of such trademarks imply any affiliation with or endorsement of this book by such owners.

The publisher's policy is to use permanent paper from mills that operate a sustainable forestry policy. Furthermore, the publisher ensures that the text paper and cover boards used have met acceptable environmental accreditation standards.

Printed and bound in China.

For more information regarding Larsen and Keller Education and its products, please visit the publisher's website www.larsen-keller.com

Table of Contents

Preface

Electromagnetism is a branch of physics that studies the fundamental interaction of electromagnetic force that arises between electrically charged particles. It studies light, electric and magnetic fields. Electricity and magnetism are different manifestations of electromagnetic phenomena and the description of each, their generation and how each is affected by the other are described by the Maxwell's equations. This book provides comprehensive insights into the field of electromagnetism. It presents this complex subject in the most comprehensible and easy to understand language. For someone with an interest and eye for detail, this textbook covers the most significant topics in the field of electromagnetism.

A foreword of all chapters of the book is provided below:

Chapter 1, Electromagnetism is a field of physics. It is concerned with the study of the electromagnetic force that occurs in electrically charged particles. The study of electromagnetism involves an in-depth examination of electricity and magnetism, which have been elaborately covered in this chapter; **Chapter 2**, An electromagnetic field is generated by electrically charged systems, which affects other charged objects in the vicinity. Electromagnetic radiation refers to electromagnetic waves that propagate through space-time and carry electromagnetic radiant energy. The topics elaborated in this chapter on the effects of electromagnetic fields and radiation and the mathematical description of electromagnetic fields will help in providing a comprehensive understanding of electromagnetic field and electromagnetic radiation; **Chapter 3**, Electrodynamics is a field of theoretical physics, which is concerned with the study of the interaction between electrical charges and currents. This chapter has been carefully written to provide an easy understanding of the varied aspects of Lorentz force, electromotive force, magnetomotive force and Maxwell's equations for an extensive understanding of electrodynamics; **Chapter 4**, Electrostatics is the study of electric charges that are at rest. Coulomb's law describes the forces of attraction and repulsion exerted by charges on each other. All the diverse concepts central to the field of electrostatics, such as electric field, electric potential, electric flux, Gauss's law, etc. have been dealt within this chapter; **Chapter 5**, The study of magnetic fields produced by steady currents falls under the domain of magnetostatics. In order to completely understand magnetostatics, it is necessary to understand the basics of magnetic field, magnetic moment, magnetization, magnetic flux, Biot-Savart law, Gauss's law for magnetism and Ampère's circuital law, which have been carefully analyzed in this chapter.

At the end, I would like to thank all the people associated with this book devoting their precious time and providing their valuable contributions to this book. I would also like to express my gratitude to my fellow colleagues who encouraged me throughout the process.

Ophelia Burgess

1

Electromagnetism: An Introduction

Electromagnetism is a field of physics. It is concerned with the study of the electromagnetic force that occurs in electrically charged particles. The study of electromagnetism involves an in-depth examination of electricity and magnetism, which have been elaborately covered in this chapter.

Electricity

Electricity, phenomenon associated with stationary or moving electric charges. Electric charge is a fundamental property of matter and is borne by elementary particles. In electricity the particle involved is the electron, which carries a charge designated, by convention, as negative. Thus, the various manifestations of electricity are the result of the accumulation or motion of numbers of electrons.

Going Atomic

To understand the fundamentals of electricity, we need to begin by focusing in on atoms, one of the basic building blocks of life and matter. Atoms exist in over a hundred different forms as chemical elements like hydrogen, carbon, oxygen, and copper. Atoms of many types can combine to make molecules, which build the matter we can physically see and touch.

Atoms are *tiny*, stretching at a max to about 300 picometers long (that's $3x10^{-10}$ or 0.0000000003 meters). A copper penny (if it actually were made of 100% copper) would have $3.2x10^{22}$ atoms (32,000,000,000,000,000,000,000 atoms) of copper inside it.

Even the atom isn't small enough to explain the workings of electricity. We need to dive down one more level and look in on the building blocks of atoms: protons, neutrons, and electrons.

Building Blocks of Atoms

An atom is built with a combination of three distinct particles: electrons, protons, and neutrons. Each atom has a center nucleus, where the protons and neutrons are densely packed together. Surrounding the nucleus are a group of orbiting electrons.

Every atom must have at least one proton in it. The number of protons in an atom is important, because it defines what chemical element the atom represents. For example, an atom with just one proton is hydrogen, an atom with 29 protons is copper, and an atom with 94 protons is plutonium. This count of protons is called the atom's atomic number.

The proton's nucleus-partner, neutrons, serve an important purpose; they keep the protons in the nucleus and determine the isotope of an atom. They're not critical to our understanding of electricity.

A very simple atom model. It's not to scale but helpful for understanding how an atom is built. A core nucleus of protons and neutrons is surrounded by orbiting electrons.

Electrons are critical to the workings of electricity (notice a common theme in their names?) In its most stable, balanced state, an atom will have the same number of electrons as protons. As in the Bohr atom model below, a nucleus with 29 protons (making it a copper atom) is surrounded by an equal number of electrons.

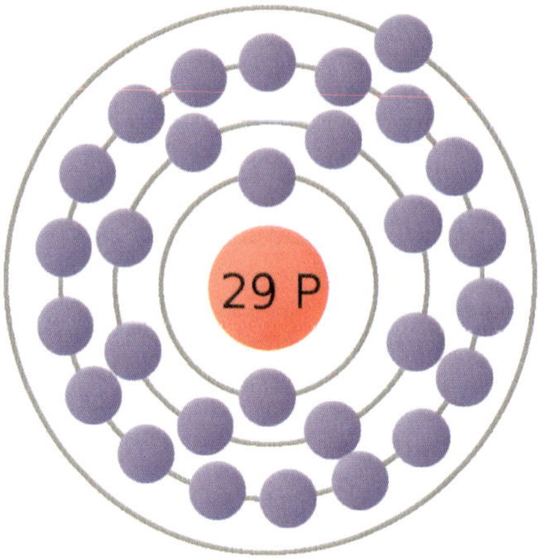

As our understanding of atoms has evolved, so too has our method for modeling them. The Bohr model is a very useful atom model as we explore electricity.

The atom's electrons aren't all forever bound to the atom. The electrons on the outer orbit of the atom are called valence electrons. With enough outside force, a valence electron can escape orbit of the atom and become free. Free electrons allow us to move charge, which is what electricity is all about.

Flowing Charges

Electricity is defined as the flow of electric charge. Charge is a property of matter–just like mass, volume, or density. It is measurable. Just as you can quantify how much mass something has, you can measure how much charge it has. The key concept with charge is that it can come in two types: positive (+) or negative (-).

In order to move charge we need charge carriers, and that's where our knowledge of atomic particles–specifically electrons and protons–comes in handy. Electrons always carry a negative charge, while protons are always positively charged. Neutrons (true to their name) are neutral, they have no charge. Both electrons and protons carry the same amount of charge, just a different type.

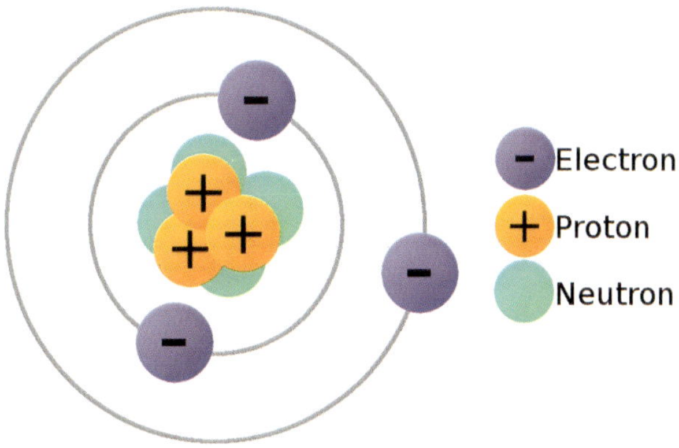

A lithium atom (3 protons) model with the charges labeled.

The charge of electrons and protons is important, because it provides us the means to exert a force on them.

Electrostatic Force

Electrostatic force (also called Coulomb's law) is a force that operates between charges. It states that charges of the same type repel each other, while charges of opposite types are attracted together. Opposites attract, and likes repel.

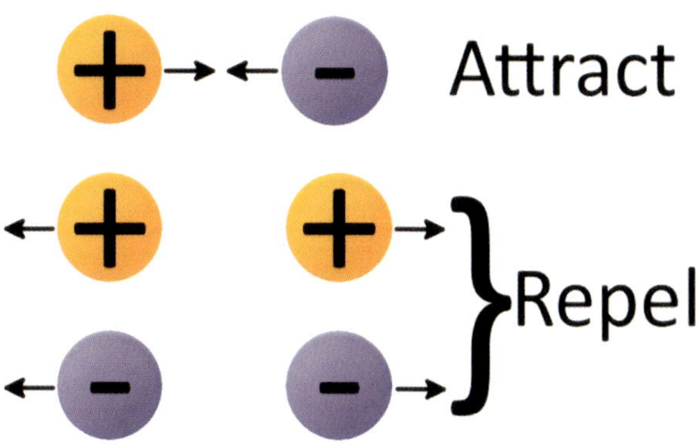

The amount of force acting on two charges depends on how far they are from each other. The closer two charges get, the greater the force (either pushing together, or pulling away) becomes.

Thanks to electrostatic force, electrons will push away other electrons and be attracted to protons. This force is part of the "glue" that holds atoms together, but it's also the tool we need to make electrons (and charges) flow.

Making Charges Flow

We now have all the tools to make charges flow. Electrons in atoms can act as our charge carrier, because every electron carries a negative charge. If we can free an electron from an atom and force it to move, we can create electricity.

Consider the atomic model of a copper atom, one of the preferred elemental sources for charge flow. In its balanced state, copper has 29 protons in its nucleus and an equal number of electrons orbiting around it. Electrons orbit at varying distances from the nucleus of the atom. Electrons closer to the nucleus feel a much stronger attraction to the center than those in distant orbits. The outermost electrons of an atom are called the valence electrons, these require the least amount of force to be freed from an atom.

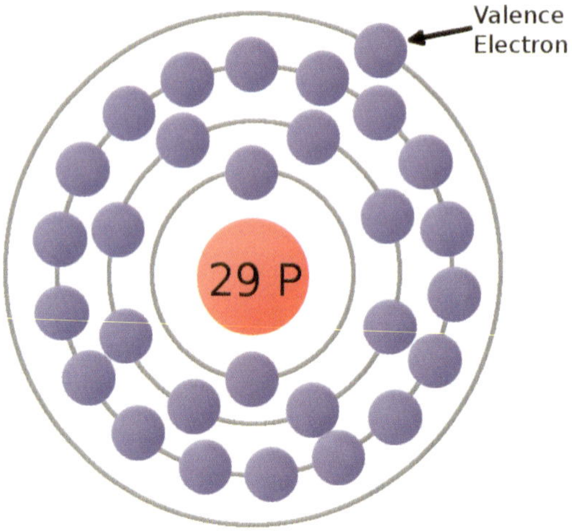

This is a copper atom diagram: 29 protons in the nucleus, surrounded by bands of circling electrons. Electrons closer to the nucleus are hard to remove while the valence (outer ring) electron requires relatively little energy to be ejected from the atom.

Using enough electrostatic force on the valence electron–either pushing it with another negative charge or attracting it with a positive charge–we can eject the electron from orbit around the atom creating a free electron.

Now consider a copper wire: matter filled with countless copper atoms. As our free electron is floating in a space between atoms, it's pulled and prodded by surrounding charges in that space. In this chaos the free electron eventually finds a new atom to latch on to; in doing so, the negative charge of that electron ejects another valence electron from the atom. Now a new electron is drifting through free space looking to do the same thing. This chain effect can continue on and on to create a flow of electrons called electric current.

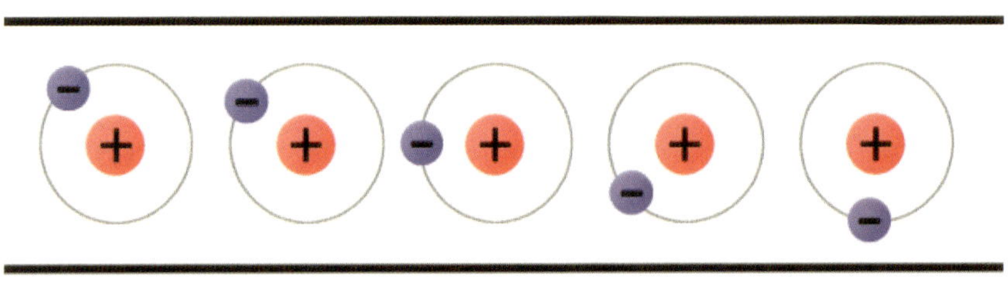

A very simplified model of charges flowing through atoms to make current.

Conductivity

Some elemental types of atoms are better than others at releasing their electrons. To get the best possible electron flow we want to use atoms which don't hold very tightly to their valence electrons. An element's conductivity measures how tightly bound an electron is to an atom.

Elements with high conductivity, which have very mobile electrons, are called conductors. These are the types of materials we want to use to make wires and other components which aid in electron flow. Metals like copper, silver, and gold are usually our top choices for good conductors.

Elements with low conductivity are called insulators. Insulators serve a very important purpose: they prevent the flow of electrons. Popular insulators include glass, rubber, plastic, and air.

Static or Current Electricity

Before we get much further, let's discuss the two forms electricity can take: static or current. In working with electronics, current electricity will be much more common, but static electricity is important to understand as well.

Static Electricity

Static electricity exists when there is a build-up of opposite charges on objects separated by an insulator. Static (as in "at rest") electricity exists until the two groups of opposite charges can find a path between each other to balance the system out.

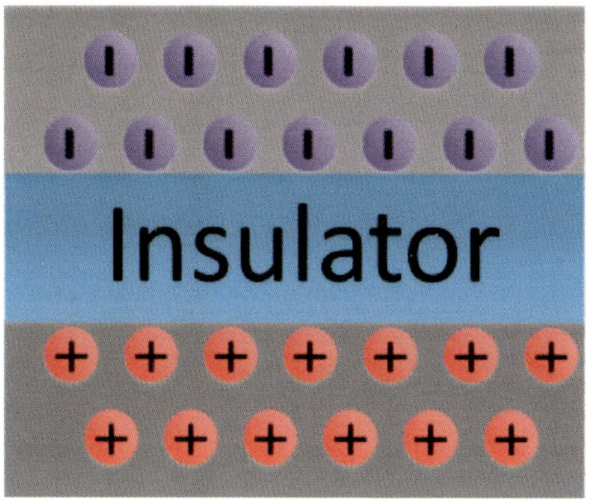

When the charges do find a means of equalizing, a static discharge occurs. The attraction of the charges becomes so great that they can flow through even the best of insulators (air, glass, plastic, rubber, etc.). Static discharges can be harmful depending on what medium the charges travel through and to what surfaces the charges are transferring. Charges equalizing through an air gap can result in a visible shock as the traveling electrons collide with electrons in the air, which become excited and release energy in the form of light.

One of the most dramatic examples of static discharge is lightning. When a cloud system gathers enough charge relative to either another group of clouds or the earth's ground, the charges will try to equalize. As the cloud discharges, massive quantities of positive (or sometimes negative) charges run through the air from ground to cloud causing the visible effect we're all familiar with.

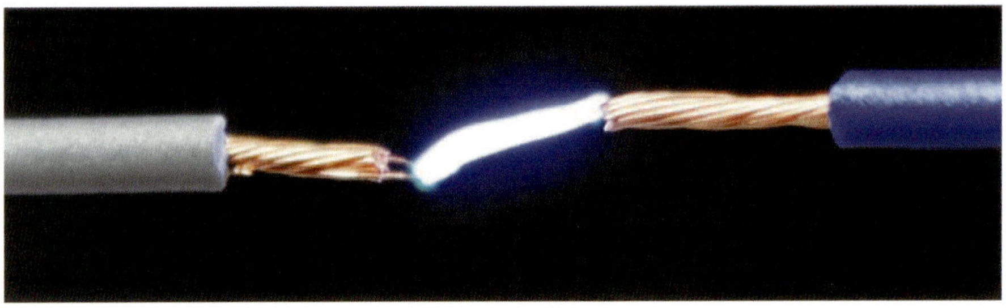

Spark gap igniters are used to create a controlled static discharge. Opposite charges build up on each of the conductors until their attraction is so great charges can flow through the air.

Static electricity also familiarly exists when we rub balloons on our head to make our hair stand up, or when we shuffle on the floor with fuzzy slippers and shock the family cat (accidentally, of course). In each case, friction from rubbing different types of materials transfers electrons. The object losing electrons becomes positively charged, while the object gaining electrons becomes negatively charged. The two objects become attracted to each other until they can find a way to equalize.

Working with electronics, we generally don't have to deal with static electricity. When we do, we're usually trying to protect our sensitive electronic components from being subjected to a static discharge. Preventative measures against static electricity include wearing ESD (electrostatic discharge) wrist straps, or adding special components in circuits to protect against very high spikes of charge.

Current Electricity

Current electricity is the form of electricity which makes all of our electronic gizmos possible. This form of electricity exists when charges are able to constantly flow. As opposed to static electricity where charges gather and remain at rest, current electricity is dynamic, charges are always on the move.

Circuits

In order to flow, current electricity requires a circuit: a closed, never-ending loop of conductive material. A circuit could be as simple as a conductive wire connected end-to-end, but useful circuits usually contain a mix of wire and other components which control the flow of electricity. The only rule when it comes to making circuits is they can't have any insulating gaps in them.

If you have a wire full of copper atoms and want to induce a flow of electrons through it, *all* free electrons need somewhere to flow in the same general direction. Copper is a great conductor, perfect for making charges flow. If a circuit of copper wire is broken, the charges can't flow through the air, which will also prevent any of the charges toward the middle from going anywhere.

On the other hand, if the wire were connected end-to-end, the electrons all have a neighboring atom and can all flow in the same general direction.

We now understand *how* electrons can flow, but how do we get them flowing in the first place? Then, once the electrons are flowing, how do they produce the energy required to illuminate light bulbs or spin motors? For that, we need to understand electric fields.

Electric Fields

We have a handle on how electrons flow through matter to create electricity. That's all there is to electricity. Well, almost all. Now we need a source to induce the flow of electrons. Most often that source of electron flow will come from an electric field.

A Field

A *field* is a tool we use to model physical interactions which don't involve any observable contact. Fields can't be seen as they don't have a physical appearance, but the effect they have is very real.

We're all subconsciously familiar with one field in particular: Earth's gravitational field, the effect of a massive body attracting other bodies. Earth's gravitational field can be modeled with a set of vectors all pointing into the center of the planet; regardless of where you are on the surface, you'll feel the force pushing you towards it.

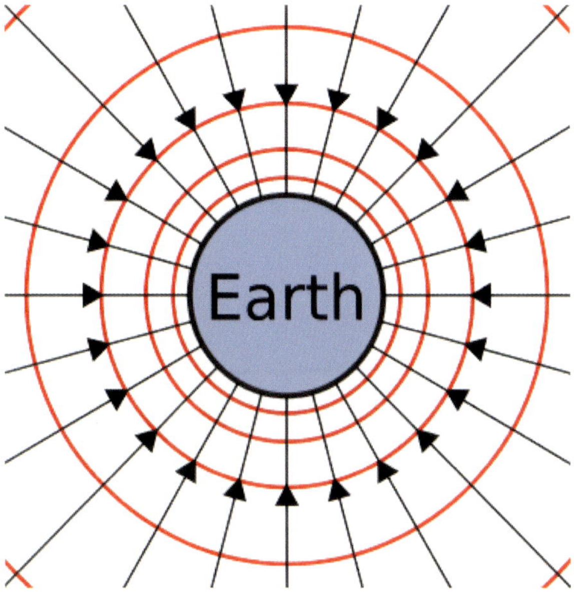

The strength or intensity of fields isn't uniform at all points in the field. The further you are from the source of the field the less effect the field has. The magnitude of Earth's gravitational field decreases as you get further away from the center of the planet.

As we go on to explore electric fields in particular remember how Earth's gravitational field works, both fields share many similarities. Gravitational fields exert a force on objects of mass, and electric fields exert a force on objects of charge.

Electric Fields

Electric fields (e-fields) are an important tool in understanding how electricity begins and continues to flow. Electric fields describe the pulling or pushing force in a space between charges. Compared to Earth's gravitational field, electric fields have one major difference: while Earth's field generally only attracts other objects of mass (since everything is *so* significantly less massive), electric fields push charges away just as often as the attract them.

The direction of electric fields is always defined as the direction a positive test charge would move if it was dropped in the field. The test charge has to be infinitely small, to keep its charge from influencing the field.

We can begin by constructing electric fields for solitary positive and negative charges. If you dropped a positive test charge near a negative charge, the test charge would be attracted towards the negative charge. So, for a single, negative charge we draw our electric field arrows pointing inward at all directions. That same test charge dropped near another positive charge would result in an outward repulsion, which means we draw arrows going out of the positive charge.

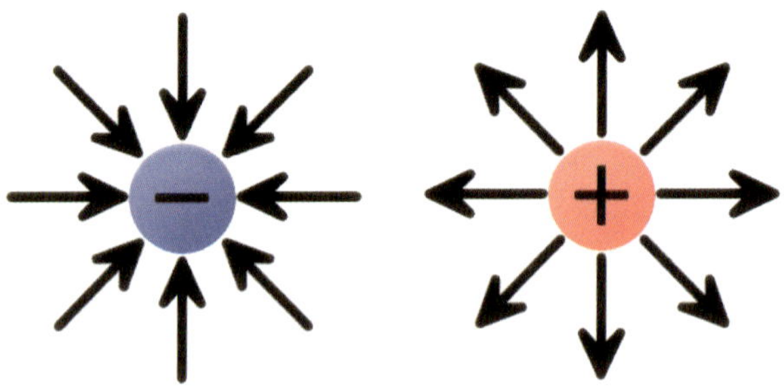

The electric fields of single charges. A negative charge has an inward electric field because it attracts positive charges. The positive charge has an outward electric field, pushing away like charges.

Groups of electric charges can be combined to make more complete electric fields.

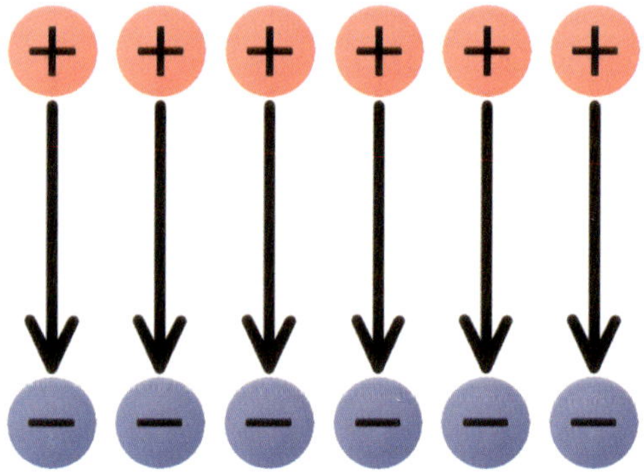

The uniform e-field above points away from the positive charges, towards the negatives. Imagine a tiny positive test charge dropped in the e-field; it should follow the direction of the arrows. As we've seen, electricity usually involves the flow of electrons–negative charges–which flow against electric fields.

Electric fields provide us with the pushing force we need to induce current flow. An electric field in a circuit is like an electron pump: a large source of negative charges that can propel electrons, which will flow through the circuit towards the positive lump of charges.

Electric Potential Energy

When we harness electricity to power our circuits, gizmos, and gadgets, we're really transforming energy. Electronic circuits must be able to store energy and transfer it to other forms like heat, light, or motion. The stored energy of a circuit is called electric potential energy.

Potential Energy

To understand potential energy we need to understand energy in general. Energy is defined as the ability of an object to do *work* on another object, which means moving that object some distance. Energy comes in many forms, some we can see (like mechanical) and others we can't (like chemical or electrical). Regardless of what form it's in, energy exists in one of two states: kinetic or potential.

An object has kinetic energy when it's in motion. The amount of kinetic energy an object has depends on its mass and speed. Potential energy, on the other hand, is a stored energy when an object is at rest. It describes how much work the object could do if set into motion. It's an energy we can generally control. When an object is set into motion, its potential energy transforms into kinetic energy.

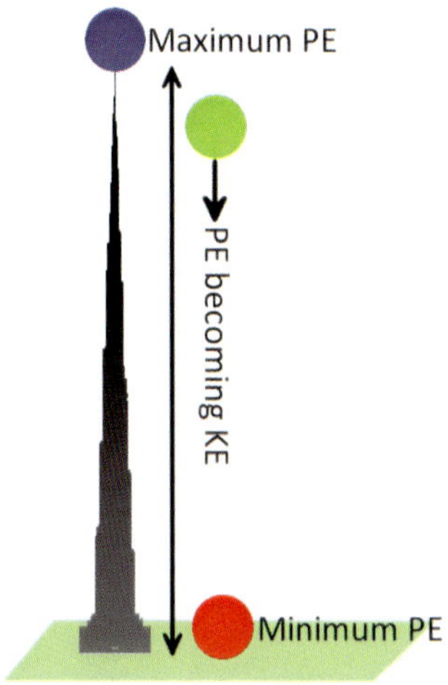

Let's go back to using gravity as an example. A bowling ball sitting motionless at the top of Khalifa tower has a lot of potential (stored) energy. Once dropped, the ball–pulled by the gravitational field–accelerates towards the ground. As the ball accelerates, potential energy is converted into kinetic energy (the energy from motion). Eventually all of the ball's energy is converted from potential to kinetic, and then passed on to whatever it hits. When the ball is on the ground, it has a very low potential energy.

Electric Potential Energy

Just like mass in a gravitational field has gravitational potential energy, charges in an electric field have an electric potential energy. A charge's electric potential energy describes how much stored energy it has, when set into motion by an electrostatic force, that energy can become kinetic, and the charge can do work.

Like a bowling ball sitting at the top of a tower, a positive charge in close proximity to another positive charge has a high potential energy; left free to move, the charge would be repelled away from the like charge. A positive test charge placed near a negative charge would have low potential energy, analogous to the bowling ball on the ground.

To instill anything with potential energy, we have to do work by moving it over a distance. In the case of the bowling ball, the work comes from carrying it up 163 floors, against the field of gravity. Similarly, work must be done to push a positive charge against the arrows of an electric field (either towards another positive charge, or away from a negative charge). The further up the field the charge goes, the more work you have to do. Likewise, if you try to pull a negative charge *away* from a positive charge–against an electric field–you have to do work.

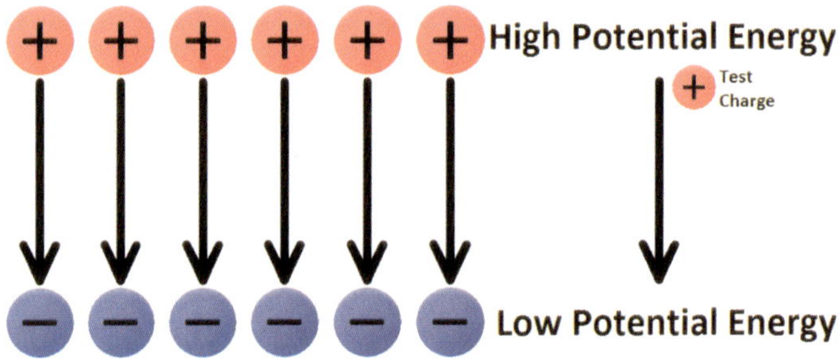

For any charge located in an electric field its electric potential energy depends on the type (positive or negative), amount of charge, and its position in the field. Electric potential energy is measured in units of joules (*J*).

Electric Potential

Electric potential builds upon electric potential *energy* to help define how much energy is stored in electric fields. It's another concept which helps us model the behavior of electric fields. Electric potential is *not* the same thing as electric potential energy.

At any point in an electric field the electric potential is the amount of electric potential energy divided by the amount of charge at that point. It takes the charge quantity out of the equation and leaves us with an idea of how much potential energy specific areas of the electric field may provide. Electric potential comes in units of joules per coulomb (*J/C*), which we define as a volt (V).

In any electric field there are two points of electric potential that are of significant interest to us. There's a point of high potential, where a positive charge would have the highest possible potential energy, and there's a point of low potential, where a charge would have the lowest possible potential energy.

One of the most common terms we discuss in evaluating electricity is voltage. A voltage is the difference in potential between two points in an electric field. Voltage gives us an idea of just how much pushing force an electric field has.

With potential and potential energy under our belt we have all of the ingredients necessary to make current electricity.

Magnetism

Magnetism is one aspect of the combined electromagnetic force. It refers to physical phenomena arising from the force caused by magnets, objects that produce fields that attract or repel other objects.

A magnetic field exerts a force on particles in the field due to the Lorentz force. The motion of electrically charged particles gives rise to magnetism. The force acting on an electrically charged particle in a magnetic field depends on the magnitude of the charge, the velocity of the particle, and the strength of the magnetic field.

All materials experience magnetism, some more strongly than others. Permanent magnets, made from materials such as iron, experience the strongest effects, known as ferromagnetism. With rare exception, this is the only form of magnetism strong enough to be felt by people.

The Nature of Magnetism

Magnets can be found in a natural state in the form of a magnetic ore, with the two main types being Magnetite also called "iron oxide", (FE_3O_4) and Lodestone, also called "leading stone". If these two natural magnets are suspended from a piece of string, they will take up a position in-line with the Earth's magnetic field always pointing north.

A good example of this effect is the needle of a compass. For most practical applications these natural occurring magnets can be disregarded as their magnetism is very low and because nowadays, man-made artificial magnets can be produced in many different shapes, sizes and magnetic strengths.

There are basically two forms of magnetism, "Permanent Magnets" and "Temporary Magnets", with the type being used dependant upon its application. There are many different types of materials available to make magnets such as iron, nickel, nickel alloys, chromium and cobalt and in their natural state some of these elements such as nickel and cobalt show very poor magnetic quantities on their own.

However, when mixed or "alloyed" together with other materials such as iron or aluminium peroxide they become very strong magnets producing unusual names such as "alcomax", "hycomax", "alni" and "alnico".

Magnetic material in the non-magnetic state has its molecular structure in the form of loose magnetic chains or individual tiny magnets loosely arranged in a random pattern. The overall effect of this type of arrangement results in zero or very weak magnetism as this haphazard arrangement of each molecular magnet tends to neutralise its neighbour.

When the material is Magnetised this random arrangement of the molecules changes and the tiny unaligned and random molecular magnets become "lined-up" in such a way that they produce a series magnetic arrangement. This idea of the molecular alignment of ferromagnetic materials is known as Weber's Theory.

Magnetic Molecule Alignment of a Piece of Iron and a Magnet

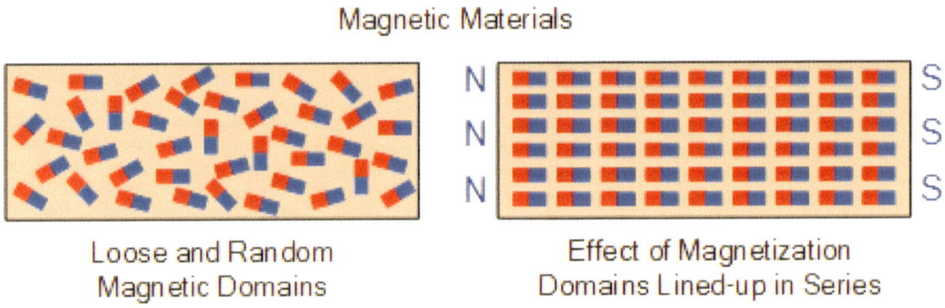

Weber's theory is based on the fact that all atoms have magnetic properties due to the spinning action of the atoms electrons. Groups of atoms join together so that their magnetic fields are all rotating in the same direction. Magnetic materials are composed of groups of tiny magnets at a molecular level around the atoms, and a magnetised material will have most of its tiny magnets lined up in one direction only to produce a north pole in one direction and a south pole in the other direction.

Likewise, a material that has its tiny molecular magnets pointing in all directions will have its molecular magnets neutralised by its neighbouring magnet, thereby neutralising any magnetic effect. These areas of molecular magnets are called "domains".

Any magnetic material will produce a magnetic field itself which depends on the degree of alignment of magnetic domains in the material set up by orbital and spinning electrons. This degree of alignment can be specified by a quantity known as magnetisation, M.

In an unmagnetised material, $M = 0$, but some of the domains remain aligned over small regions in the material once the magnetic field is removed. The effect of applying a magnetising force to the material is to align some of the domains to produce a non-zero magnetisation value.

Once the magnetising force has been removed, the magnetism within the material will either remain or decay away quict quickly depending on the magnetic material being used. This ability of a material to retain its magnetism is called Retentivity.

Materials which are required to retain their magnetism will have a fairly high retentivity and as such are used to make permanent magnets, while those materials required to lose their magnetism quickly such as soft iron cores for relays and solenoids will have a very low retentivity.

Magnetic Flux

All magnets, no matter what their shape, have two regions called magnetic poles with the magnetism both in and around a magnetic circuit producing a definite chain of organised and balanced pattern of invisible lines of flux around it. These lines of flux are collectively referred to as the "magnetic field" of the magnet. The shape of this magnetic field is more intense in some parts than others with the area of the magnet that has the greatest magnetism being called "poles". At each end of a magnet is a pole.

These lines of flux (called a vector field) can not be seen by the naked eye, but they can be seen visually by using iron fillings sprinkled onto a sheet of paper or by using a small compass to trace them out. Magnetic poles are always present in pairs, there is always a region of the magnet called the North-pole and there is always an opposite region called the South-pole.

Magnetic fields are always shown visually as lines of force that give a definite pole at each end of the material where the flux lines are more dense and concentrated. The lines which go to make up a magnetic field showing the direction and intensity are called Lines of Force or more commonly "Magnetic Flux" and are given the Greek symbol, Phi (Φ) as shown below.

Lines of Force From a Bar Magnets Magnetic Field

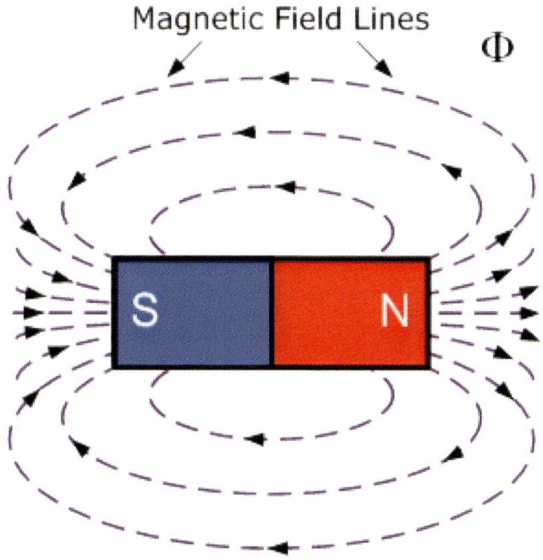

As shown above, the magnetic field is strongest near to the poles of the magnet were the lines of flux are more closely spaced. The general direction for the magnetic flux flow is from the North (N) to the South (S) pole. In addition, these magnetic lines form closed loops that leave at the north pole of the magnet and enter at the south pole. Magnetic poles are always in pairs.

However, magnetic flux does not actually flow from the north to the south pole or flow anywhere for that matter as magnetic flux is a static region around a magnet in which the magnetic force exists. In other words magnetic flux does not flow or move it is just there and is not influenced by gravity. Some important facts emerge when plotting lines of force:

- Lines of force NEVER cross.
- Lines of force are CONTINUOUS.
- Lines of force always form individual CLOSED LOOPS around the magnet.
- Lines of force have a definite DIRECTION from North to South.
- Lines of force that are close together indicate a STRONG magnetic field.
- Lines of force that are farther apart indicate a WEAK magnetic field.

Magnetic forces attract and repel like electric forces and when two lines of force are brought close together the interaction between the two magnetic fields causes one of two things to occur:

- When adjacent poles are the same, (north-north or south-south) they REPEL each other.
- When adjacent poles are not the same, (north-south or south-north) they ATTRACT each other.

This effect is easily remembered by the famous expression that "opposites attract" and this interaction of magnetic fields can be easily demonstrated using iron fillings to show the lines of force around a magnet. The effect upon the magnetic fields of the various combinations of poles as like poles repel and unlike poles attract can be seen below.

Magnetic Field of Like and Unlike Poles

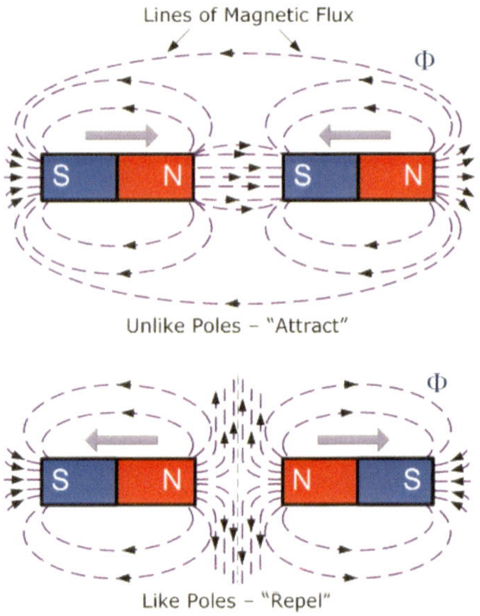

When plotting magnetic field lines with a compass it will be seen that the lines of force are produced in such a way as to give a definite pole at each end of the magnet where the lines of force leave the North pole and re-enter at the South pole. Magnetism can be destroyed by heating or hammering the magnetic material, but cannot be destroyed or isolated by simply breaking the magnet into two pieces.

So if you take a normal bar magnet and break it into two pieces, you do not have two halves of a magnet but instead each broken piece will somehow have its own North pole and a South pole. If you take one of those pieces and break it into two again, each of the smaller pieces will have a North pole and a South pole and so on. No matter how small the pieces of the magnet become, each piece will still have a North pole and a South pole, crazy.

Then in order for us to make use of magnetism in electrical or electronic calculations, it is necessary to define what are the various aspects of magnetism.

The Magnitude of Magnetism

We now know that the lines of force or more commonly the magnetic flux around a magnetic material is given the Greek symbol, Phi, (Φ) with the unit of flux being the Weber, (Wb) after Wilhelm Eduard Weber. But the number of lines of force within a given unit area is called the "Flux Density" and since flux (Φ) is measured in (Wb) and area (A) in metres squared, (m^2), flux density is therefore measured in Webers/Metre^2 or (Wb/m^2) and is given the symbol B.

However, when referring to flux density in magnetism, flux density is given the unit of the Tesla after Nikola Tesla so therefore one Wb/m^2 is equal to one Tesla, $1Wb/m^2 = 1T$. Flux density is proportional to the lines of force and inversely proportional to area so we can define Flux Density as:

Magnetic Flux Density

$$\text{Magnetic Flux Density, (tesla)} = \frac{\text{Magnetic Flux, (weber)}}{\text{Area, (m2)}}$$

The symbol for magnetic flux density is B and the unit of magnetic flux density is the Tesla, T.

$$B = \frac{\Phi}{A} \text{ in Teslas}$$

It is important to remember that all calculations for flux density are done in the same units, e.g., flux in webers, area in m2 and flux density in Teslas.

Magnetism Example No1

The amount of flux present in a round magnetic bar was measured at 0.013 webers. If the material has a diameter of 12cm, calculate the flux density.

The cross sectional area of the magnetic material in m2 is given as:

$$\text{Diameter} = 12\text{cm}$$

$$\therefore \text{Area} = \pi r^2$$

$$A = 3.142 \times 0.06^2 = 0.0113\text{m}^2$$

The magnetic flux is given as 0.013 webers, therefore the flux density can be calculated as:

$$B = \frac{\Phi}{A} = \frac{0.013}{0.0113} = 1.15\text{T}$$

So the flux density is calculated as 1.15 Teslas.

Measuring Magnetism

The strength of the field around a magnet depends on how close you get: it's strongest very near the magnet and falls off quickly as you move away. (That's why a small magnet on your table has to be quite close to things to attract them.) We measure the strength of magnetic fields in units called gauss and tesla (the modern SI unit, named for Croatian-born US electricity pioneer Nikola Tesla). It's interesting to note that the strength of Earth's magnetic field is very weak—about 100–1000 times weaker than that of a typical bar or fridge magnet. On Earth, gravity, not magnetism, is the force that sticks you to the floor. We'd notice Earth's magnetism much more if its gravity weren't so very strong.

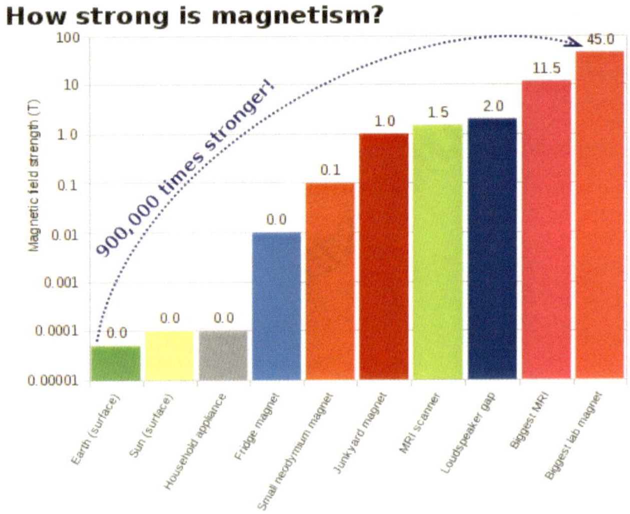

Comparing the strength of some "everyday" sources of magnetism. Please note that the vertical scale is logarithmic: each step up the scale means the strength of the magnetic field has increased ten times. The main thing to note here is how very weak Earth's magnetism is (the green block on the extreme left) compared to everything else we routinely encounter (never mind the giant magnets used in hospitals and labs). The record-breaking lab magnet shown on the extreme right is about 900,000 times stronger than Earth's magnetic field.

Electromagnet

The Homer Simpson or Mickey Mouse magnet that holds things to your refrigerator is a permanent magnet: it keeps hold of its magnetism all the time. Not all magnets work this way. You can make a temporary magnet by passing electricity through a coil of wire wrapped around an iron nail (a device you'll sometimes see referred to as a solenoid). Switch on the current and the nail becomes a magnet; switch it off again and the magnetism disappears. (This is the basic idea behind an electric chime doorbell: you make an electromagnet when you press the button, which pulls a hammer onto the chime bar—ding-dong.) Temporary magnets like this are called electromagnets—magnets worked by electricity—and they hint at a deeper connection between electricity and magnetism that we'll come on to in a moment.

Like permanent magnets, temporary electromagnets come in different sizes and strengths. You can make an electromagnet powerful enough to pick up paperclips with a single 1.5-volt battery. Use a much bigger voltage to make a bigger electric current and you can build an electromagnet powerful enough to pick up a car. That's how scrapyard electromagnets work. The strength of an electromagnet depends on two main things: the size of the electric current you use and the number of times you coil the wire. Increase either or both of these and you get a more powerful electromagnet.

Uses of Magnets

Maybe you think magnets are interesting; maybe you think they're boring! What use are they, you might ask, apart from in childish magic tricks and scrapyards?

You might be surprised just how many things around you work by magnetism or electromagnetism. Every electric appliance with an electric motor in it (everything from your electric toothbrush to your lawn mower) uses magnets to turn electricity into motion. Motors use electricity to generate temporary magnetism in wire coils. The magnetic field thus produced pushes against the fixed field of a permanent magnet, spinning the inside part of the motor around at high speed. You can harness this spinning motion to drive all kinds of machines.

There are magnets in your refrigerator holding the door closed. Magnets read and write data (digital information) on your computer's hard drive and on cassette tapes in old-fashioned personal stereos. More magnets in your hi-fi loudspeakers or headphones help to turn stored music back into sounds you can hear. If you're sick with a serious internal illness, you might have a type of body scan called NMR (nuclear magnetic resonance), which draws the world beneath your skin using patterns of magnetic fields. Magnets are used to recycle your metal trash (steel food cans are strongly magnetic but aluminum drinks cans are not, so a magnet is an easy way to separate the two different metals).

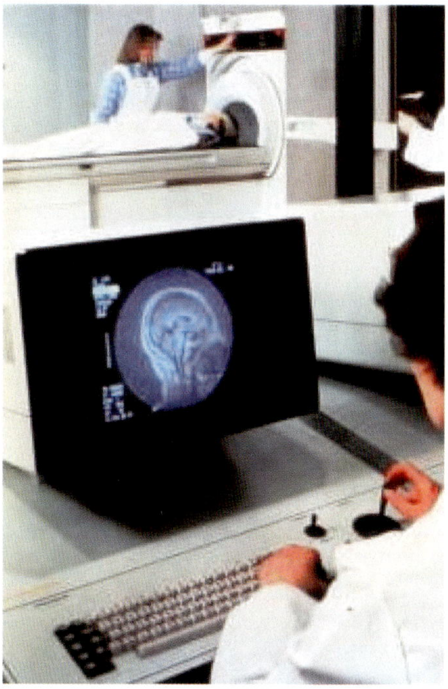

An NMR scan like this builds up a detailed image of a patient's body (or, in this case, their head) on a computer screen using the magnetic activity of atoms in their body tissue. You can see the patient going into the scanner at the top and the image of their head on the screen below.

Materials Which are Magnetic

Iron is the king of magnetic materials—the metal we all think of when we think of magnets. Most other common metals (such as copper, gold, silver, and aluminum) are, at first sight, nonmagnetic and most nonmetals (including paper, wood, plastic, concrete, glass, and textiles such as cotton and wool) are nonmagnetic too. But iron's not the only magnetic metal. Nickel, cobalt, and elements that belong to a part of the Periodic Table (the orderly arrangement chemists use to describe all the known chemical elements) known as the rare-Earth metals (notably samarium and neodymium) also make good magnets. Some of the best magnets are alloys (mixtures) of these elements with one another and with other elements. Ferrites (compounds made of iron, oxygen, and other elements) also make superb magnets. Lodestone (which is also called magnetite) is an example of a ferrite that's commonly found inside Earth (it has the chemical formula $FeO \cdot Fe_2O_3$).

Materials like iron turn into good temporary magnets when you put a magnet near them, but tend to lose some or all of their magnetism when you take the magnet away again. We say these materials are magnetically soft. By contrast, alloys of iron and the rare-Earth metals retain most of their magnetism even when you remove them from a magnetic field, so they make good permanent magnets. We call those materials magnetically hard.

Is it true to say that all materials are either magnetic or nonmagnetic? People used to think that but scientists now know that the materials we consider to be nonmagnetic are also affected by magnetism, though extremely weakly. The extent to which a material can be magnetized is called its susceptibility.

Induced and Permanent Atomic Magnetic Dipoles

Whether a substance is paramagnetic or diamagnetic is determined primarily by the presence or absence of free magnetic dipole moments (i.e., those free to rotate) in its constituent atoms. When there are no free moments, the magnetization is produced by currents of the electrons in their atomic orbits. The substance is then diamagnetic, with a negative susceptibility independent of both field strength and temperature.

In matter with free magnetic dipole moments, the orientation of the moments is normally random and, as a result, the substance has no net magnetization. When a magnetic field is applied, the dipoles are no longer completely randomly oriented; more dipoles point with the field than against the field. When this results in a net positive magnetization in the direction of the field, the substance has a positive susceptibility and is classified as paramagnetic.

The forces opposing alignment of the dipoles with the external magnetic field are thermal in origin and thus weaker at low temperatures. The excess number of dipoles pointing with the field is determined by (mB/kT), where mB represents the magnetic energy and kT the thermal energy. When the magnetic energy is small compared to the thermal energy, the excess number of dipoles pointing with the field is proportional to the field and inversely proportional to the absolute temperature, corresponding to Curie's law. When the value of (mB/kT) is large enough to align nearly all the dipoles with the field, the magnetization approaches a saturation value.

There is a third category of matter in which intrinsic moments are not normally present but appear under the influence of an external magnetic field. The intrinsic moments of conduction electrons in metals behave this way. One finds a small positive susceptibility independent of temperature comparable with the diamagnetic contribution, so that the overall susceptibility of a metal may be positive or negative. The molar susceptibility of elements is shown in figure.

In addition to the forces exerted on atomic dipoles by an external magnetic field, mutual forces exist between the dipoles. Such forces vary widely for different substances. Below a certain transition temperature depending on the substance, they produce an ordered arrangement of the orientations of the atomic dipoles even in the absence of an external field. The mutual forces tend to align neighbouring dipoles either parallel or antiparallel to one another. Parallel alignment of atomic dipoles throughout large volumes of the substance results in ferromagnetism, with a permanent magnetization on a macroscopic scale. On the other hand, if equal numbers of atomic dipoles are aligned in opposite directions and the dipoles are of the same size, there is no permanent macroscopic magnetization, and this is known as antiferromagnetism. If the atomic dipoles are of different magnitudes and those pointing in one direction are all different in size from those pointing in the opposite direction, there exists permanent magnetization on a macroscopic scale in an effect known as ferrimagnetism. In all cases, the material behaves as a paramagnet above the characteristic transition temperature; it acquires a macroscopic magnetic moment only when an external field is applied.

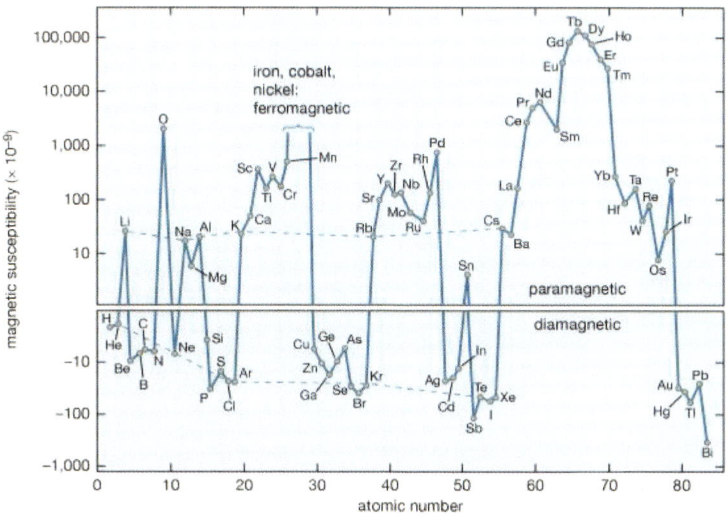

magnetic susceptibility of chemical elements: The susceptibility of a kilogram mole of the elements. Broken lines connect the alkali metals (paramagnetic) and the rare gases of the atmosphere (diamagnetic)

Diamagnetism

When an electron moving in an atomic orbit is in a magnetic field *B*, the force exerted on the electron produces a small change in the orbital motion; the electron orbit precesses about the direction of *B*. As a result, each electron acquires an additional angular momentum that contributes to the magnetization of the sample. The susceptibility χ is given by

$$x = -\mu_0 N\left(\frac{e^2}{6m}\right)\sum r^2$$

where $\Sigma < r^2 >$ is the sum of the mean square radii of all electron orbits in each atom, *e* and *m* are the charge and mass of the electron, and *N* is the number of atoms per unit volume. The negative sign of this susceptibility is a direct consequence of Lenz's law. When *B* is switched on, the change in motion of each orbit is equivalent to an induced circulating electric current in such a direction that its own magnetic flux opposes the change in magnetic flux through the orbit; i.e., the induced magnetic moment is directed opposite to *B*.

Since the magnetization *M* is proportional to the number *N* of atoms per unit volume, it is sometimes useful to give the susceptibility per mole, χ_{mole}. For a kilogram mole (the molecular weight in kilograms), the numerical value of the molar susceptibility is

$$X_{mole} = -3.55 \times 10^{12} \sum (r)^2$$

For an atom, the mean value of $\Sigma < r^2 >$ is about 10^{-21} square metre and χ_{mole} has values of 10^{-9} to 10^{-10}; the atomic number Z equals the number of electrons in each atom. The quantity $\Sigma < r^2 >$ for each atom, and therefore the diamagnetic susceptibility, is essentially independent of temperature. It is also not affected by the surroundings of the atom.

A different kind of diamagnetism occurs in superconductors. The conduction electrons are spread out over the entire metal, and so the induced magnetic moment is governed by the size of the superconducting sample rather than by the size of the individual constituent atoms (a very large effective) $< r^2 >$. The diamagnetism is so strong that the magnetic field is kept out of the superconductor.

Paramagnetism

Paramagnetism occurs primarily in substances in which some or all of the individual atoms, ions, or molecules possess a permanent magnetic dipole moment. The magnetization of such matter depends on the ratio of the magnetic energy of the individual dipoles to the thermal energy. This dependence can be calculated in quantum theory and is given by the Brillouin function, which depends only on the ratio (B/T). At low magnetic fields, the magnetization is linearly proportional to the field and reaches its maximum saturation value when the magnetic energy is much greater than the thermal energy. Figure below shows the dependence of the magnetic moment per ion in units of Bohr magnetons as a function of B/T. (One Bohr magneton equals $9.274009994 \times 10^{-24}$ ampere times square metre.)

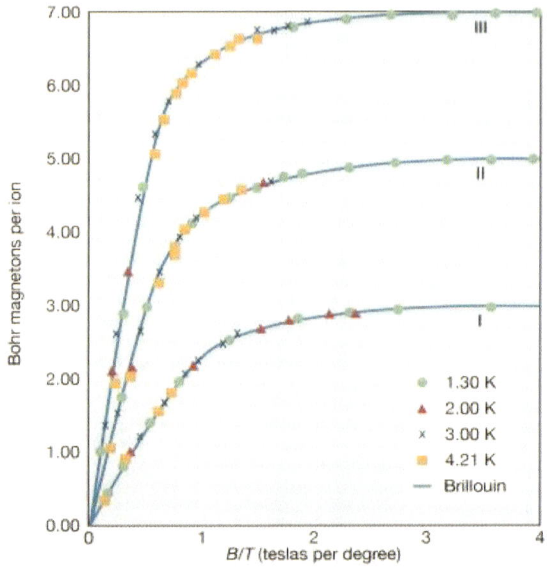

magnetization of paramagnetic substances: The approach to saturation in the magnetization of a paramagnetic substance following a Brillouin curve. Curves I, II, and III refer to ions of chromium, potassium alum, iron ammonium alum, and gadolinium sulfate octahydrate for which $g = 2$ and $j = 3/2$, $5/2$, and $7/2$, respectively

In substances that have a nuclear magnetic dipole moment, there is a further contribution to susceptibility. The size of the nuclear magnetic moment is only about one-thousandth that of an atom. Per kilogram mole, χ_n is on the order of $10^{-8}/T$; in solid hydrogen this just exceeds the electronic diamagnetism of 1 K.

Curie's law should hold when mB is much smaller than kT, provided that no other forces act on the atomic dipoles. In many solids, the presence of internal forces may cause the susceptibility to vary in a complicated way. If the forces orient the dipoles parallel to each other, the behaviour is ferromagnetic. The forces may orient the dipoles so that the normal state has no free moment. If the force is sufficiently weak, a small magnetic field can reorient the dipoles, resulting in a net magnetization. This type of paramagnetism occurs for conduction electrons in a metal. In normal metals, each occupied electron state has two electrons with opposite spin orientation. This is a consequence of the Pauli exclusion principle of quantum mechanics, which permits no greater occupancy of the energetically favoured states. In the presence of a magnetic field, however, it is energetically more favourable for some of the electrons to move to higher states. With only single electrons in these states, the electron moments can be oriented along the field. The resulting paramagnetic susceptibility is independent of temperature. The net susceptibility is independent of temperature. The net susceptibility of a metal can be of either sign, since the diamagnetic and paramagnetic contributions are of comparable magnitudes.

Ferromagnetism

A ferromagnetic substance contains permanent atomic magnetic dipoles that are spontaneously oriented parallel to one another even in the absence of an external field. The magnetic repulsion between two dipoles aligned side by side with their moments in the same direction makes it difficult to understand the phenomenon of ferromagnetism. It is known that within a ferromagnetic material, there is a spontaneous alignment of atoms in large clusters. A new type of interaction, a quantum mechanical effect known as the exchange interaction, is involved.

Causes of Magnetism

In the early 20th century, before scientists properly understood the structure of atoms and how they work, they came up with an easy-to-understand idea called the domain theory to explain magnetism. A few years later, when they understood atoms better, they found the domain theory still worked but could itself be explained, at a deeper level, by the theory of atoms. All the different aspects of magnetism we observe can be explained, ultimately, by talking about either domains, electrons in atoms, or both. Let's look at the two theories in turn.

Explaining Magnetism with the Domain Theory

Imagine a factory somewhere that makes little bar magnets and ships them out to schools for their science lessons. Picture a guy called Dave who has to drive their truck, transporting lots of cardboard boxes, each one with a magnet inside it, to a different school. Dave doesn't have time to worry which way the boxes are stacked, so he piles them inside his truck any old how. The magnet inside one box could be pointing north while the one next to it is pointing south, east, or west. Overall, the magnets are all jumbled up so, even though magnetic fields leak out of each box, they all cancel one another out.

The same factory employs another truck driver called Bill who couldn't be more different. He likes everything tidy, so he loads his truck a different way, stacking all the boxes neatly so they line up exactly the same way. Can you see what will happen? The magnetic field from one box will align with the field from all the other boxes effectively turning the truck into one giant magnet. The cab will be like a giant north pole and the back of the truck a huge south pole.

What happens inside these two trucks is what happens on a tiny scale inside magnetic materials. According to the domain theory, something like an iron bar contains lots of tiny pockets called domains. Each domain is a bit like a box with a magnet inside. See where we're heading? The iron bar is just like the truck. Normally, all its onboard "boxes" are arranged randomly and there's no overall magnetism: the iron is not magnetized. But arrange all the boxes in order, make them all face the same way, and you get an overall magnetic field: hey presto, the bar is magnetized. When you bring a magnet up to an unmagnetized iron bar and stroke it systematically and repeatedly up and down, what you're doing is rearranging all the magnetic "boxes" (domains) inside so they point the same way.

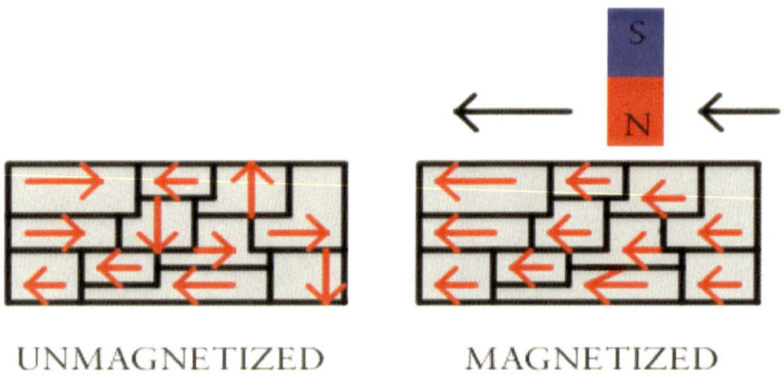

Domain theory explains what happens inside materials when they are magnetized. In an unmagnetized material (left), the domains are randomly arranged so there is no overall magnetic field. When you magnetize a material (right), by stroking a bar magnet over it repeatedly in the same direction, the domains rearrange so their magnetic fields align, producing a combined magnetic field in the same direction.

This theory explains how magnetism can arise, but can it explain some of the other things we know about magnets? If you chop a magnet in half, we know you get two magnets, each with a north and south pole. That makes sense according to the domain theory. If you cut a magnet in half, you get a smaller magnet that's still packed with domains, and these can be arranged north-south just like in the original magnet. What about the way magnetism disappears when you hit a magnet or heat it? That can be explained too. Imagine the van full of orderly boxes again. Drive it erratically, at really high speed, and it's a bit like shaking or hammering it. All the boxes will jumble up so they face different ways and the overall magnetism will disappear. Heating a magnet agitates it internally and jumbles up the boxes in much the same way.

Explaining Magnetism with the Atomic Theory

The domain theory is easy enough to understand, but it's not a complete explanation. We know that iron bars aren't full of boxes packed with little magnets—and, if you think about, trying to explain a magnet by saying it's full of smaller magnets isn't really an explanation at all, because it immediately prompts the question: what are the smaller magnets made of? Fortunately, there's another theory we can turn to.

Back in the 19th century, scientists discovered they could use electricity to make magnetism and magnetism to make electricity. James Clerk Maxwell said that the two phenomena were really different aspects of the same thing—electromagnetism—like two sides of the same piece of paper. Electromagnetism was a brilliant idea, but it was more of a description than an explanation: it showed how things were rather than explaining why they were that way. It wasn't until the 20th century, when later scientists came to understand the world inside atoms, that the explanation for electromagnetism finally appeared.

We know everything is made of atoms and that atoms are made up of a central lump of matter called the nucleus. Minute particles called electrons move around the nucleus in orbit, a bit like satellites in the sky above us, but they also spin on their axis at the same time (just like spinning tops). We know electrons carry electric currents (flows of electricity) when they move through materials such as metals. Electrons are, in a sense, tiny particles of electricity. Now back in the 19th century, scientists knew that moving electricity made magnetism. In the 20th century, it became clear that magnetism was caused by electrons moving inside atoms and creating magnetic fields all around them. Domains are actually groups of atoms in which spinning electrons produce an overall magnetic field pointing one way or another.

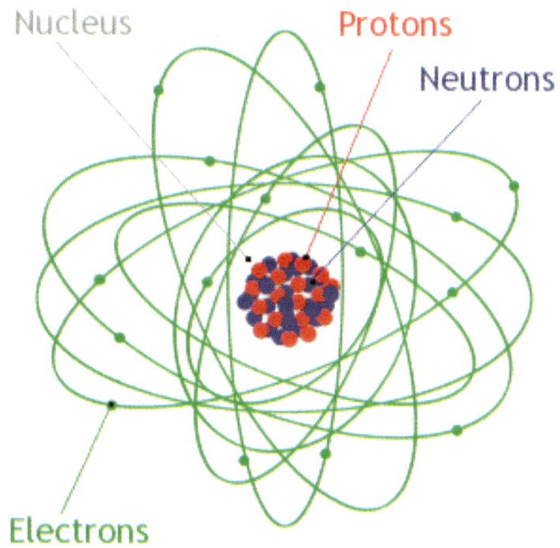

Magnetism is caused by electrons orbiting and spinning inside atoms. Note that this picture is not drawn to scale: most of an atom is empty space and the electrons are actually much further from the nucleus than here.

Like the domain theory, atomic theory can explain many of the things we know about magnets, including paramagnetism (the way magnetic materials line up with magnetic fields). Most of the electrons in an atom exist in pairs that spin in opposite directions, so the magnetic effect of one electron in a pair cancels out the effect of its partner. But if an atom has some unpaired electrons (iron atoms have four), these produce net magnetic fields that line up with one another and turn the whole atom into a mini magnet. When you put a paramagnetic material such as iron in a magnetic field, the electrons change their motion to produce a magnetic field that lines up with the field outside.

What about diamagnetism? In diamagnetic materials, there are no unpaired electrons, so this doesn't happen. The atoms have little or no overall magnetism and are less affected by outside magnetic fields. However, the electrons orbiting inside them are electrically charged particles and, when they move in a magnetic field, they behave like any other electrically charged particles in a magnetic field and experience a force. That changes their orbits very slightly, producing some net magnetism that opposes the very thing that causes it (according to the classic bit of electromagnetic theory know as Lenz's law, which is related to the law of conservation of energy). As a result, the weak magnetic field they produce *opposes* the magnetic field that causes it—which is exactly what we see when diamagnetic materials try to "fight" the magnetic field they're placed in.

Electromagnetism

Electromagnetism is a branch of Physics, deals with the electromagnetic force that occurs between electrically charged particles. Electromagnetic force is one of the four fundamental forces and exhibits electromagnetic fields such as magnetic fields, electric fields and light. It is the basic reason electrons bound to the nucleus and responsible for complete structure of nucleus.

Before the invention of electromagnetism, people or scientists used to think magnetism and electricity are two different topics. The view has changed after James Clerk Maxwell published *A Treatise on Electricity and Magnetism* in the year 1873. The publication states that the interaction of positive and negative charges are mediated by one force. This observation laid foundation to Electromagnetism. Later many scientists like Michael Faraday, Oliver Heaviside and Heinrich Hertz contributed their ideas in electromagnetism.

Electromagnetism is a process where magnetic field is created by introducing current in the conductor. When a conductor is electrically charged it generates magnetic lines of force of conductor. For example, if current i.e., positive charges moving in a wire, it produces magnetic field along the wire and the direction of magnetic lines, and force can be determined using Right Hand Rule.

Magnetic Field Around a Conductor

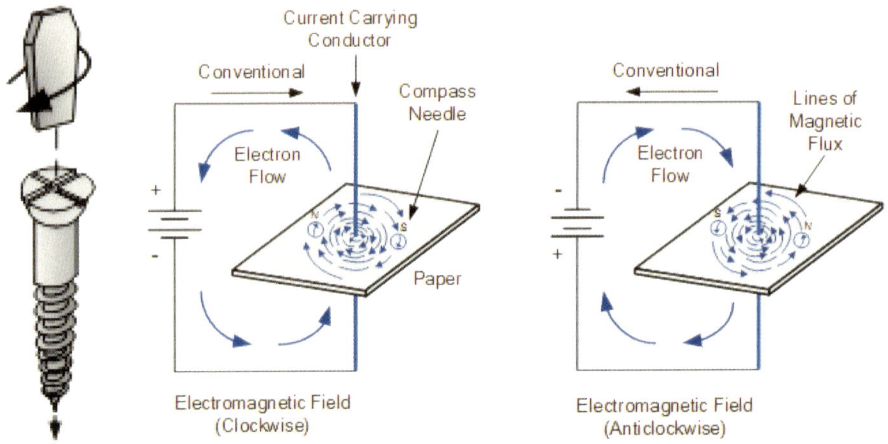

A simple way to determine the direction of the magnetic field around the conductor is to consider screwing an ordinary wood screw into a sheet of paper. As the screw enters the paper the rotational action is CLOCKWISE and the only part of the screw that is visible above the paper is the screw head.

If the wood screw is of the pozidriv or philips type head design, the cross on the head will be visible and it is this cross that is used to indicate current flowing "into" the paper and away from the observer.

Likewise, the action of removing the screw is the reverse, anticlockwise. As the current enters from the top it therefore leaves the underside of the paper and the only part of the wood screw that is visible from below is the tip or point of the screw and it is this point which is used to indicate current flowing "out of" the paper and towards the observer.

Then the physical action of screwing the wood screw in and out of the paper indicates the direction of the current in the conductor and therefore, the direction of rotation of the electromagnetic field around it as shown below. This concept is known generally as the Right Hand Screw Action.

The Right Hand Screw Action

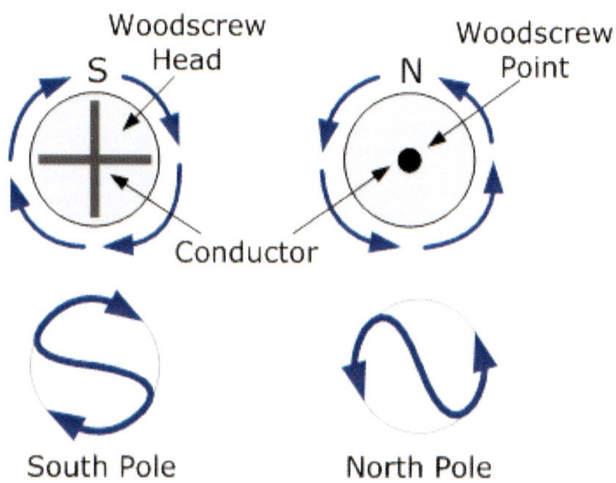

A magnetic field implies the existence of two poles, a north and a south. The polarity of a current carrying conductor can be established by drawing the capital letters S and N and then adding arrow heads to the free end of the letters as shown above giving a visual representation of the magnetic field direction.

Another more familiar concept which determines both the direction of current flow and the resulting direction of the magnetic flux around the conductor is called the "Left Hand Rule".

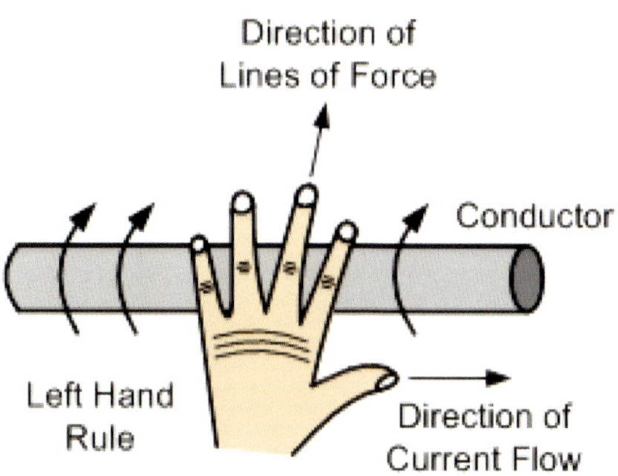

Left Hand Rule of Electromagnetism

The recognised direction of a magnetic field is from its north pole to its south pole. This direction can be deduced by holding the current carrying conductor in your left hand with the thumb extended pointing in the direction of the *electron flow* from negative to positive.

The position of the fingers laid across and around the conductor will now be pointing in the direction of the generated magnetic lines of force as shown.

If the direction of the electron flowing through the conductor is reversed, the left hand will need to be placed onto the other side of the conductor with the thumb pointing in the new direction of the electron current flow.

Also as the current is reversed the direction of the magnetic field produced around the conductor will also be reversed because as we have said previously, the direction of the magnetic field depends upon the direction of current flow.

This "Left Hand Rule" can also be used to determine the magnetic direction of the poles in an electromagnetic coil. This time, the fingers point in the direction of the electron flow from negative to positive while the extended thumb indicating the direction of the north pole. There is a variation on this rule called the "right hand rule" which is based on so-called conventional current flow, (positive to negative).

Consider when a single straight piece of wire is bent into the form of a single loop as shown below. Although the electric current is flowing in the same direction through the whole length of the wire conductor, it will be flowing in opposite directions through the paper. This is because the current leaves the paper on one side and enters the paper on the other therefore a clockwise field and an anticlockwise field are produced next to each other across the sheet of paper.

The resulting space between these two conductors becomes an "intensified" magnetic field with the lines of force spreading out in such a way that they assume the form of a bar magnet generating a distinctive north and south pole at the point of intersection.

Electromagnetism Around a Loop

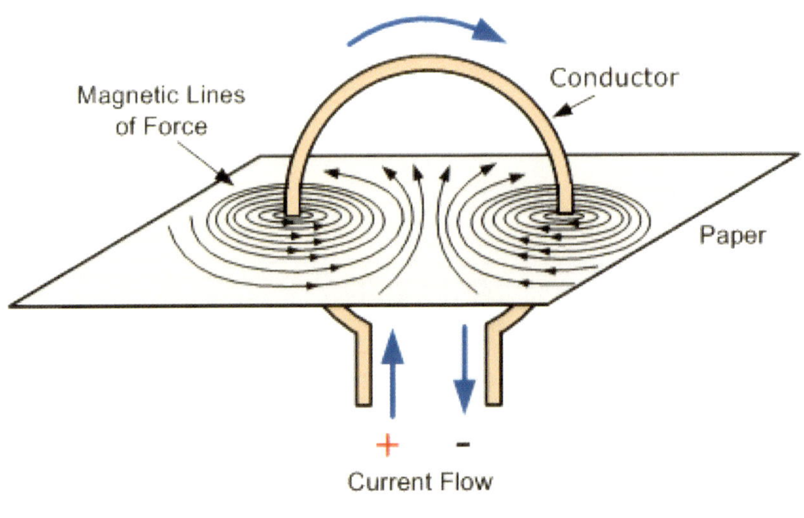

Lines of Force Around the Loop

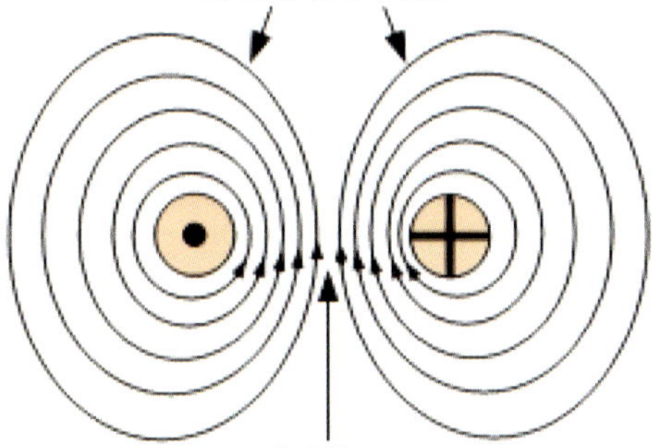

The current flowing through the two parallel conductors of the loop are in opposite directions as the current through the loop exits the left hand side and returns on the right hand side. This results in the magnetic field around each conductor inside the loop being in the "SAME" direction to each other.

The resulting lines of force generated by the current flowing through the loop oppose each other in the space between the two conductors where the two like poles meet thereby deforming the lines of force around each conductor as shown.

However, the distortion of the magnetic flux inbetween the two conductors results in an intensity of the magnetic field at the middle junction were the lines of force become closer together. The resulting interaction between the two like fields produces a mechanical force between the two conductors as they try to repel away from each other. In an electrical machine this repelling of these two magnetic fields produces motion.

However, as the conductors cannot move, the two magnetic fields therefore help each other by generating a north and a south pole along this line of interaction. This results in the magnetic field being strongest in the middle between the two conductors. The intensity of the magnetic field around the conductor is proportional to the distance from the conductor and by the amount of current flowing through it.

The magnetic field generated around a straight length of current-carrying wire is very weak even with a high current passing through it. However, if several loops of the wire are wound together along the same axis producing a coil of wire, the resultant magnetic field will become even more concentrated and stronger than that of just a single loop. This produces an electromagnetic coil more commonly called a Solenoid.

Then every length of wire has the effect of electromagnetism around itself when an electrical current flows through it. The direction of the magnetic field being dependant upon the direction of the flow of current. We can increase the strength of the generated magnetic field by forming the length of wire into a coil.

Explanation with an Example

Permanent Magnetic speakers commonly used in TV's and Radios are perfect example of

Electromagnetic devices. Let's see the operation of these devices which are based on the principle of electromagnetism. See the picture below.

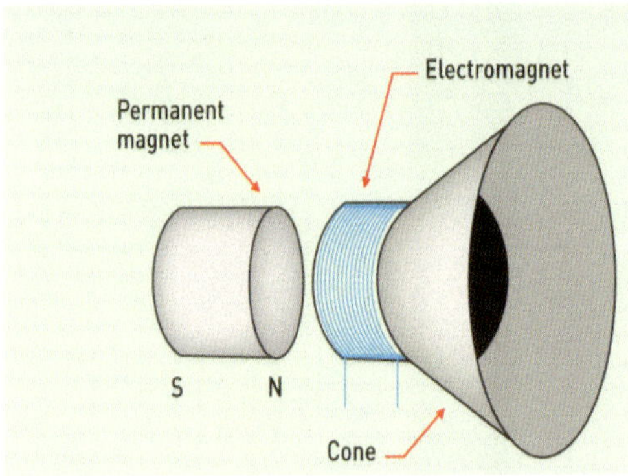

In order to convert electrical waves into audible sound, the speakers are designed. A metal coil is attached to a permanent magnet and when current passes through the coil it generates magnetic field. The newly formed magnetic field is repelled by the permanent magnetic field resulting in the vibrations. These vibrations are amplified by the cone like structure causing the sound. This is how speakers works based on electromagnetism.

Electromagnetic Induction

We have seen what happens when a conductor is electrically charged. Now, lets see what happens if we place conductor in between magnetic field.

When a conductor is placed or moved through the magnetic field it generates voltage i.e., electricity. This principle is called Electromagnetic Induction. The voltages generated will be based on the speed of conductor moving through the electric field. Faster the speed of conductor, greater the induced electricity or voltage.

Faraday's Law

Faraday's Law states that whenever there is relative motion between magnetic field and conductor, the flux linkage changes and this change in flux induces a voltage across the coil.

Explanation with an Example

DC Generator works on the principle of Faraday's Law of Electromagnetic Induction. It is a system that converts mechanical energy into electrical energy.

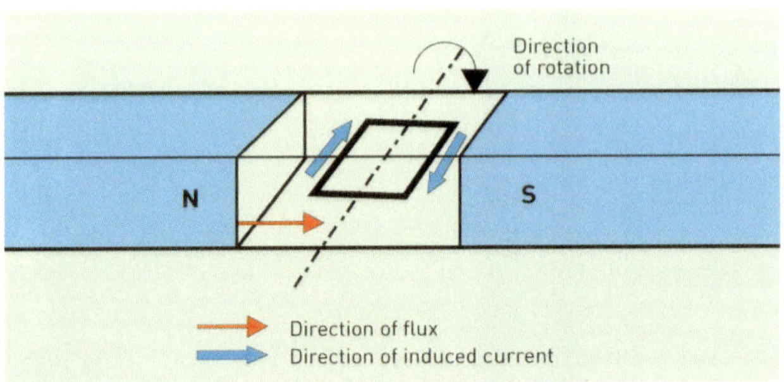

In the above figure, A rectangular conductor width sides is placed in between a magnetic field. When the rectangular conductor rotates in between magnetics, it cuts magnetic field thereby causing Electromagnetic field (e m f).

Lines of Force Around an Electromagnet

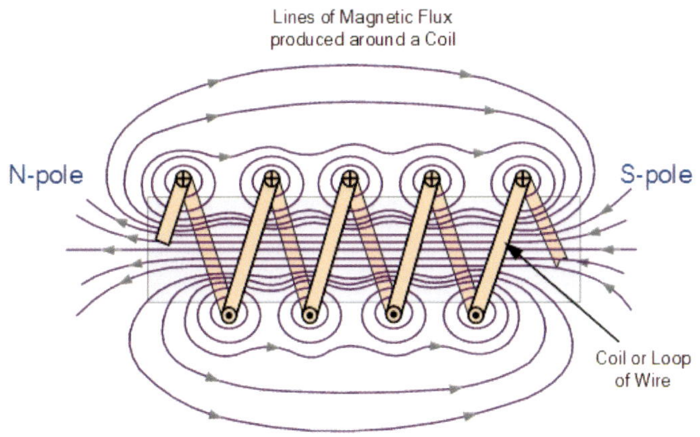

The magnetic field that this produces is stretched out in a form of a bar magnet giving a distinctive north and south pole with the flux being proportional to the amount of current flowing in the coil. If additional layers of wire are wound upon the same coil with the same current flowing, the magnetic field strength will be increased.

It can be seen from this therefore that the amount of flux available in any given magnetic circuit is directly proportional to the current flowing through it and the number of turns of wire within the coil. This relationship is called Magneto Motive Force or m.m.f. and is defined as:

$$\text{Magneto Motive Force, (m.m.f.)} = I \times N \text{ ampere turns}$$

Magneto Motive Force is expressed as a current, I flowing through a coil of N turns. The magnetic field strength of an electromagnet is therefore determined by the *ampere turns* of the coil with the more turns of wire in the coil the greater will be the strength of the magnetic field.

The Magnetic Strength of the Electromagnet

We now know that were two adjacent conductors are carrying current, magnetic fields are set up according to the direction of the current flow. The resulting interaction of the two fields is such that a mechanical force is experienced by the two conductors.

When the current is flowing in the same direction (the same side of the coil) the field between the two conductors is weak causing a force of attraction as shown above. Likewise, when the current is flowing in opposite directions the field between them becomes intensified and the conductors are repelled.

The intensity of this field around the conductor is proportional to the distance from it with the strongest point being next to the conductor and progressively getting weaker further away from the conductor. In the case of a single straight conductor, the current flowing and the distance from it are factors which govern the intensity of the field.

The formula therefore for calculating the "Magnetic Field Strength", H sometimes called "Magnetising Force" of a long straight current carrying conductor is derived from the current flowing through it and the distance from it.

Magnetic Field Strength for Electromagnets

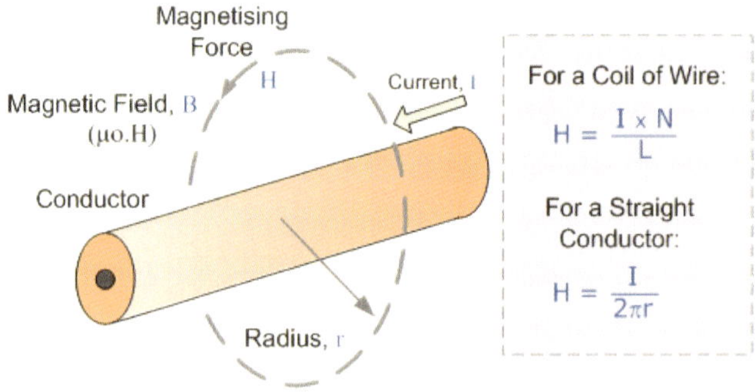

Where:

- H – is the strength of the magnetic field in ampere-turns/metre, At/m
- N – is the number of turns of the coil
- I – is the current flowing through the coil in amps, A
- L – is the length of the coil in metres, m

Then to summarise, the strength or intensity of a coils magnetic field depends on the following factors.

- The number of turns of wire within the coil.
- The amount of current flowing in the coil.
- The type of core material.

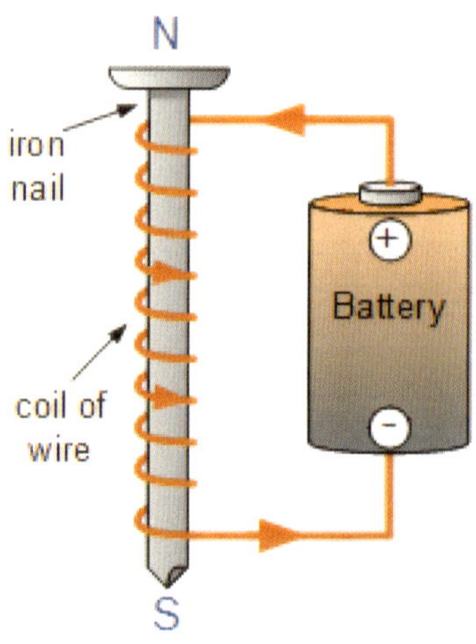

The magnetic field strength of the electromagnet also depends upon the type of core material being used as the main purpose of the core is to concentrate the magnetic flux in a well defined and predictable path. So far only air cored (hollow) coils have been considered but the introduction of other materials into the core (the centre of the coil) has a very large controlling effect on the strength of the magnetic field.

Electromagnet Using a Nail

If the material is non-magnetic for example wood, for calculation purposes it can be regarded as free space as they have very low values of permeability. If however, the core material is made from a Ferromagnetic material such as iron, nickel, cobalt or any mixture of their alloys, a considerable difference in the flux density around the coil will be observed.

Ferromagnetic materials are those which can be magnetised and are usually made from soft iron, steel or various nickel alloys. The introduction of this type of material into a magnetic circuit has the effect of concentrating the magnetic flux making it more concentrated and dense and amplifies the magnetic field created by the current in the coil.

We can prove this by wrapping a coil of wire around a large soft-iron nail and connecting it to a battery as shown. This simple classroom experiment allows us to pick-up a large quantity of clips or pins and we can make the electromagnet stronger by adding more turns to the coil. This degree of intensity of the magnetic field either by a hollow air core or by introducing ferromagnetic materials into the core is called Magnetic Permeability.

Permeability of Electromagnets

If cores of different materials with the same physical dimensions are used in the electromagnet, the strength of the magnet will vary in relation to the core material being used. This variation in the magnetic strength is due to the number of flux lines passing through the central core. if the magnetic material has a high permeability then the flux lines can easily be created and pass through the central core and permeability (μ) and it is a measure of the ease by which the core can be magnetised.

The numerical constant given for the permeability of a vacuum is given as: $\mu_o = 4.\pi.10^{-7}$ H/m with the relative permeability of free space (a vacuum) generally given a value of one. It is this value that is used as a reference in all calculations dealing with permeability and all materials have their own specific values of permeability.

The problem with using just the permeability of different iron, steel or alloy cores is that the calculations involved can become very large so it is more convenient to define the materials by their relative permeability.

Relative Permeability, symbol μ_r is the product of μ (absolute permeability) and μ_0 the permeability of free space and is given as.

Relative Permeability

$$\mu_r = \frac{\mu}{\mu_0} = \frac{\text{Flux Density in the Material}}{\text{Flux Density in a Vacuum}}$$

Materials that have a permeability slightly less than that of free space (a vacuum) and have a weak, negative susceptibility to magnetic fields are said to be Diamagnetic in nature such as: water, copper, silver and gold. Those materials with a permeability slightly greater than that of free space and themselves are only slightly attracted by a magnetic field are said to be Paramagnetic in nature such as: gases, magnesium, and tantalum.

Electromagnet Example No1

The absolute permeability of a soft iron core is given as 80 milli-henries/m (80.10^{-3}). Calculate the equivalent relative permeability value.

$$\mu_r = \frac{\mu}{\mu_0} = \frac{80 \times 10^{-3}}{4 \times \pi \times 10^{-7}} = 63,654 \text{ or } 64 \times 10^3$$

When ferromagnetic materials are used in the core the use of relative permeability to define the field strength gives a better idea of the strength of the magnetic field for the different types of materials used. For example, a vacuum and air have a relative permeability of one and for an iron core it is around 500, so we can say that the field strength of an iron core is 500 times stronger than an equivalent hollow air coil and this relationship is much easier to understand than 0.628×10^{-3} H/m, ($500.4.\pi.10^{-7}$).

While, air may have a permeability of just one, some ferrite and permalloy materials can have a permeability of 10,000 or more. However, there are limits to the amount of magnetic field strength that can be obtained from a single coil as the core becomes heavily saturated as the magnetic flux increases and this is looked at in the next tutorial about B-H curves and Hysteresis.

Application of Electromagnetism

Household Appliances

Electromagnetism serves as a basic principle of working for many of the home appliances in household applications. These applications include lighting, kitchen appliances, air conditioning systems, etc.

- The most dominant use of power in homes as well as commercial buildings is lighting systems. These lighting systems used numerous fluorescent lighting fixtures. Ballasts used in the fluorescent lamps use electromagnetism principle so that at the time of turn ON of the light it produces a high voltage.

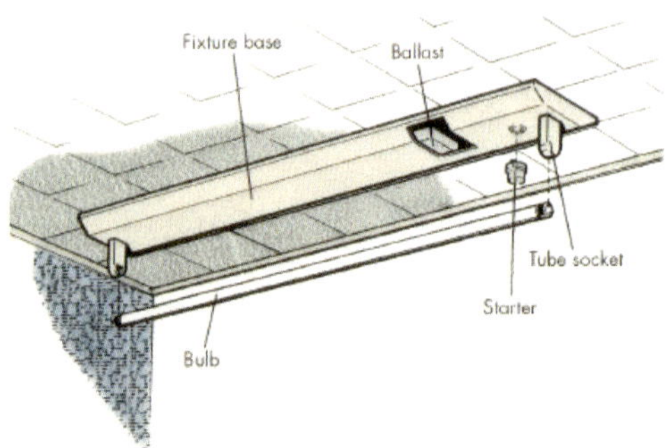

fluorescent lighting fixtures

- Electric fans, blowers and other cooling systems use electric motors. These motors work on the principle of electromagnetic induction which is the branch of electromagnetism. In any electrical appliance, electric motor is moved by the magnetic field produced by the electric current according to the Lorenz force principle. These motors are vary in size, rating and cost based on the application.

Electric fan

- Kitchen appliances like induction cookers, microwave ovens, electric mixers and grinders, bread toasters, etc. use the electromagnetism for their operation.
- Alarming systems use electrical bells which works based on electromagnetic principle. In these bells, the sound is produced by electromagnetic coils which move the striker against the bell. As long as the coil is energized, iron striker gets attracted by it, hence it strikes the bell.
- The electromagnets will be demagnetized when the striker get contact with bell and by the spring tension striker comes back to its original position and again electrical contact will be made once again. This process repeats until the switch gets opened.

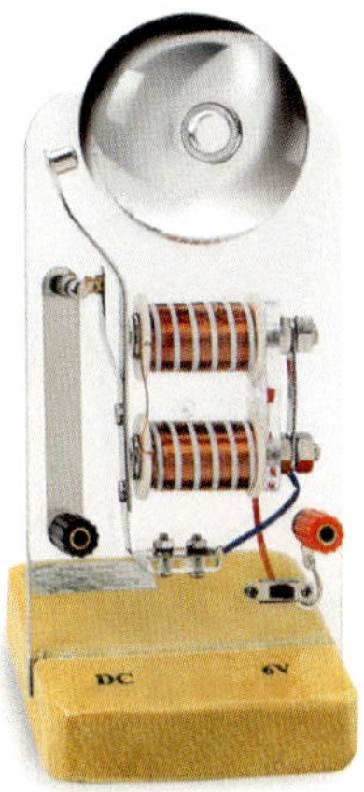

Electric Bell

- Security systems use locking systems for doors which are generally magnetic locking systems. These systems are unlocked either by a magnetic card swiping or having a security code.
- Magnetic card reader on the doors reads the number of keys stored in the magnetic tape of the card. When key stored in memory matches with data on the card, then the door opens.

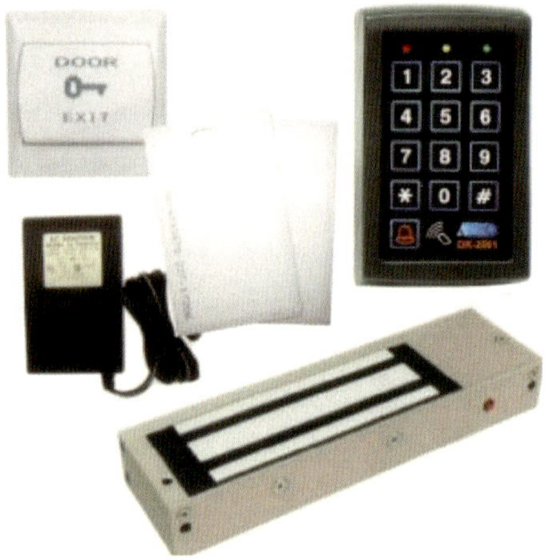

Security systems

- Entertainment system like television, radio or stereo systems use loudspeaker. This device consists of electromagnet which is attached to the membrane or cone surrounded by the magnetic flux produced by the permanent magnet.
- When the current through the electromagnet is varied, electromagnet and membrane of the speaker get moved back and forth. If the current is varied at the same frequencies of the sound waves, results a vibration of the speaker which will further create sound waves.

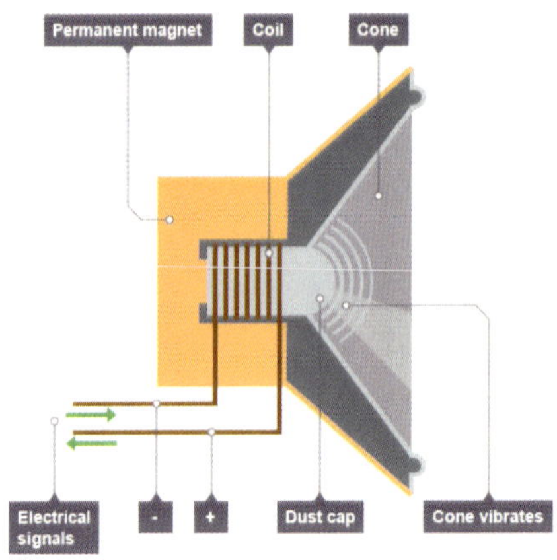

Electric Speaker

Industrial Applications

Almost all of the instruments or devices used in industries are based on the electromagnetism. Materials used in constructing such devices include iron, cobalt, nickel, etc which naturally responds to the magnetic fields.

Starting from small control instruments to the large power equipments, the electromagnetism is used at least at one stage of their working.

- Generators and motors dominate in most of the industries which are the primary power source and driving systems respectively. Generators convert the mechanical to electrical energy whereas the motors convert electrical energy to mechanical energy.

- Generators supply the electrical energy in the time of mains power interruption and most of the cases, these are driven by the IC engines. There are different classes of motors which are employed in industries. These are used for cranes, hoists, lifts, conveyor systems, etc.

Industrial Machines

- Various sensors and actuating devices are works based on electromagnetism. Electromagnetic sensors include Hall-effect sensors, magnetoresistive sensors, fluxgate sensors, etc. These sensors convert the physical quantity such as flow, pressure, level, proximity, etc into an electrical signal.
- Actuators are the final control elements which drives the load at specific conditions. These actuator devices include solenoid valves, relays, motors, etc. and all these works on the principle of electromagnetism.

Sensors and actuators

Magnetic Levitation Trains

This is the modern technology of transportation systems that use the concept of electromagnetism. These are called as high speed trains which use powerful electromagnets to develop the speed.

These trains will float over a guide way using the basic principles of magnets such as electromagnetic suspension (EMS) and electrodynamic suspension (EDS). In EMS, electromagnets employed on the train body are attracted to the iron rails.

These magnets wrap around the guided rails and attractive force between the guideways and magnets lifts the train upwards. In EDS, train is levitated by the repulsive force in the conductive guideways by induced currents.

A guideway is nothing but an arrangement of specially designed magnetic coils and tracks at regular intervals. Along this guideway a maglev train is suspended by the phenomenon of the magnetic levitation with no supports other than magnetic fields.

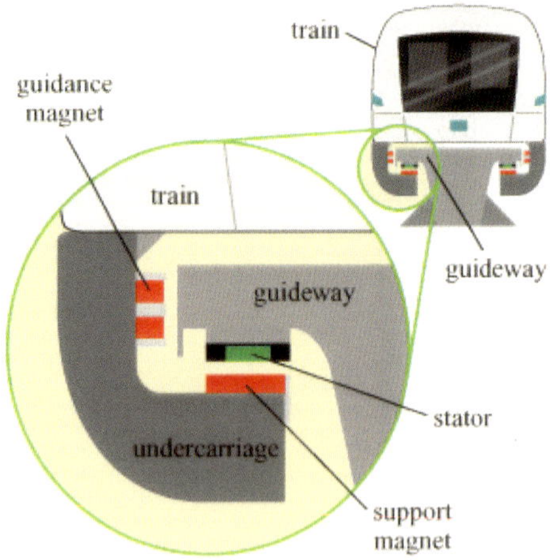

Magnetic Levitation Trains

The above figure shows the EMS magnetic levitation system in which electromagnets are attached to the body of the train and are powered by the batteries in the train. The guideway consists of electromagnetic coils built on the surface of the track.

These coils are constructed with core material and coil windings. When the electromagnets are energized, the train will be levitated by the attractive force between the electromagnets and the coils.

Hence this type of systems needs a large electrical power source, electromagnetic coils lining a guideway and guidance magnets attached to the underside of the train.

Communication System

It is the process of transmitting information from a source to a receiver. This transmission of energy over long distances is carried out through electromagnetic waves at high frequencies. These waves are also called as microwaves or high frequency radio waves.

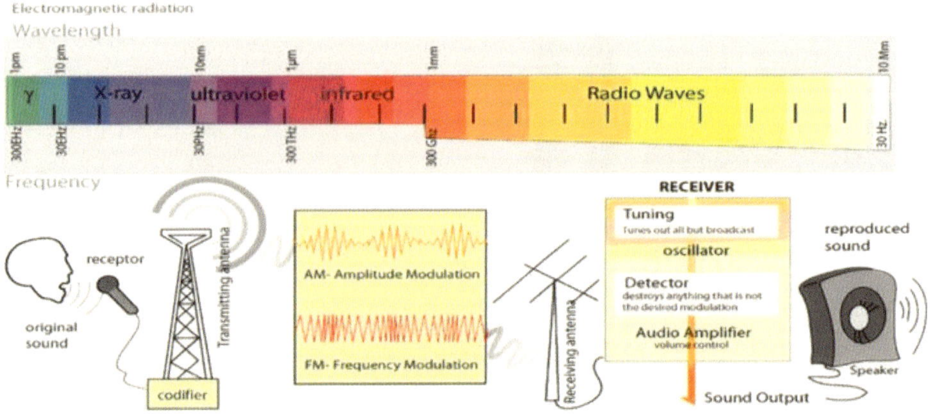

Communication System

Suppose in case of mobile phones, sound energy is converted into electromagnetic energy. By using radio transmitters, this electromagnetic energy is transferred to the receiver. At the receiver these electromagnetic waves are again transformed back into sound energy.

Depending on the nature of the baseband signal, communication systems can be analog or digital communication systems. Depending on the nature of transmitting signal, this system is categorized as baseband and carrier communication systems.

Electromagnetic fields produced by the time varying sources are propagated through a waveguide or transmission line. An electromagnetic wave radiation is formed when these electromagnetic fields are propagated away from the sources without any connection or conducting medium to the sources.

An electromagnetic spectrum consists of different range of all possible wavelength or frequencies of electromagnetic radiation. These frequency ranges include low frequency, ultra low frequency, medium frequency, high frequency, ultra high frequency, super high frequency etc.

Medical System

Nowadays electromagnetic fields play a key role in advanced medical equipments such as hyperthermia treatments for cancer, implants and magnetic resonance imaging (MRI).

RF range frequencies are mostly used in medical applications. In MRI scans, sophisticated equipment works based on the electromagnetism can scan minute details of the human body.

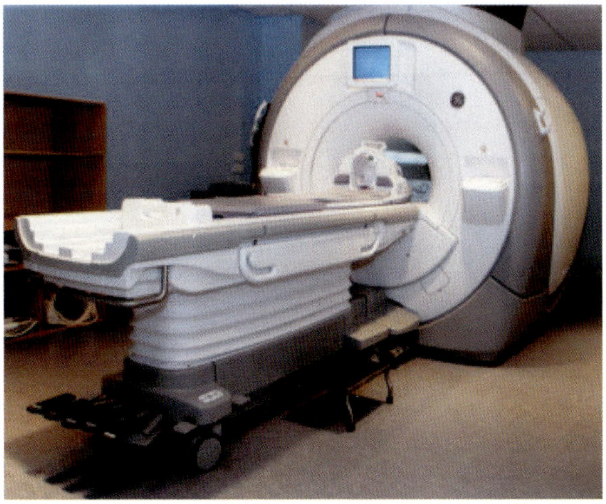

Medical system

The electromagnetic therapy is an alternate form of medicine which claims to treat disease by applying pulsed electromagnetic fields or electromagnetic radiation to the body. This type of treatments is used for wide range of ailments such as nervous disorders, diabetes, spinal cord injuries, ulcers, asthma, etc.

Many of the medical equipments such as scanners, x-ray equipments and other equipments uses electromagnetism principle for their functioning.

References

- Electricity, science: britannica.com, Retrieved 28 March 2018
- What-is-electricity, tutorials: learn.sparkfun.com, Retrieved 30 June 2018
- Magnetism-38059: livescience.com, Retrieved 31 March 2018
- Magnetism, electromagnetism: electronics-tutorials.ws, Retrieved 21 May 2018
- Induced-and-permanent-atomic-magnetic-dipoles, magnetism, science: Retrieved 31 March 2018
- Applications-of-electromagnetism: electronicshub.org, Retrieved 01 May 2018

2

Electromagnetic Field and Radiation

An electromagnetic field is generated by electrically charged systems, which affects other charged objects in the vicinity. Electromagnetic radiation refers to electromagnetic waves that propagate through space-time and carry electromagnetic radiant energy. The topics elaborated in this chapter on the effects of electromagnetic fields and radiation and the mathematical description of electromagnetic fields will help in providing a comprehensive understanding of electromagnetic field and electromagnetic radiation.

Electromagnetic Field

An electromagnetic field, sometimes referred to as an EM field, is generated when charged particles, such as electrons, are accelerated. All electrically charged particles are surrounded by electric fields. Charged particles in motion produce magnetic fields. When the velocity of a charged particle changes, an EM field is produced.

Electromagnetic fields were first discovered in the 19th century, when physicists noticed that electric arcs (sparks) could be reproduced at a distance, with no connecting wires in between. This led scientists to believe that it was possible to communicate over long distances without wires. The first radio transmitters made use of electric arcs. These "spark transmitters" and the associated receivers were as exciting to people in the early 20th century as the Internet is today. This was the beginning of what we now call wireless communication.

Electromagnetic fields are typically generated by alternating current (AC) in electrical conductors. The frequencyof the AC can range from one cycle in thousands of years (at the low extreme) to trillions or quadrillions of cycles per second(at the high extreme). The standard unit of EM frequency is the hertz, abbreviated Hz.Larger units are often used. A frequency of 1,000 Hz is onekilohertz(kHz); a frequency of 1,000 kHz is one megahertz (MHz); a frequency of 1,000 MHz is one gigahertz (GHz).

The Basics of Wavelength and Frequency

What makes the various forms of electromagnetic fields so different.

One of the main characteristics which defines an electromagnetic field (EMF) is its frequency or its corresponding wavelength. Fields of different frequencies interact with the body in different ways. One can imagine electromagnetic waves as series of very regular waves that travel at an enormous speed, the speed of light. The frequency simply describes the number of oscillations or cycles per second, while the term wavelength describes the distance between one wave and the next. Hence wavelength and frequency are inseparably intertwined: the higher the frequency the shorter the wavelength.

A simple analogy should help to illustrate the concept: Tie a long rope to a door handle and keep hold of the free end. Moving it up and then down slowly will generate a single big wave; more rapid motion will generate a whole series of small waves. The length of the rope remains constant, therefore, the more waves you generate (higher frequency) the smaller will be the distance between them (shorter wavelength).

Difference between Non-ionizing Electromagnetic Fields and Ionising Radiation

Wavelength and frequency determine another important characteristic of electromagnetic fields: Electromagnetic waves are carried by particles called quanta. Quanta of higher frequency (shorter wavelength) waves carry more energy than lower frequency (longer wavelength) fields. Some electromagnetic waves carry so much energy per quantum that they have the ability to break bonds between molecules. In the electromagnetic spectrum, gamma rays given off by radioactive materials, cosmic rays and X-rays carry this property and are called 'ionizing radiation'. Fields whose quanta are insufficient to break molecular bonds are called 'non-ionizing radiation'. Man-made sources of electromagnetic fields that form a major part of industrialized life - electricity, microwaves and radiofrequency fields – are found at the relatively long wavelength and low frequency end of the electromagnetic spectrum and their quanta are unable to break chemical bonds.

Electromagnetic Fields at Low Frequencies

Electric fields exist whenever a positive or negative electrical charge is present. They exert forces on other charges within the field. The strength of the electric field is measured in volts per metre (V/m). Any electrical wire that is charged will produce an associated electric field. This field exists even when there is no current flowing. The higher the voltage, the stronger the electric field at a given distance from the wire.

Electric fields are strongest close to a charge or charged conductor, and their strength rapidly diminishes with distance from it. Conductors such as metal shield them very effectively. Other materials, such as building materials and trees, provide some shielding capability. Therefore, the electric fields from power lines outside the house are reduced by walls, buildings, and trees. When power lines are buried in the ground, the electric fields at the surface are hardly detectable.

Magnetic fields arise from the motion of electric charges. The strength of the magnetic field is measured in amperes per meter (A/m); more commonly in electromagnetic field research, scientists specify a related quantity, the flux density (in microtesla, µT) instead. In contrast to electric fields, a magnetic field is only produced once a device is switched on and current flows. The higher the current, the greater the strength of the magnetic field.

Like electric fields, magnetic fields are strongest close to their origin and rapidly decrease at greater distances from the source. Magnetic fields are not blocked by common materials such as the walls of buildings.

Electric fields	Magnetic fields
1. Electric fields arise from voltage.	1. Magnetic fields arise from current flows.
2. Their strength is measured in Volts per metre (V/m)	2. Their strength is measured in amperes per meter (A/m). Commonly, EMF investigators use a related measure, flux density (in microtesla (μT) or millitesla (mT) instead.
3. An electric field can be present even when a device is switched off.	3. Magnetic fields exist as soon as a device is switched on and current flows.
4. Field strength decreases with distance from the source.	4. Field strength decreases with distance from the source.
5. Most building materials shield electric fields to some extent.	5. Magnetic fields are not attenuated by most materials.

Electric Fields

Plugging a wire into an outlet creates electric fields in the air surrounding the appliance. The higher the voltage the stronger the field produced. Since the voltage can exist even when no current is flowing, the appliance does not have to be turned on for an electric field to exist in the room surrounding it.

Magnetic Fields

Magnetic fields are created only when the electric current flows. Magnetic fields and electric fields then exist together in the room environment. The greater the current the stronger the magnetic field. High voltages are used for the transmission and distribution of electricity whereas relatively low voltages are used in the home. The voltages used by power transmission equipment vary little from day to day, currents through a transmission line vary with power consumption.

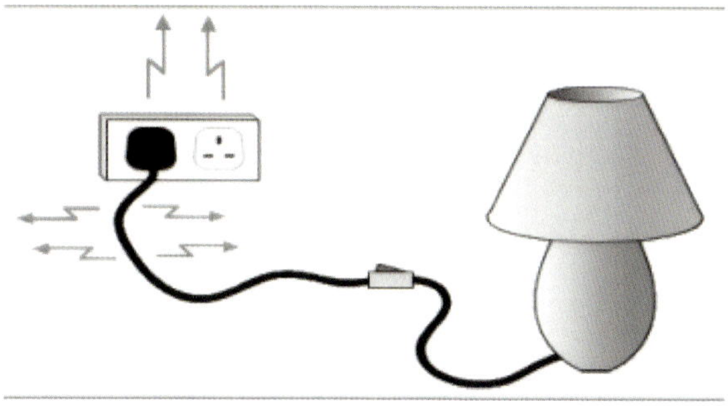

Electric fields around the wire to an appliance only cease to exist when the appliance is unplugged or switched off at the wall. They will still exist around the cable behind the wall.

Difference Between Static Fields Differ and Time-varying Fields

A static field does not vary over time. A direct current (DC) is an electric current flowing in one direction only. In any battery-powered appliance the current flows from the battery to the appliance and then back to the battery. It will create a static magnetic field. The earth's magnetic field is also a static field. So is the magnetic field around a bar magnet which can be visualized by observing the pattern that is formed when iron filings are sprinkled around it.

In contrast, time-varying electromagnetic fields are produced by alternating currents (AC). Alternating currents reverse their direction at regular intervals. In most European countries electricity changes direction with a frequency of 50 cycles per second or 50 Hertz. Equally, the associated electromagnetic field changes its orientation 50 times every second. North American electricity has a frequency of 60 Hertz.

Main Sources of Low, Intermediate and High Frequency Fields

The time-varying electromagnetic fields produced by electrical appliances are an example of extremely low frequency (ELF) fields. ELF fields generally have frequencies up to 300 Hz. Other technologies produce intermediate frequency (IF) fields with frequencies from 300 Hz to 10 MHz and radiofrequency (RF) fields with frequencies of 10 MHz to 300 GHz. The effects of electromagnetic fields on the human body depend not only on their field level but on their frequency and energy. Our electricity power supply and all appliances using electricity are the main sources of ELF fields; computer screens, anti-theft devices and security systems are the main sources of IF fields; and radio, television, radar and cellular telephone antennas, and microwave ovens are the main sources of RF fields. These fields induce currents within the human body, which if sufficient can produce a range of effects such as heating and electrical shock, depending on their amplitude and frequency range. (However, to produce such effects, the fields outside the body would have to be very strong, far stronger than present in normal environments.)

Electromagnetic Fields at High Frequencies

Mobile telephones, television and radio transmitters and radar produce RF fields. These fields are used to transmit information over long distances and form the basis of telecommunications as well as radio and television broadcasting all over the world. Microwaves are RF fields at high frequencies in the GHz range. In microwaves ovens, we use them to quickly heat food.

At radio frequencies, electric and magnetic fields are closely interrelated and we typically measure their levels as power densities in watts per square metre (W/m^2).

Electromagnetic Field Structure

The structure of electromagnetic field can be classified in two distinct methods: Continuous Structure and Discrete Structure.

Continuous Structure

Magnetic and electric fields are produced by an even motion of charged objects. The oscillating charges that produce magnetic and electric fields can be observed in a continuous, smooth and wavelike motion. The energy is observed as being continuous transfer between two electromagnetic field locations.

Discrete Structure

Several experiments has revealed that in a particular circumstance the electromagnetic energy flows in a fixed frequency or a more harsh way in case of a photon. The Planck's link the photon energy to its frequency. The equation is given below.

$$E=hv$$

Here,

E = Photon Energy

h = Planck's Constant

v = Frequency of Photon

Perception and Behavioral Effects of Electromagnetic Fields

Some animals respond to extremely low levels of electromagnetic fields (EMF), usually at frequencies ranging from DC to extremely-low-frequency (ELF) and usually with specialized receptors. Although the responses have been described and can be demonstrated at will, the mechanisms are not understood. Perceptual and behavioral responses to very low levels of EMFs at low frequency and above have not been reported in humans, and there are no mechanisms at DC or ELF that might imply as yet unreported responses, although they cannot be excluded. There is a considerable literature on perception and behavioral responses to stimulation at magnitudes consistent with direct electrical stimulation of nervous tissues by induced currents in the tissue, but they are limited to the frequency ranges that stimulate excitable membranes.

In the radiofrequency (RF) range of concern for assessment of GWEN sites, there have been many reports of sensory perception. The responses may be organized in the following categories:

- Stimulation of nervous structures by electric and magnetic fields and associated currents in the body. Above a threshold that is frequency-dependent, these currents are perceived as a painful stimulus that increases with current intensity.

  Electroencephalographic (EEG) activity in cats and rabbits has been reported to be altered by exposure to amplitude-modulated RF. Exposure to 147-MHz fields, amplitude-modulated between 1 and 25 Hz, altered the ability of cats to produce selected EEG rhythms. Changes in EEG frequency spectrum were also observed in rabbits chronically exposed to 1-10 MHz fields that were amplitude-modulated at 14-16 Hz. Other studies have shown small changes in EEG patterns, particularly desynchronization, in rats and rabbits after exposure to 12.95-GHz field at 1 W/kg. Some later studies failed to find an effect. The lowest levels used, in the above studies are 10,000 times that which would be encountered near GWEN installations.

- Shocks and burns. When the human body is in an EMF of suitable frequency and intensity and it makes contact with a conducting body in the same field, an electrical current is produced that can cause perceptible electrical shock, muscular contractions, burns, and possible death.
- Heating. If enough RF power is absorbed in human tissue, especially skin, it can raise the tissue temperature and cause a sensation of warming that will be due to thermal stimulation of temperature receptors. Thermal perception of absorbed RF energy is frequency-dependent: the threshold energy decreases as the frequency increases. There is a delay in the perception of warmth after the start of irradiation; the delay may vary from 5 sec or more at GHz frequencies to as little as 1 sec for infrared radiation. Justesen et al. compared thermal perception in human volunteers who were irradiated on the forearm in a 100-cm^2 area with far-infrared radiation or 2.45-GHz microwave radiation. The thresholds of perception for a 10-sec exposure were 1.7 mW/cm^2 for far infrared radiation and 26.7 mW/cm^2 for microwave radiation.
- Auditory perception. A special effect has been reported in which microwave RF emitted in the form of very short pulses (1-20 μsec) is perceived by humans and animals as clicks or other sounds. This perception could well result when thermal absorption leads to thermoelastic expansion of tissues and fluids in the head and is sensed by auditory receptors. If the energy flux in the pulse exceeds about 40 μJ/cm^2, delivered in a few microseconds, auditory perceptions occur.

  The auditory perception of pulsed microwave fields was first reported in 1947 and has been studied extensively. Frey reported on controlled experimental exposures at frequencies of 0.2-8.9 GHz and pulse widths of 1-1,000 μsec. He found that, depending on the characteristics of the field, sensations were perceived as buzzing, ticking, hissing, or knocking sounds. Sound was perceived at all frequencies up to 8.9 GHz. Guy et al. demonstrated that the threshold for auditory perception was four times higher at 3.75 kHz in subjects with neurosensory deficits compared with normal subjects, thus indicating that the effect was in the acoustic elements involved in hearing.
- Behavioral changes. Epidemiologic studies of groups of people occupationally or environmentally exposed to electromagnetic fields in the RF and ELF range have yielded perceptual and behavioral responses, including fatigue, difficulty in concentrating, and increased frequency of headaches. A number of researchers have used disruption of behavior patterns, such as work stoppage, to study the effect of RF fields on animals, including rodents and monkeys. Several carrier frequencies, field zones, and modulation characteristics were used. A relatively narrow range of threshold of specific absorption rates (SARs), about 4-9 W/kg, was found. Lebovitz examined the effect of repeated exposures to a pulsed 1,300-MHz field on behavioral performance in rats. He exposed animals to SARs of 1.6, 3.6, or 6.7 W/kg for 3 h/day, 5 d/wk for 6-9 wk and found that rates of lever-pressing for food were slightly reduced at the highest SAR. However, the ability of the rats to discriminate improved as a positive function of SAR when lever-pressing was not reinforced by the presence of food. DeLorge used a different experimental paradigm and showed a disruption in performance of rats at an SAR of 2.5 W/kg when they were exposed to a pulsed 1,300-MHz field. Behavioral studies performed by Hjeresen et al. showed that

rats placed in a shuttlebox tended to remain in the side shielded from pulsed RF fields. Because the rats also tended to avoid pulsed sound waves, the investigators suggested that the rats' avoidance of the RF fields might be related to the hearing of the pulsed fields. The response to a pulsed field is stronger than that to a continuous field. For example, Carroll et al.18 found that rats exposed to an intense field that was not pulse-modulated did not readily learn to escape from it. It appears that hearing the pulses is a more effective cue for escaping than is the warming that results from a continuous field.

Studies with ELF fields have suggested that behavior can be influenced by exposure to either magnetic or electric fields. Persinger reported that prenatal exposure of rats to 0.5-Hz, 0.05-to 3-mT fields, resulted in changes in juvenile or adult rats' emotionality and ability to perform a conditioned-suppression test. Frey found that prenatal exposure of rats to a 60-Hz field at 3.5 kV/m caused changes in open-field activity.

McGivern et al. studied male rats that had been exposed prenatally on days 15-21 of gestation. Exposures were for 15 min twice a day, to a 15-Hz, 800-μT pulsed magnetic field. Exposed animals showed a significant reduction in scent marking, compared with sham-exposed or caged controls. Exposed males had larger seminal vesicles, prostates, and epididymides than did control males.

Transient neurobehavioral changes in rats exposed prenatally and postnatally to a 60-Hz electric field (65-kV/m effective field) were reported by Sikov et al. Exposed animals showed significantly more motility than did controls. Development of righting reflex and negative geotropism was also delayed in exposed rats; the percentage of pups that failed to show these behaviors was increased on day 14 of postnatal life, but not on day 21.

Lovely et al. exposed gravid Sprague-Dawley rats to a 60-kV/m, 60-Hz electric field and then tested the offspring at the age of 90 days in three tasks: shuttlebox avoidance, a residential maze, and a preference-avoidance test. No differences were noted between exposed and sham-exposed animals.

Changes in learning have been reported in rats exposed to a combination of 60-Hz electric fields (30 kV/m and 0.1 mT) and magnetic fields (10 kV/m and 0.033 mT) throughout gestation and during the first 8 days of postnatal life. Both acquisition and extinction of a schedule-controlled response were affected in the exposed animals. In contrast, other studies failed to find an effect of exposure to ELF fields on behavior.

Exposure to 60-Hz electric fields (30 or 60 kV/m) has been reported to affect the social behavior of baboons. The investigators used a number of measures of social behavior, but found that passive affinity, tension, and stereotypy performance were significantly increased in exposed groups. The authors suggested that the changes might indicate a stress response to the fields.

Although there is evidence that exposure of experimental animals to electric or magnetic fields can influence neurobehavioral function, there is a paucity of direct observations at the 175-kHz frequency and at the ultra-high frequencies of 200400 MHz used in the GWEN system. Moreover, magnitudes of the exposure usually required to produce an effect are substantially higher than those likely to be encountered as a result of operation of the GWEN system. It therefore seems unlikely that electromagnetic fields from GWEN will affect neurosensory or neurobehavioral function in persons living around GWEN sites.

Applications of Electromagnetic Field Theory

Electromagnetic Field Theory has applications in analysing and designing of communication system like:

- Satellite Communication
- TV Communication
- Wireless Communication
- Mobile Communication
- Microwave Communication
- This theory is also used in analysis and designing of antenna, transmission
- Lines and wave guides.

Besides the above applications, EMFT is used in:

- Bio-medical systems
- Weather forecast radars
- Remote sensing radars
- Radio astronomy radars
- Plasmas
- Electric motors
- Radiation therapy
- Surface hardening
- Annealing
- Soldering
- Dielectric heating
- Joining and Sealing
- Reducing acidity in vegetables to improve taste
- Lasers
- Masers

Electromagnetic Radiation

Electromagnetic radiation is energy that is propagated through free space or through a material medium in the form of electromagnetic waves, such as radio waves, visible light, and gamma rays. The term also refers to the emission and transmission of such radiant energy.

The Scottish physicist James Clerk Maxwell was the first to predict the existence of electromagnetic waves. In 1864 he set forth his electromagnetic theory, proposing that light--including various other forms of radiant energy--is an electromagnetic disturbance in the form of waves. In 1887 Heinrich Hertz, a German physicist, provided experimental confirmation by producing the first man-made electromagnetic waves and investigating their properties. Subsequent studies resulted in a broader understanding of the nature and origin of radiant energy.

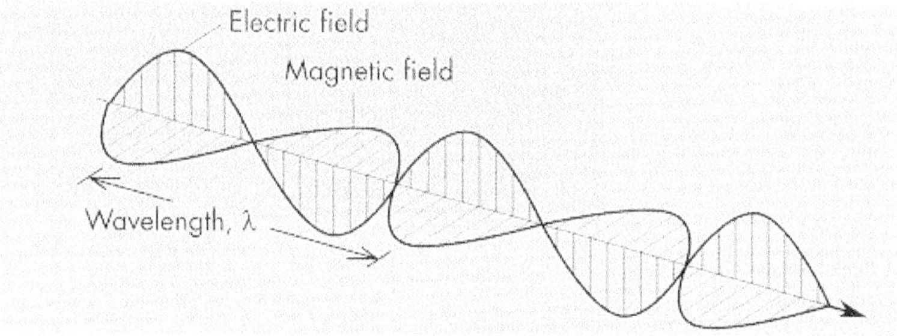

It has been established that time-varying electric fields can induce magnetic fields and that time-varying magnetic fields can in like manner induce electric fields. Because such electric and magnetic fields generate each other, they occur jointly, and together they propagate as electromagnetic waves. An electromagnetic wave is a transverse wave in that the electric field and the magnetic field at any point and time in the wave are perpendicular to each other as well as to the direction of propagation. In free space (i.e., a space that is absolutely devoid of matter and that experiences no intrusion from other fields or forces), electromagnetic waves always propagate with the same speed--that of light (299,792,458 m per second, or 186,282 miles per second)--independent of the speed of the observer or of the source of the waves.

Electromagnetic radiation has properties in common with other forms of waves such as reflection, refraction, diffraction, and interference. Moreover, it may be characterized by the frequency with which it varies over time or by its wavelength. Electromagnetic radiation, however, has particle-like properties in addition to those associated with wave motion. It is quantized in that for a given frequency , its energy occurs as an integer times h , in which h is a fundamental constant of nature known as Planck's constant. A quantum of electromagnetic energy is called a photon. Visible light and other forms of electromagnetic radiation may be thought of as a stream of photons, with photon energy directly proportional to frequency.

Electromagnetic radiation spans an enormous range of frequencies or wavelengths, as is shown by the electromagnetic spectrum. Customarily, it is designated by fields, waves, and particles in increasing magnitude of frequencies--radio waves, microwaves, infrared rays, visible light, ultraviolet light, X rays, and gamma rays. The corresponding wavelengths are inversely proportional, and both the frequency and wavelength scales are logarithmic.

Electromagnetic radiation of different frequencies interacts with matter differently. A vacuum is the only perfectly transparent medium, and all material media absorb strongly some regions of the electromagnetic spectrum. For example, molecular oxygen ($O2$), ozone ($O3$), and molecular nitrogen ($N2$) in the Earth's atmosphere are almost perfectly transparent to infrared rays of all frequencies, but they strongly absorb ultraviolet light, X rays, and gamma rays. The frequency (or energy equal to hv) of X rays is substantially higher than that of visible light, and so X rays are able to penetrate many materials that do not transmit light. Moreover, absorption of X rays by a molecular system can cause chemical reactions to occur. When X rays are absorbed in a gas, for instance, they eject photoelectrons from the gas, which in turn ionize its molecules. If these processes occur in living tissue, the photoelectrons emitted from the organic molecules destroy the cells of the tissue. Gamma rays, though generally of somewhat higher frequency than X rays, have basically the same nature. When the energy of gamma rays is absorbed in matter, its effect is virtually indistinguishable from the effect produced by X rays.

There are many sources of electromagnetic radiation, both natural and man-made. Radio waves, for example, are produced by cosmic objects such as pulsars and quasars and by electronic circuits. Sources of ultraviolet radiation include mercury vapour lamps and high-intensity lights, as well as the Sun. The latter also generates X rays, as do certain types of particle accelerators and electronic devices.

Waves and Their Characteristics

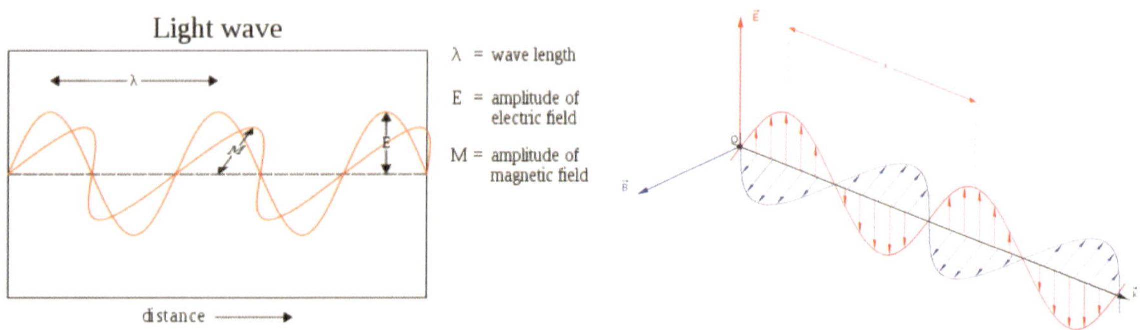

Electromagnetic Waves

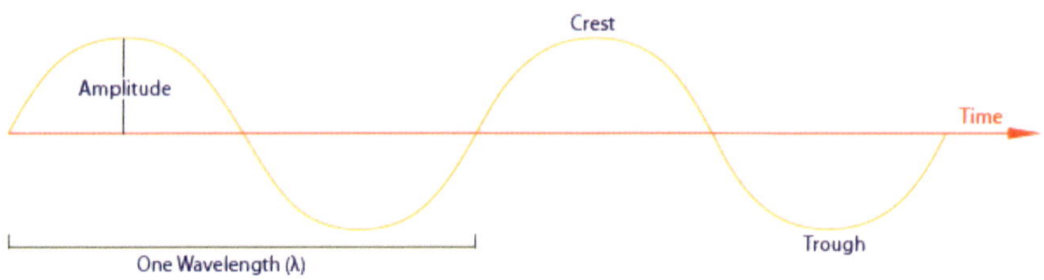

An EM Wave

Amplitude

Amplitude is the distance from the maximum vertical displacement of the wave to the middle of the wave. This measures the magnitude of oscillation of a particular wave. In short, the amplitude is basically the height of the wave. Larger amplitude means higher energy and lower amplitude means lower energy. Amplitude is important because it tells you the intensity or brightness of a wave in comparison with other waves.

Wavelength

Wavelength (λ) is the distance of one full cycle of the oscillation. Longer wavelength waves such as radio waves carry low energy; this is why we can listen to the radio without any harmful consequences. Shorter wavelength waves such as x-rays carry higher energy that can be hazardous to our health. Consequently lead aprons are worn to protect our bodies from harmful radiation when we undergo x-rays. This wavelength frequently relationship is characterized by:

$$c = \lambda v$$

where

- c is the speed of light,
- λ is wavelength, and
- v is frequency.

Shorter wavelength means greater frequency, and greater frequency means higher energy. Wavelengths are important in that they tell one what type of wave one is dealing with.

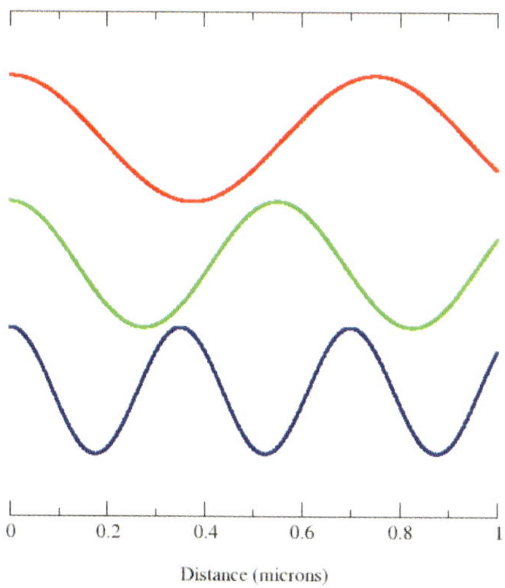

Different Wavelengths and Frequencies: Wavelength tells you the type of light and Amplitude tells you about the intensity of the light

Frequency

Frequency is defined as the number of cycles per second, and is expressed as sec^{-1} or Hertz (Hz). Frequency is directly proportional to energy and can be express as:

$$E = hv$$

where

- E is energy,
- h is Planck's constant, (h= 6.62607×10^{-34} J), and
- v is frequency.

Period

Period (T) is the amount of time a wave takes to travel one wavelength; it is measured in seconds (s).

Velocity

The velocity of wave in general is expressed as:

$$\text{velocity} = \lambda v$$

For Electromagnetic wave, the velocity in vacuum is 2.99 x10^8 m/s or 186,282 miles/second.

Theory

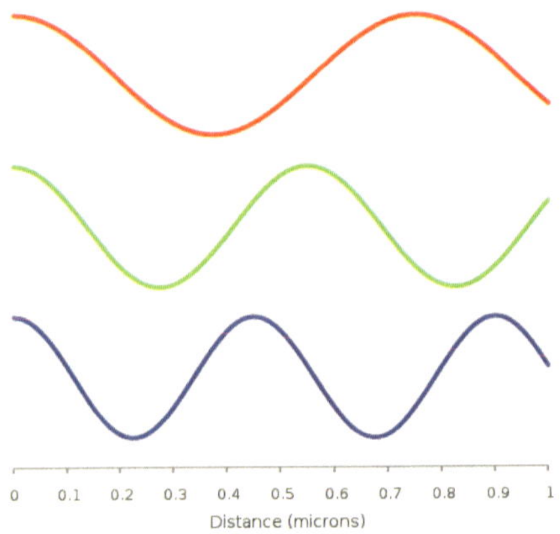

Shows the relative wavelengths of the electromagnetic waves of three different colours of light (blue, green, and red) with a distance scale in micrometers along the x-axis.

Maxwell's Equations

Maxwell derived a wave form of the electric and magnetic equations, thus uncovering the wave-like nature of electric and magnetic fields and their symmetry. Because the speed of EM waves predicted by the wave equation coincided with the measured speed of light, Maxwell concluded that light itself is an EM wave. Maxwell's equations were confirmed by Heinrich Hertz through experiments with radio waves.

According to Maxwell's equations, a spatially varying electric field is always associated with a magnetic field that changes over time. Likewise, a spatially varying magnetic field is associated with specific changes over time in the electric field. In an electromagnetic wave, the changes in the electric field are always accompanied by a wave in the magnetic field in one direction, and vice versa. This relationship between the two occurs without either type field causing the other; rather, they occur together in the same way that time and space changes occur together and are interlinked in special relativity. In fact, magnetic fields may be viewed as relativistic distortions of electric fields, so the close relationship between space and time changes here is more than an analogy. Together, these fields form a propagating electromagnetic wave, which moves out into space and need never again affect the source. The distant EM field formed in this way by the acceleration of a charge carries energy with it that "radiates" away through space, hence the term.

Near and Far Fields

In electromagnetic radiation (such as microwaves from an antenna, shown here) the term applies only to the parts of the electromagnetic field that radiate into infinite space and decrease in t by an inverse-square law of power, so that the total radiation energy that crosses through an imaginary spherical surface is the same, no matter how far away from the antenna the spherical surface is drawn. Electromagnetic radiation thus includes the far field part of the electromagnetic field around a transmitter. A part of the "near-field" close to the transmitter, forms part of the changing electromagnetic field, but does not count as electromagnetic radiation.

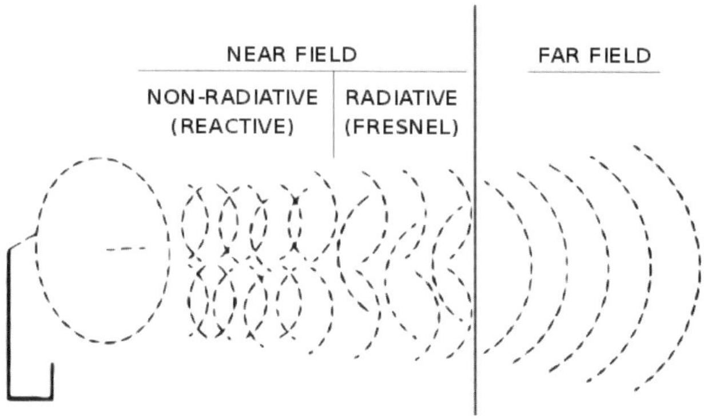

Maxwell's equations established that some charges and currents ("sources") produce a local type of electromagnetic field near them that does *not* have the behaviour of EMR. Currents directly produce a magnetic field, but it is of a magnetic dipole type that dies out with distance from the current. In a similar manner, moving charges pushed apart in a conductor by a changing electrical potential (such as in an antenna) produce an electric dipole type electrical field, but this also declines with distance. These fields make up the near-field near the EMR source. Neither of these behaviours are responsible for EM radiation. Instead, they cause electromagnetic field behaviour that only efficiently transfers power to a receiver very close to the source, such as the magnetic induction inside a transformer, or the feedback behaviour that happens close to the coil of a metal detector. Typically, near-fields have a powerful effect on their own sources, causing an increased "load" (decreased electrical reactance) in the source or transmitter, whenever energy is withdrawn from the EM field by a receiver. Otherwise, these fields do not "propagate" freely out into space, carrying their energy away without distance-limit, but rather oscillate, returning their energy to the transmitter if it is not received by a receiver.

By contrast, the EM far-field is composed of *radiation* that is free of the transmitter in the sense that (unlike the case in an electrical transformer) the transmitter requires the same power to send these changes in the fields out, whether the signal is immediately picked up or not. This distant part of the electromagnetic field *is* "electromagnetic radiation" (also called the far-field). The far-fields propagate (radiate) without allowing the transmitter to affect them. This causes them to be independent in the sense that their existence and their energy, after they have left the transmitter, is completely independent of both transmitter and receiver. Due to conservation of energy, the amount of power passing through any spherical surface drawn around the source is the same. Because such a surface has an area proportional to the square of its distance from the source, the power density of EM radiation always decreases with the inverse square of distance from the source; this is called the inverse-square law. This is in contrast to dipole parts of the EM field close to the source (the near-field), which varies in power according to an inverse cube power law, and thus does *not* transport a conserved amount of energy over distances, but instead fades with distance, with its energy (as noted) rapidly returning to the transmitter or absorbed by a nearby receiver (such as a transformer secondary coil).

The far-field (EMR) depends on a different mechanism for its production than the near-field, and upon different terms in Maxwell's equations. Whereas the magnetic part of the near-field is due to currents in the source, the magnetic field in EMR is due only to the local change in the electric field. In a similar way, while the electric field in the near-field is due directly to the charges and charge-separation in the source, the electric field in EMR is due to a change in the local magnetic field. Both processes for producing electric and magnetic EMR fields have a different dependence on distance than do near-field dipole electric and magnetic fields. That is why the EMR type of EM field becomes dominant in power "far" from sources. The term "far from sources" refers to how far from the source (moving at the speed of light) any portion of the outward-moving EM field is located, by the time that source currents are changed by the varying source potential, and the source has therefore begun to generate an outwardly moving EM field of a different phase.

A more compact view of EMR is that the far-field that composes EMR is generally that part of the EM field that has traveled sufficient distance from the source, that it has become completely disconnected from any feedback to the charges and currents that were originally responsible for it. Now independent of the source charges, the EM field, as it moves farther away, is dependent only upon the accelerations of the charges that produced it. It no longer has a strong connection to the direct fields of the charges, or to the velocity of the charges (currents).

In the Liénard–Wiechert potential formulation of the electric and magnetic fields due to motion of a single particle (according to Maxwell's equations), the terms associated with acceleration of the particle are those that are responsible for the part of the field that is regarded as electromagnetic radiation. By contrast, the term associated with the changing static electric field of the particle and the magnetic term that results from the particle's uniform velocity, are both associated with the electromagnetic near-field, and do not comprise EM radiation.

Properties

Electrodynamics is the physics of electromagnetic radiation, and electromagnetism is the physical phenomenon associated with the theory of electrodynamics. Electric and magnetic fields obey the properties of superposition. Thus, a field due to any particular particle or time-varying electric or magnetic field contributes to the fields present in the same space due to other causes. Further, as they are vector fields, all magnetic and electric field vectors add together according to vector addition. For example, in optics two or more coherent lightwaves may interact and by constructive or destructive interference yield a resultant irradiance deviating from the sum of the component irradiances of the individual lightwaves.

Since light is an oscillation it is not affected by traveling through static electric or magnetic fields in a linear medium such as a vacuum. However, in nonlinear media, such as some crystals, interactions can occur between light and static electric and magnetic fields — these interactions include the Faraday effect and the Kerr effect.

In refraction, a wave crossing from one medium to another of different density alters its speed and direction upon entering the new medium. The ratio of the refractive indices of the media determines the degree of refraction, and is summarized by Snell's law. Light of composite wavelengths (natural sunlight) disperses into a visible spectrum passing through a prism, because of the wavelength-dependent refractive index of the prism material (dispersion); that is, each component wave within the composite light is bent a different amount.

EM radiation exhibits both wave properties and particle properties at the same time. Both wave and particle characteristics have been confirmed in many experiments. Wave characteristics are more apparent when EM radiation is measured over relatively large timescales and over large distances while particle characteristics are more evident when measuring small timescales and distances. For example, when electromagnetic radiation is absorbed by matter, particle-like properties will be more obvious when the average number of photons in the cube of the relevant wavelength is much smaller than. It is not too difficult to experimentally observe non-uniform deposition of energy when light is absorbed, however this alone is not evidence of "particulate" behavior. Rather, it reflects the quantum nature of *matter*. Demonstrating that the light itself is quantized, not merely its interaction with matter, is a more subtle affair.

Some experiments display both the wave and particle natures of electromagnetic waves, such as the self-interference of a single photon. When a single photon is sent through an interferometer, it passes through both paths, interfering with itself, as waves do, yet is detected by a photomultiplier or other sensitive detector only once.

A quantum theory of the interaction between electromagnetic radiation and matter such as electrons is described by the theory of quantum electrodynamics.

Electromagnetic waves can be polarized, reflected, refracted, diffracted or interfere with each other.

Wave Model

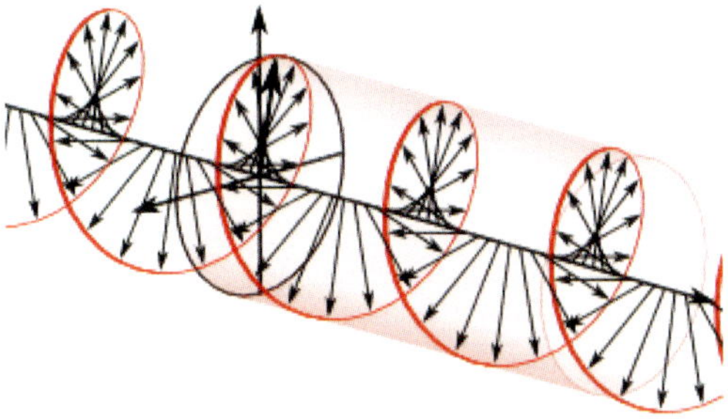

Representation of the electric field vector of a wave of circularly polarized electromagnetic radiation.

Electromagnetic radiation is a transverse wave, meaning that its oscillations are perpendicular to the direction of energy transfer and travel. The electric and magnetic parts of the field stand in a fixed ratio of strengths in order to satisfy the two Maxwell equations that specify how one is produced from the other. These E and B fields are also in phase, with both reaching maxima and minima at the same points in space. A common misconception is that the E and B fields in electromagnetic radiation are out of phase because a change in one produces the other, and this would produce a phase difference between them as sinusoidal functions (as indeed happens in electromagnetic induction, and in the near-field close to antennas). However, in the far-field EM radiation which is described by the two source-free Maxwell curl operator equations, a more correct description is that a time-change in one type of field is proportional to a space-change in the other. These derivatives require that the E and B fields in EMR are in-phase.

An important aspect of light's nature is its frequency. The frequency of a wave is its rate of oscillation and is measured in hertz, the SI unit of frequency, where one hertz is equal to one oscillation per second. Light usually has multiple frequencies that sum to form the resultant wave. Different frequencies undergo different angles of refraction, a phenomenon known as dispersion.

A wave consists of successive troughs and crests, and the distance between two adjacent crests or troughs is called the wavelength. Waves of the electromagnetic spectrum vary in size, from very long radio waves the size of buildings to very short gamma rays smaller than atom nuclei. Frequency is inversely proportional to wavelength, according to the equation:

$$v = f\lambda$$

where υ is the speed of the wave (c in a vacuum, or less in other media), f is the frequency and λ is the wavelength. As waves cross boundaries between different media, their speeds change but their frequencies remain constant.

Electromagnetic waves in free space must be solutions of Maxwell's electromagnetic wave equation. Two main classes of solutions are known, namely plane waves and spherical waves. The plane waves may be viewed as the limiting case of spherical waves at a very large (ideally infinite) distance from the source. Both types of waves can have a waveform which is an arbitrary time function (so long as it is sufficiently differentiable to conform to the wave equation). As with any time function, this can be decomposed by means of Fourier analysis into its frequency spectrum, or individual sinusoidal components, each of which contains a single frequency, amplitude and phase. Such a component wave is said to be *monochromatic*. A monochromatic electromagnetic wave can be characterized by its frequency or wavelength, its peak amplitude, its phase relative to some reference phase, its direction of propagation and its polarization.

Interference is the superposition of two or more waves resulting in a new wave pattern. If the fields have components in the same direction, they constructively interfere, while opposite directions cause destructive interference. An example of interference caused by EMR is electromagnetic interference (EMI) or as it is more commonly known as, radio-frequency interference (RFI). Additionally, multiple polarization signals can be combined (i.e. interfered) to form new states of polarization, which is known as parallel polarization state generation.

The energy in electromagnetic waves is sometimes called radiant energy.

Particle Model and Quantum Theory

An anomaly arose in the late 19th century involving a contradiction between the wave theory of light and measurements of the electromagnetic spectra that were being emitted by thermal radiators known as black bodies. Physicists struggled with this problem, which later became known as the ultraviolet catastrophe, unsuccessfully for many years. In 1900, Max Planck developed a new theory of black-body radiation that explained the observed spectrum. Planck's theory was based on the idea that black bodies emit light (and other electromagnetic radiation) only as discrete bundles or packets of energy. These packets were called quanta. Later, Albert Einstein proposed that light quanta be regarded as real particles. Later the particle of light was given the name photon, to correspond with other particles being described around this time, such as the electron and proton. A photon has an energy, *E*, proportional to its frequency, *f*, by

$$E = hf = \frac{hc}{\lambda}$$

where *h* is Planck's constant, λ is the wavelength and *c* is the speed of light. This is sometimes known as the Planck–Einstein equation. In quantum theory the energy of the photons is thus directly proportional to the frequency of the EMR wave.

Likewise, the momentum *p* of a photon is also proportional to its frequency and inversely proportional to its wavelength:

$$p = \frac{E}{c} = \frac{hf}{c} = \frac{h}{\lambda}.$$

The source of Einstein's proposal that light was composed of particles (or could act as particles in some circumstances) was an experimental anomaly not explained by the wave theory: the photoelectric effect, in which light striking a metal surface ejected electrons from the surface, causing an electric current to flow across an applied voltage. Experimental measurements demonstrated that the energy of individual ejected electrons was proportional to the *frequency*, rather than the *intensity*, of the light. Furthermore, below a certain minimum frequency, which depended on the particular metal, no current would flow regardless of the intensity. These observations appeared to contradict the wave theory, and for years physicists tried in vain to find an explanation. In 1905, Einstein explained this puzzle by resurrecting the particle theory of light to explain the observed effect. Because of the preponderance of evidence in favor of the wave theory, however, Einstein's ideas were met initially with great skepticism among established physicists. Eventually Einstein's explanation was accepted as new particle-like behavior of light was observed, such as the Compton effect.

As a photon is absorbed by an atom, it excites the atom, elevating an electron to a higher energy level (one that is on average farther from the nucleus). When an electron in an excited molecule or atom descends to a lower energy level, it emits a photon of light at a frequency corresponding to the energy difference. Since the energy levels of electrons in atoms are discrete, each element and each molecule emits and absorbs its own characteristic frequencies. Immediate photon emission is called fluorescence, a type of photoluminescence. An example is visible light emitted from fluorescent paints, in response to ultraviolet (blacklight). Many other fluorescent emissions are known in spectral bands other than visible light. Delayed emission is called phosphorescence.

Wave–particle Duality

The modern theory that explains the nature of light includes the notion of wave–particle duality. More generally, the theory states that everything has both a particle nature and a wave nature, and various experiments can be done to bring out one or the other. The particle nature is more easily discerned using an object with a large mass. A bold proposition by Louis de Broglie in 1924 led the scientific community to realize that electrons also exhibited wave–particle duality.

Wave and Particle Effects of Electromagnetic Radiation

Together, wave and particle effects fully explain the emission and absorption spectra of EM radiation. The matter-composition of the medium through which the light travels determines the nature of the absorption and emission spectrum. These bands correspond to the allowed energy levels in the atoms. Dark bands in the absorption spectrum are due to the atoms in an intervening medium between source and observer. The atoms absorb certain frequencies of the light between emitter and detector/eye, then emit them in all directions. A dark band appears to the detector, due to the radiation scattered out of the beam. For instance, dark bands in the light emitted by a distant star are due to the atoms in the star's atmosphere. A similar phenomenon occurs for emission, which is seen when an emitting gas glows due to excitation of the atoms from any mechanism, including heat. As electrons descend to lower energy levels, a spectrum is emitted that represents the jumps between the energy levels of the electrons, but lines are seen because again emission happens only at particular energies after excitation. An example is the emission spectrum of nebulae. Rapidly moving electrons are most sharply accelerated when they encounter a region of force, so they are responsible for producing much of the highest frequency electromagnetic radiation observed in nature.

These phenomena can aid various chemical determinations for the composition of gases lit from behind (absorption spectra) and for glowing gases (emission spectra). Spectroscopy (for example) determines what chemical elements comprise a particular star. Spectroscopy is also used in the determination of the distance of a star, using the red shift.

Propagation Speed

When any wire (or other conducting object such as an antenna) conducts alternating current, electromagnetic radiation is propagated at the same frequency as the current. In many such situations it is possible to identify an electrical dipole moment that arises from separation of charges due to the exciting electrical potential, and this dipole moment oscillates in time, as the charges move back and forth. This oscillation at a given frequency gives rise to changing electric and magnetic fields, which then set the electromagnetic radiation in motion.

At the quantum level, electromagnetic radiation is produced when the wavepacket of a charged particle oscillates or otherwise accelerates. Charged particles in a stationary state do not move, but a superposition of such states may result in a transition state that has an electric dipole moment that oscillates in time. This oscillating dipole moment is responsible for the phenomenon of radiative transition between quantum states of a charged particle. Such states occur (for example) in atoms when photons are radiated as the atom shifts from one stationary state to another.

As a wave, light is characterized by a velocity (the speed of light), wavelength, and frequency. As particles, light is a stream of photons. Each has an energy related to the frequency of the wave given by Planck's relation $E = hf$, where E is the energy of the photon, $h = 6.626 \times 10^{-34}$ J·s is Planck's constant, and f is the frequency of the wave.

One rule is obeyed regardless of circumstances: EM radiation in a vacuum travels at the speed of light, *relative to the observer*, regardless of the observer's velocity. (This observation led to Einstein's development of the theory of special relativity.)

In a medium (other than vacuum), velocity factor or refractive index are considered, depending on frequency and application. Both of these are ratios of the speed in a medium to speed in a vacuum.

Special Theory of Relativity

By the late nineteenth century, various experimental anomalies could not be explained by the simple wave theory. One of these anomalies involved a controversy over the speed of light. The speed of light and other EMR predicted by Maxwell's equations did not appear unless the equations were modified in a way first suggested by FitzGerald and Lorentz, or else otherwise that speed would depend on the speed of observer relative to the "medium" (called luminiferous aether) which supposedly "carried" the electromagnetic wave (in a manner analogous to the way air carries sound waves). Experiments failed to find any observer effect. In 1905, Einstein proposed that space and time appeared to be velocity-changeable entities for light propagation and all other processes and laws. These changes accounted for the constancy of the speed of light and all electromagnetic radiation, from the viewpoints of all observers—even those in relative motion.

Electromagnetic Spectrum

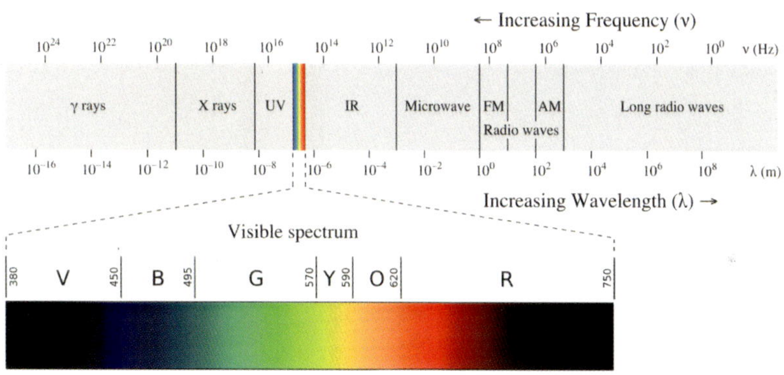

Electromagnetic spectrum with visible light highlighted

CLASS	FREQUENCY	WAVELENGTH	ENERGY
γ	300 EHz	1 pm	1.24 MeV
HX	30 EHz	10 pm	124 keV
	3 EHz	100 pm	12.4 keV
SX	300 PHz	1 nm	1.24 keV
EUV	30 PHz	10 nm	124 eV
NUV	3 PHz	100 nm	12.4 eV
NIR	300 THz	1 μm	1.24 eV
MIR	30 THz	10 μm	124 meV
FIR	3 THz	100 μm	12.4 meV
EHF	300 GHz	1 mm	1.24 meV
SHF	30 GHz	1 cm	124 μeV
UHF	3 GHz	1 dm	12.4 μeV
VHF	300 MHz	1 m	1.24 μeV
HF	30 MHz	10 m	124 neV
MF	3 MHz	100 m	12.4 neV
LF	300 kHz	1 km	1.24 neV
VLF	30 kHz	10 km	124 peV
VF/ULF	3 kHz	100 km	12.4 peV
SLF	300 Hz	1 Mm	1.24 peV
ELF	30 Hz	10 Mm	124 feV
	3 Hz	100 Mm	12.4 feV

Legend:

- γ = Gamma rays
- HX = Hard X-rays
- SX = Soft X-Rays
- EUV = Extreme-ultraviolet
- NUV = Near-ultraviolet

Visible light (colored bands)

- NIR = Near-infrared
- MIR = Mid-infrared
- FIR = Far-infrared
- EHF = Extremely high frequency (microwaves)
- SHF = Super-high frequency (microwaves)
- UHF = Ultrahigh frequency (radio waves)
- VHF = Very high frequency (radio)
- HF = High frequency (radio)
- MF = Medium frequency (radio)
- LF = Low frequency (radio)
- VLF = Very low frequency (radio)
- VF = Voice frequency
- ULF = Ultra-low frequency (radio)
- SLF = Super-low frequency (radio)
- ELF = Extremely low frequency(radio)

EM radiation (the designation 'radiation' excludes static electric and magnetic and near fields) is classified by wavelength into radio, microwave, infrared, visible, ultraviolet, X-rays and gamma rays. Arbitrary electromagnetic waves can be expressed by Fourier analysis in terms of sinusoidal monochromatic waves, which in turn can each be classified into these regions of the EMR spectrum.

For certain classes of EM waves, the waveform is most usefully treated as *random*, and then spectral analysis must be done by slightly different mathematical techniques appropriate to random or stochastic processes. In such cases, the individual frequency components are represented in terms of their *power* content, and the phase information is not preserved. Such a representation is called the power spectral density of the random process. Random electromagnetic radiation requiring this kind of analysis is, for example, encountered in the interior of stars, and in certain other very wideband forms of radiation such as the Zero point wave field of the electromagnetic vacuum.

The behavior of EM radiation depends on its frequency. Lower frequencies have longer wavelengths, and higher frequencies have shorter wavelengths, and are associated with photons of higher energy. There is no fundamental limit known to these wavelengths or energies, at either end of the spectrum, although photons with energies near the Planck energy or exceeding it (far too high to have ever been observed) will require new physical theories to describe.

Soundwaves are not electromagnetic radiation. At the lower end of the electromagnetic spectrum, about 20 Hz to about 20 kHz, are frequencies that might be considered in the audio range. However, electromagnetic waves cannot be directly perceived by human ears. Sound waves are instead the oscillating compression of molecules. To be heard, electromagnetic radiation must be converted to pressure waves of the fluid in which the ear is located (whether the fluid is air, water or something else).

Interactions as a Function of Frequency

When EM radiation interacts with matter, its behavior changes qualitatively as its frequency changes.

Radio and Microwave

At radio and microwave frequencies, EMR interacts with matter largely as a bulk collection of charges which are spread out over large numbers of affected atoms. In electrical conductors, such induced bulk movement of charges (electric currents) results in absorption of the EMR, or else separations of charges that cause generation of new EMR (effective reflection of the EMR). An example is absorption or emission of radio waves by antennas, or absorption of microwaves by water or other molecules with an electric dipole moment, as for example inside a microwave oven. These interactions produce either electric currents or heat, or both.

Infrared

Like radio and microwave, infrared also is reflected by metals (as is most EMR into the ultraviolet). However, unlike lower-frequency radio and microwave radiation, Infrared EMR commonly interacts with dipoles present in single molecules, which change as atoms vibrate at the ends of a single chemical bond. It is consequently absorbed by a wide range of substances, causing them to increase in temperature as the vibrations dissipate as heat. The same process, run in reverse, causes bulk substances to radiate in the infrared spontaneously.

Visible Light

As frequency increases into the visible range, photons have enough energy to change the bond structure of some individual molecules. It is not a coincidence that this happens in the "visible range," as the mechanism of vision involves the change in bonding of a single molecule (retinal) which absorbs light in the rhodopsin in the retina of the human eye. Photosynthesis becomes possible in this range as well, for similar reasons, as a single molecule of chlorophyll is excited by a single photon. Animals that detect infrared make use of small packets of water that change temperature, in an essentially thermal process that involves many photons. For this reason, infrared, microwaves and radio waves are thought to damage molecules and biological tissue only by bulk heating, not excitation from single photons of the radiation.

Visible light is able to affect a few molecules with single photons, but usually not in a permanent or damaging way, in the absence of power high enough to increase temperature to damaging levels. However, in plant tissues that conduct photosynthesis, carotenoids act to quench electronically excited chlorophyll produced by visible light in a process called non-photochemical quenching, in order to prevent reactions that would otherwise interfere with photosynthesis at high light levels. Limited evidence indicate that some reactive oxygen species are created by visible light in skin, and that these may have some role in photoaging, in the same manner as ultraviolet A.

Ultraviolet

As frequency increases into the ultraviolet, photons now carry enough energy (about three electron volts or more) to excite certain doubly bonded molecules into permanent chemical rearrangement. In DNA, this causes lasting damage. DNA is also indirectly damaged by reactive oxygen species produced by ultraviolet A (UVA), which has energy too low to damage DNA directly. This is why ultraviolet at all wavelengths can damage DNA, and is capable of causing cancer, and (for UVB) skin burns (sunburn) that are far worse than would be produced by simple heating (temperature increase) effects. This property of causing molecular damage that is out of proportion to heating effects, is characteristic of all EMR with frequencies at the visible light range and above. These properties of high-frequency EMR are due to quantum effects that permanently damage materials and tissues at the molecular level.

At the higher end of the ultraviolet range, the energy of photons becomes large enough to impart enough energy to electrons to cause them to be liberated from the atom, in a process called photoionisation. The energy required for this is always larger than about 10 electron volts (eV) corresponding with wavelengths smaller than 124 nm (some sources suggest a more realistic cutoff of 33 eV, which is the energy required to ionize water). This high end of the ultraviolet spectrum with energies in the approximate ionization range, is sometimes called "extreme UV." Ionizing UV is strongly filtered by the Earth's atmosphere).

X-rays and Gamma Rays

Electromagnetic radiation composed of photons that carry minimum-ionization energy, or more, (which includes the entire spectrum with shorter wavelengths), is therefore termed ionizing radiation. (Many other kinds of ionizing radiation are made of non-EM particles). Electromagnetic-type ionizing radiation extends from the extreme ultraviolet to all higher frequencies and shorter wavelengths, which means that all X-rays and gamma rays qualify. These are capable of the most severe types of molecular damage, which can happen in biology to any type of biomolecule, including mutation and cancer, and often at great depths below the skin, since the higher end of the X-ray spectrum, and all of the gamma ray spectrum, penetrate matter.

Atmosphere and Magnetosphere

Most UV and X-rays are blocked by absorption first from molecular nitrogen, and then (for wavelengths in the upper UV) from the electronic excitation of dioxygen and finally ozone at the mid-range of UV. Only 30% of the Sun's ultraviolet light reaches the ground, and almost all of this is well transmitted.

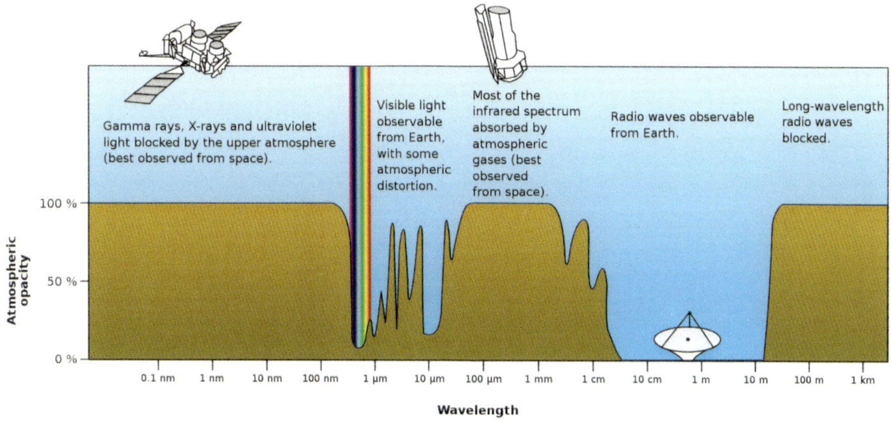

Rough plot of Earth's atmospheric absorption and scattering (or opacity) of various wavelengths of electromagnetic radiation

Visible light is well transmitted in air, as it is not energetic enough to excite nitrogen, oxygen, or ozone, but too energetic to excite molecular vibrational frequencies of water vapor.

Absorption bands in the infrared are due to modes of vibrational excitation in water vapor. However, at energies too low to excite water vapor, the atmosphere becomes transparent again, allowing free transmission of most microwave and radio waves.

Finally, at radio wavelengths longer than 10 meters or so (about 30 MHz), the air in the lower atmosphere remains transparent to radio, but plasma in certain layers of the ionosphere begins to interact with radio waves. This property allows some longer wavelengths (100 meters or 3 MHz) to be reflected and results in shortwave radio beyond line-of-sight. However, certain ionospheric effects begin to block incoming radiowaves from space, when their frequency is less than about 10 MHz (wavelength longer than about 30 meters).

Types and Sources, Classed by Spectral Band

Radio Waves

Radio waves have the least amount of energy and the lowest frequency. When radio waves impinge upon a conductor, they couple to the conductor, travel along it and induce an electric current on the conductor surface by moving the electrons of the conducting material in correlated bunches of charge. Such effects can cover macroscopic distances in conductors (such as radio antennas), since the wavelength of radiowaves is long.

Microwaves

Microwaves are a form of electromagnetic radiation with wavelengths ranging from as long as one meter to as short as one millimeter; with frequencies between 300 MHz (0.3 GHz) and 300 GHz.

Visible Light

Natural sources produce EM radiation across the spectrum. EM radiation with a wavelength between approximately 400 nm and 700 nm is directly detected by the human eye and perceived as visible light. Other wavelengths, especially nearby infrared (longer than 700 nm) and ultraviolet (shorter than 400 nm) are also sometimes referred to as light.

Thermal and Electromagnetic Radiation as a form of Heat

The basic structure of matter involves charged particles bound together. When electromagnetic radiation impinges on matter, it causes the charged particles to oscillate and gain energy. The ultimate fate of this energy depends on the context. It could be immediately re-radiated and appear as scattered, reflected, or transmitted radiation. It may get dissipated into other microscopic motions within the matter, coming to thermal equilibrium and manifesting itself as thermal energy, or even kinetic energy, in the material. With a few exceptions related to high-energy photons (such as fluorescence, harmonic generation, photochemical reactions, the photovoltaic effect for ionizing radiations at far ultraviolet, X-ray and gamma radiation), absorbed electromagnetic radiation simply deposits its energy by heating the material. This happens for infrared, microwave and radio wave radiation. Intense radio waves can thermally burn living tissue and can cook food. In addition to infrared lasers, sufficiently intense visible and ultraviolet lasers can easily set paper afire.

Ionizing radiation creates high-speed electrons in a material and breaks chemical bonds, but after these electrons collide many times with other atoms eventually most of the energy becomes thermal energy all in a tiny fraction of a second. This process makes ionizing radiation far more dangerous per unit of energy than non-ionizing radiation. This caveat also applies to UV, even though almost all of it is not ionizing, because UV can damage molecules due to electronic excitation, which is far greater per unit energy than heating effects.

Infrared radiation in the spectral distribution of a black body is usually considered a form of heat, since it has an equivalent temperature and is associated with an entropy change per unit of thermal energy. However, "heat" is a technical term in physics and thermodynamics and is often confused with thermal energy. Any type of electromagnetic energy can be transformed into thermal energy in interaction with matter. Thus, *any* electromagnetic radiation can "heat" (in the sense of increase the thermal energy temperature of) a material, when it is absorbed.

The inverse or time-reversed process of absorption is thermal radiation. Much of the thermal energy in matter consists of random motion of charged particles, and this energy can be radiated away from the matter. The resulting radiation may subsequently be absorbed by another piece of matter, with the deposited energy heating the material.

The electromagnetic radiation in an opaque cavity at thermal equilibrium is effectively a form of thermal energy, having maximum radiation entropy.

Biological Effects

Bioelectromagnetics is the study of the interactions and effects of EM radiation on living organisms. The effects of electromagnetic radiation upon living cells, including those in humans, depends upon the radiation's power and frequency. For low-frequency radiation (radio waves to visible light) the best-understood effects are those due to radiation power alone, acting through heating when radiation is absorbed. For these thermal effects, frequency is important only as it affects penetration into the organism (for example, microwaves penetrate better than infrared). It is widely accepted that low frequency fields that are too weak to cause significant heating could not possibly have any biological effect.

Despite the commonly accepted results, some research has been conducted to show that weaker *non-thermal* electromagnetic fields, (including weak ELF magnetic fields, although the latter does not strictly qualify as EM radiation), and modulated RF and microwave fields have biological effects. Fundamental mechanisms of the interaction between biological material and electromagnetic fields at non-thermal levels are not fully understood.

The World Health Organization has classified radio frequency electromagnetic radiation as Group 2B - possibly carcinogenic. This group contains possible carcinogens such as lead, DDT, and styrene. For example, epidemiological studies looking for a relationship between cell phone use and brain cancer development, have been largely inconclusive, save to demonstrate that the effect, if it exists, cannot be a large one.

At higher frequencies (visible and beyond), the effects of individual photons begin to become important, as these now have enough energy individually to directly or indirectly damage biological molecules. All UV frequences have been classed as Group 1 carcinogens by the World Health Organization. Ultraviolet radiation from sun exposure is the primary cause of skin cancer.

Thus, at UV frequencies and higher (and probably somewhat also in the visible range), electromagnetic radiation does more damage to biological systems than simple heating predicts. This is most obvious in the "far" (or "extreme") ultraviolet. UV, with X-ray and gamma radiation, are referred to as ionizing radiation due to the ability of photons of this radiation to produce ions and free radicals in materials (including living tissue). Since such radiation can severely damage life at energy levels that produce little heating, it is considered far more dangerous (in terms of damage-produced per unit of energy, or power) than the rest of the electromagnetic spectrum.

Use as Weapon

The heat ray is an application of EMR that makes use of microwave frequencies to create an unpleasant heating effect in the upper layer of the skin. A publicly known heat ray weapon called the Active Denial System was developed by the US military as an experimental weapon to deny the enemy access to an area. A death ray is a weapon that delivers heat ray electromagnetic energy at levels that injure human tissue. The inventor of the death ray, Harry Grindell Matthews, claims to have lost sight in his left eye while developing his death ray weapon based on a primitive microwave magnetron from the 1920s (note that a typical microwave oven induces a tissue damaging cooking effect inside the oven at about 2 kV/m).

Derivation from Electromagnetic Theory

Electromagnetic waves were predicted by the classical laws of electricity and magnetism, known as Maxwell's equations. Inspection of Maxwell's equations without sources (charges or currents) results in nontrivial solutions of changing electric and magnetic fields. Beginning with Maxwell's equations in free space:

$$\nabla \cdot \mathrm{E} = 0$$

$$\nabla \times E = -\frac{\partial B}{\partial t}$$

$$\nabla \cdot B = 0$$

$$\nabla \times B = \mu_0 \epsilon_0 \frac{\partial E}{\partial t}$$

where

∇ is a vector differential operator.

In Cartesian coordinate system ∇ is defined in terms of partial derivative operators as $\nabla = \tilde{y}\frac{\partial}{\partial x} + \hat{\jmath}\frac{\partial}{\partial y} + \hat{k}\frac{\partial}{\partial z}$.

One solution,

$$E = B = 0,$$

is trivial.

For a more useful solution, we utilize vector identities, which work for any vector, as follows:

$$\nabla \times (\nabla \times A) = \nabla(\nabla \cdot A) - \nabla^2 A$$

The curl of second equation:

$$\nabla \times (\nabla \times E) = \nabla \times \left(-\frac{\partial B}{\partial t}\right)$$

Evaluating the left hand side:

$$\nabla \times (\nabla \times E) = \nabla(\nabla \cdot E) - \nabla^2 E = -\nabla^2 E$$

simplifying the above by using first equation .

Evaluating the right hand side:

$$\nabla \times \left(-\frac{\partial B}{\partial t}\right) = -\frac{\partial}{\partial t}(\nabla \times B) = -\mu_0 \epsilon_0 \frac{\partial^2 E}{\partial t^2}$$

Sixth and seventh equations are equal, so this results in a vector-valued differential equation for the electric field, namely,

$$\nabla^2 E = \mu_0 \epsilon_0 \frac{\partial^2 E}{\partial t^2}$$

Applying a similar pattern results in similar differential equation for the magnetic field:

$$\nabla^2 B = \mu_0 \epsilon_0 \frac{\partial^2 B}{\partial t^2}.$$

These differential equations are equivalent to the wave equation:

$$\nabla^2 f = \frac{1}{c_0^{\,2}} \frac{\partial^2 f}{\partial t^2}$$

where

c_0 is the speed of the wave in free space and

f describes a displacement

Or more simply:

$$\Box f = 0$$

where $\Box$ is d'Alembertian:

$$\Box = \nabla^2 - \frac{1}{c_0^{\,2}} \frac{\partial^2}{\partial t^2} = \frac{\partial^2}{\partial x^2} + \frac{\partial^2}{\partial y^2} + \frac{\partial^2}{\partial z^2} - \frac{1}{c_0^{\,2}} \frac{\partial^2}{\partial t^2}$$

In the case of the electric and magnetic fields, the speed is:

$$c_0 = \frac{1}{\sqrt{\mu_0 \epsilon_0}}$$

This is the speed of light in vacuum. Maxwell's equations unified the vacuum permittivity ϵ_0, the vacuum permeability and μ_0 the speed of light itself, c_0. This relationship had been discovered by Wilhelm Eduard Weber and Rudolf Kohlrausch prior to the development of Maxwell's electrodynamics, however Maxwell was the first to produce a field theory consistent with waves traveling at the speed of light.

These are only two equations versus the original four, so more information pertains to these waves hidden within Maxwell's equations. A generic vector wave for the electric field.

$$E = E_0 f\left(\hat{k} \cdot x - c_0 t\right),$$

Here, E_0 is the constant amplitude, f is any second differentiable function, $\hat{k}$ is a unit vector in the direction of propagation, and x is a position vector. $f\left(\hat{k} \cdot x - c_0 t\right)$ is a generic solution to the wave equation. In other words,

$$\nabla^2 f\left(\hat{k} \cdot x - c_0 t\right) = \frac{1}{c_0^{\,2}} \frac{\partial^2}{\partial t^2} f\left(\hat{k} \cdot x - c_0 t\right),$$

for a generic wave traveling in the $\hat{k}$ direction.

This form will satisfy the wave equation.

$$\nabla \cdot E = \hat{k} \cdot E_0 f'\left(\hat{k} \cdot x - c_0 t\right) = 0$$

$$E \cdot \hat{k} = 0$$

The first of Maxwell's equations implies that the electric field is orthogonal to the direction the wave propagates.

$$\nabla \times E = \hat{k} \times E_0 f'\left(\hat{k} \cdot x - c_0 t\right) = -\frac{\partial B}{\partial t}$$

$$B = \frac{1}{c_0} \hat{k} \times E$$

The second of Maxwell's equations yields the magnetic field. The remaining equations will be satisfied by this choice of E, B.

The electric and magnetic field waves in the far-field travel at the speed of light. They have a special restricted orientation and proportional magnitudes, $E_0 = c_0 B_0$, which can be seen immediately from the Poynting vector. The electric field, magnetic field, and direction of wave propagation are all orthogonal, and the wave propagates in the same direction as $E \times B$. Also, E and B far-fields in free space, which as wave solutions depend primarily on these two Maxwell equations, are in-phase with each other. This is guaranteed since the generic wave solution is first order in both space and time, and the curl operator on one side of these equations results in first-order spatial derivatives of the wave solution, while the time-derivative on the other side of the equations, which gives the other field, is first-order in time, resulting in the same phase shift for both fields in each mathematical operation.

From the viewpoint of an electromagnetic wave traveling forward, the electric field might be oscillating up and down, while the magnetic field oscillates right and left. This picture can be rotated with the electric field oscillating right and left and the magnetic field oscillating down and up. This is a different solution that is traveling in the same direction. This arbitrariness in the orientation with respect to propagation direction is known as polarization. On a quantum level, it is described as photon polarization. The direction of the polarization is defined as the direction of the electric field.

More general forms of the second-order wave equations given above are available, allowing for both non-vacuum propagation media and sources. Many competing derivations exist, all with varying levels of approximation and intended applications. One very general example is a form of the electric field equation, which was factorized into a pair of explicitly directional wave equations, and then efficiently reduced into a single uni-directional wave equation by means of a simple slow-evolution approximation.

Sources of Electromagnetic Fields

Natural sources

Electromagnetic fields are present everywhere in our environment but are invisible to the human eye. Electric fields are produced by the local build-up of electric charges in the atmosphere associated with thunderstorms. The earth's magnetic field causes a compass needle to orient in a North-South direction and is used by birds and fish for navigation. The Earth's magnetic field plays an important role for all living things. Without a geomagnetic field, life on this planet would not be possible. Impressive examples such as tides show us how powerful geomagnetic influences work on water.

Because humans consist largely of water, they need an intact natural geomagnetic field. The communication of living beings with the natural geomagnetic field is influenced by artificial (electrosmog) and natural disturbance fields (earth rays). Inside buildings, distortions of the geomagnetic field are detected again and again. Reasons for this are either the use of iron and steel for the ceiling construction or natural (geopathic) fault zones and technical interferences (electro smog).

If a human being spends an extended period of time in a distorted earth magnetic field (sleeping, learning, work area, etc.) adhered biological stress arises, which has a negative effect on cell communication in the human body - the immune system is weakened.

Human-made Sources

Besides natural sources the electromagnetic spectrum also includes fields generated by human-made sources: X-rays are employed to diagnose a broken limb after a sport accident. The electricity that comes out of every power socket has associated low frequency electromagnetic fields and various kinds of higher frequency radio waves are used to transmit information – whether via TV antennas, radio stations or mobile phone base stations. Exposure to human-made electromagnetic fields (EMF) has increased over the past century. The widespread use of EMF sources has been accompanied by public debate about possible adverse effects on human health.

Mathematical Descriptions of the Electromagnetic Field

Mathematical frameworks for representing Fields and waves and expressing Maxwell's equations of electromagnetism include vector calculus, differential forms, dyadics, bivectors, tensors, quaternions, and Clifford algebras. Vector notation is by far the most widely used, particularly in applications. Of the more sophisticated notations, differential forms stand out as being close enough to vectors that most practitioners can readily understand the notation, yet at the same time offering unique visualization tools and graphical insight into the behavior of fields and waves.

Since James Clerk Maxwell's discovery of the full set of mathematical laws that govern electromagnetic fields, other mathematicians, physicists, and engineers have proposed a surprisingly large number of mathematical frameworks for representing fields and waves and working with electromagnetic theory.

Differential Forms

The scalar and vector fields used in electromagnetic theory may be represented by exterior differential forms. Differential forms are an extension of the vector concept. The use of differential forms does not necessarily replace vector analysis. The differential form representation supplies additional physical insight in addition to the conventional vector picture, and once one is familiar with the notation, it is easy to translate back and forth between vectors and differential forms.

One way to understand the difference between vector fields and differential forms is to contrast the two possible interpretations of the electric field as a force on a test charge (force picture, represented by vectors) or the change in energy experienced by the test charge as it moves through the field (energy picture, represented by differential forms). The two interpretations are not in contradiction, but complement each other in jointly providing a deep understanding of a fundamental physical property of the universe.

The energy W_{21} required to move a particle with electric charge q in an electric field with components E_x, E_y, and E_z, from point r_1 to r_2 is obtained by the integral

$$W_{21}(t) = -q\int_{r_1}^{r_2} \bar{E}.dl$$

Figure below shows the path of integration for the definition of the voltage v_{21} from node 2 to node 1. The vector path integral sums up the projection of the field vector on the vectorial path element. The contribution of the integrand over each small length of the integration path is proportional to the product of the magnitudes of field vector with the infinitesimal path element and the cosine of the angle enclosed between them.

Instead of using a vector, the electric field can be represented by a linear combination of differentials, according to

$$\varepsilon = Ex(x, y, z, t)dx + Ey(x, y, z, t)dy + Ez(x, y, z, t)dz$$

The object on the right hand side is known as a one-form, and has three independent components, similar to a vector. The key difference is that the unit vectors $\hat{x}$, $\hat{y}$, and $\hat{z}$ are replaced by differentials dx, dy and dz. The energy can be expressed in terms of the differential form as

$$W_{21}(t) = -q\int_{r_1}^{r_2} \varepsilon$$

which is simpler than the vector line integral, as there is no dot product or differential line element. In a sense, the dot product in eq. second converts the vector into a differential that can be integrated. The integrand in eq. second is already given in terms of differentials, so the dot product is not needed. In this sense, the differential form representation is more natural than the vector representation. In the differential forms notation, the electric field is represented by a one-form. A one-form is a linear combination of differentials of each coordinate. An arbitrary one-form can be written

$$A = A_1 dx + A_2 dy + A_3 dz:$$

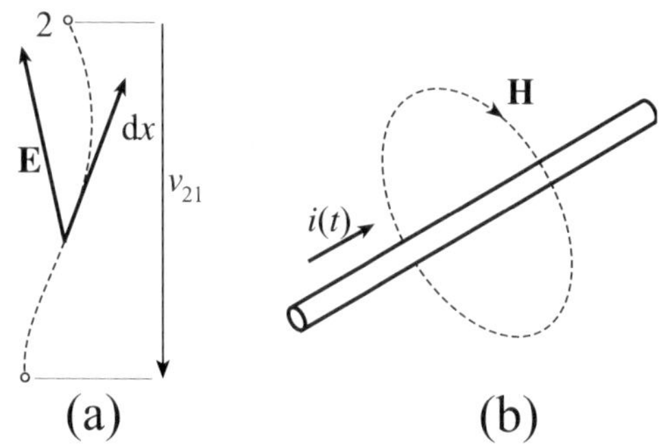

Path of integration, (a) for the definition of the voltage and (b) for the definition of the current.

The three quantities A_1, A_2, and A_3 are the components of the one-form. Two one-forms A and B can be added, so that

$$A + B = (A_1 + B_1)dx + (A_2 + B_2)dy + (A_3 + B_3)dz$$

One-forms can be integrated over paths. As shown in the introduction, we graphically represent a one-form as surfaces. The one-form d x has surfaces perpendicular to the x -axis spaced a unit distance apart. These surfaces are infinite in the y and z directions. The integral of dx over a path from the point (0, 0, 0) to (4, 0, 0) is,

$$\int_0^4 dx = 4$$

This matches the graphical representation in figure(a), since the path shown in the figure crosses four surfaces. If the path were not of integer length, we would have to imagine fractional surfaces in between the unit spaced surfaces. A path from (0, 0, 0) to (0.25, 0, 0), for example, crosses 0.25 surfaces.

We can also think of d x as a one-form in the plane. In this case, the picture becomes a series of lines perpendicular to the x -axis spaced a unit distance apart, as shown in figure(b). Graphically, integrals in the plane are similar to integrals in three dimensions: the value of a path integral is the number of lines pierced by the path.

In order to graphically integrate a one-form properly, we also have to think of the surfaces as having an orientation. The integral of the one-form -dx over a path from (0, 0, 0) to (4, 0, 0) is -4. Thus, when we count surfaces pierced by a path, we have to compare the sign of the one-form with the direction of the path in order to determine whether the surface contributes positively or negatively. The orientation of surfaces can be indicated using an arrowhead on each surface, but since the orientation is usually clear from context, to reduce clutter we do not indicate it in figures.

A more complicated one-form, such as 3d x + 5d y , has surfaces that are oblique to the coordinate axes. This one-form is shown in figure below. The greater the magnitude of the components of a one-form, the closer the surfaces are spaced. For one-forms with components that are not constant, these surfaces can be curved. The surfaces can also originate along a line or curve and extend away to infinity, or the surfaces may be infinite. In this case, the integral over a path is still the number of surfaces or fractional surfaces pierced by the path. There are one-forms that cannot be drawn as surfaces in three dimensions, but in most practical situations, the surface picture can be used.

A one-form represents a quantity which is integrated over a path, whereas vector represents a quantity with a magnitude and direction, such as displacement or velocity. Despite this difference, both types of quantities have three independent components, and can be used interchangeably in describing electromagnetic field quantities. Mathematically, vectors and differential forms are closely related. In Euclidean coordinates, we can make a correspondence between vectors and forms. The one-form A and the vector A are equivalent if they have the same components:

$$A_1 dx + A_2 dy + A_3 dz \leftrightarrow A_1\hat{x} + A_2\hat{y} + A_3\hat{z}.$$

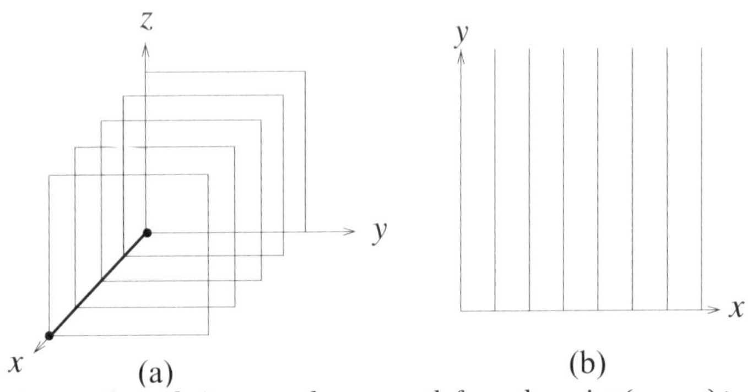

(a) (b)

(a) The one-form dx integrated over a path from the point (0, 0, 0) to (4, 0, 0). (b) The one-form dx in the plane.

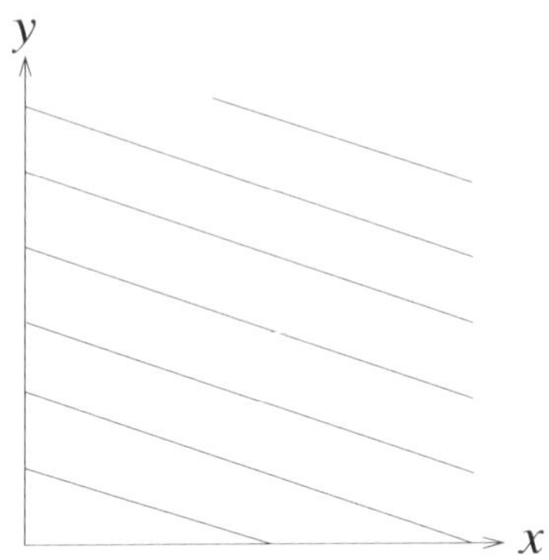

The one-form 3dx + 5dy.

We say that the one-form A and the vector A are dual. Since it is easy to convert between the differential form and vector representations, one can choose the quantity which best suits a particular problem.

The common physical interpretation of the electric field is related to the force on a point-like unit charge. This force picture yields in a natural way to the vector representation and to the visualization of the electric field via field lines. Another viewpoint is to consider the energy of a charge moved through the field. We can visualize the field via the change of the energy of a test charge moved through the field. This energy picture is more related to differential forms. Figure (a) shows the representation of the field via the surfaces of constant test charge energy or constant electric potential. For an electrostatic field the surfaces associated with the one-form $\mathcal{E}$ are equipotentials. The voltage between two points 2 and 1 is given by,

$$V_{21} = -\int_1^2 \mathcal{E}$$

For a static electric field, the surfaces of the one-form never meet or end. In general the surfaces of a one- form also may end or meet each other, which corresponds to fields produced by sources or time-varying fields.

The magnetic field intensity can be represented also by a one-form according to

$$H = H_x(x, y, z, t)d_x + H_y(x, y, z, t)dy + H_z(x, y, z, t)dz$$

The dimension of the differential form $\mathcal{E}$ is V and H has the dimension A. The coefficients Ex, E_y, and E_z have units of V/m, similar to the electric field intensity vector. The differentials dx, dy, and dz have units of length, which combines with the coefficients so that the one-form $\mathcal{E}$ has units of V. Mathematically, the differential forms $\mathcal{E}$ and H express the changes of the electric and magnetic potentials over an infinitesimal path element.

Figure below shows a situation we encounter in time variable fields. In the center of the structure the field intensity is higher than at its edges. New one-form surfaces appear as we move closer to the center

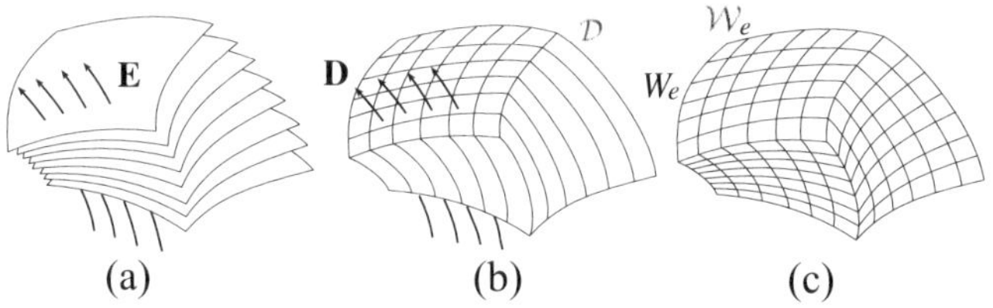

Geometric representation of (a) a one-form, (b) a two-form, (c) a three-form. The one- form is represented by surfaces, the two-form as tubes of flux or flow, and the three-form as cubes representing volumetric density.

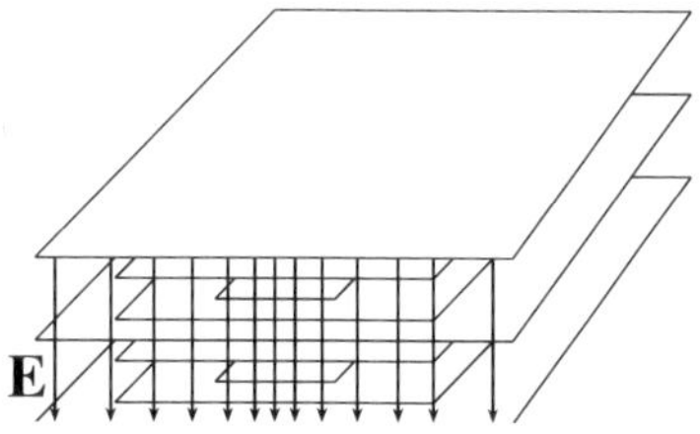

One-form with ending surfaces representing a nonconservative field.

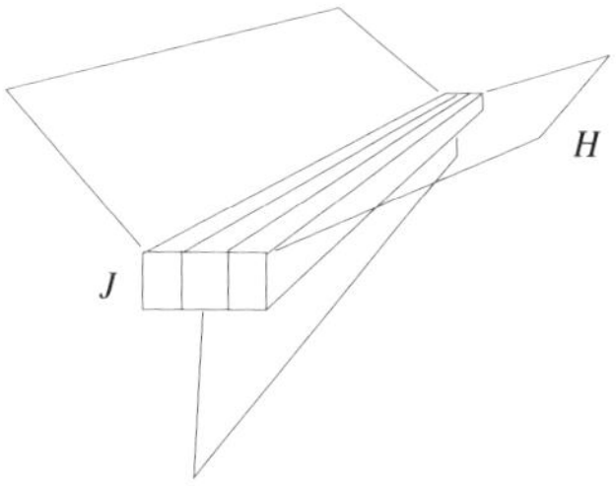

A graphical representation of the physical behavior of magnetic field and electric current: tubes of J produce surfaces of H.

of the structure. In this case the integral will depend on the path from x_1 to x_2 and we cannot assign a scalar potential to the field. This type of field is sometimes referred to as nonconservative. With the vector picture, it can be challenging to determine if a particular field is nonconservative, but with the one-form surface representation, it is visually obvious.

The magnetic field produced by a line current behaves similarly, as new one-form surfaces of H are created by the presence of an electric current. one-form surfaces radiate outward from the line current. The integration path passes through a number of one-form surfaces equal to the amount of current flowing through the path. The differential forms picture of the line current and associated magnetic field is shown in figure above. The surfaces of H emanate from the tubes of electric current J . With the vector picture, the vector field corresponding to the magnetic field intensity appear to rotate around the line current everywhere in space, but the curl of the vector field is actually zero everywhere except at the location of the current.

The differential forms picture makes it clear that the surfaces of the magnetic field intensity one-form are created by the tubes of current.

Integrating One-forms Over Paths

The laws of electromagnetics are expressed in terms of integrals of fields represented by differential forms. To apply these laws, we must be able to compute the values of integrals of differential forms. Since one-forms are by definition mathematical quantities which are integrated over paths, the process of evaluating an integral of a one-form is quite natural. The key idea is that we can replace the coordinates x, y and z (or μ, v, w in curvilinear coordinates) with a parametric equation for the integration path.

A parametric representation of an integration path has the form $[f(t), g(t), h(t)]$, so that the functions f, g, and h give the coordinates of a point on the path for each value of the parameter t. We replace the coordinates in a one-form by these functions, and then the integral can be evaluated. For a differential, when the coordinate is replaced by a function defining the path, we then take the derivative by t to produce a new differential in the variable t. For example, dx becomes $df = f'(t)dt$, where the prime denotes the derivative of the function $f(t)$ by t. The differential form now has a single differential, dt, and the integral can be performed using standard rules of calculus. Two-forms can be integrated similarly, except that the integration surface requires two parameters, and the final integrand over the parametric surface is a double integral.

Two-forms, Three-forms, and the Exterior Product

To introduce the concept of higher order differential forms, let us consider the current i flowing in x-direction through the surface A in Figure. To compute the current we have to integrate the x-component J_x of the current density over the surface A in the yz-plane

$$i = \int_A J_x \mathrm{dydz}.$$

If we integrate a current density over an area we have to consider the orientation of the area. If in Figure the current density J_x is positive, the current i also will be positive. Inverting the direction of J_x will yield a negative current. This inversion may be performed by mirroring the coordinates with respect to the yz-plane. How do we know whether a surface integral is positive or negative? The answer is: we have to define a positive orientation. A positive oriented or right-handed Cartesian coordinate system is specified as follows: if we are looking in z-direction on the xy-plane the x-axis may be rotated clockwise by 90± into the y-axis. In Figure the vector component J_x is pointing in positive orientation. In Figure the coordinate system as well as the vector field were rotated by 180± around the z-axis. Physically nothing has changed. In the left figure, the vector pointing towards the observer is positive, whereas in the right figure the vector pointing away from the observer is positive.

Exterior differential forms allow one to represent the orientation of a coordinate system in a rigorous way. We introduce the exterior product or wedge product $dy \wedge dz$ with the property

$$dy \wedge dz = -dz \wedge dy$$

An exterior differential form is the exterior product of differential forms. Exterior differential forms consisting wedge products of two differentials or sums of such products are called two-forms. We may decide either $dy \wedge dz = dydz$ or $dy \wedge dz = -dydz$. Deciding that,

$$dy \wedge dz = dydz$$

assigns to $dy \wedge dz$ the positive orientation and to $dz \wedge dy$ the negative orientation. The integral can now be written in the orientation-independent form

$$i = \int_A J_x \, dy \wedge dz$$

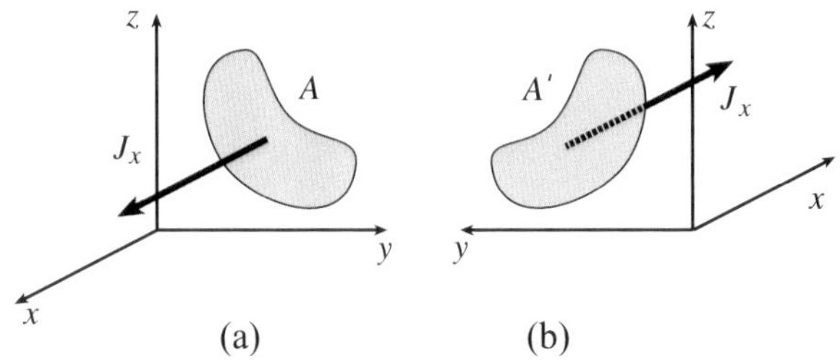

The orientation of an area.

Physically, a two-form is a quantity which is integrated over a two-dimensional surface. A quantity representing flow of a fluid, for example, has units of flow rate per area, and would be integrated over a surface to find the total flow rate through the surface. Similarly, the integral of electric flux density over a surface is the total flux through the surface, which has units of charge. A general two-form C is written as,

$$C = C_1 dy \wedge dz + C_2 dz \wedge dx + C_3 dx \wedge dy$$

For convenience, we usually use the antisymmetry of the exterior product to put the differentials of two-forms into right cyclic order.

We can use the exterior product to create differential forms of higher order as well. An exterior differential form of order p is called a *p-form* and has p differentials in each term. In n-dimensional space the order of a differential form may assume values $0,\ldots,n$. In differential form notation a clear distinction between scalars, pseudoscalars, polar vectors and axial vectors is made. Scalars are represented by zero-forms, pseudoscalars by three-forms, polar vectors by one-forms and axial vectors by two-forms. For a p-form μ and a q-form V the commutation relation for the exterior product is,

$$\mu \wedge v = (-1)^{p+q+1} v \wedge \mu$$

Figure below shows the fundamental 1-, 2- and three-forms in Cartesian coordinates.

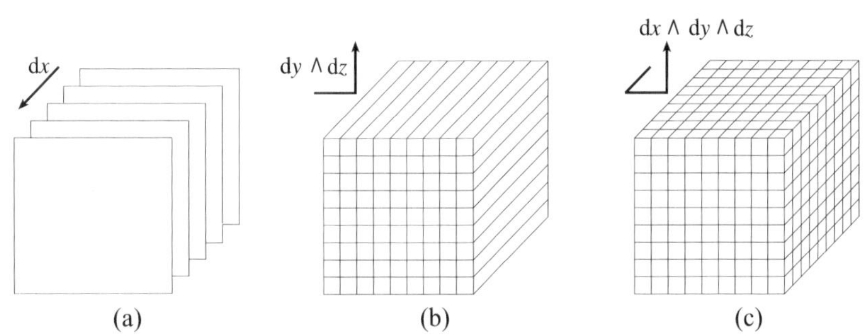

The fundamental (a) one-form, (b) two-form, (c) three-form in Cartesian coordinates.

Current Density Two-form

We can describe the flow of current in a conductor by a current density vector field J(x) = [$J_x(x)\ J_y(x)\ J_z(x)]^T$. The current flows along the current density field lines going through the boundary ∂A of the area A as shown in Figure. In the differential forms representation, the current is depicted as tubes of flow. Figure shows the tube representation of a two-form. The two-form is visualized as a bundle of tubes carrying the current. Each tube carries a unit amount of current, and the current density is inversely proportional to the cross-sectional area of the tubes. Figure shows the tube representations of the fundamental two-form $dy \wedge dz$. The tubes are made up of the one-form surfaces of dy and dz.

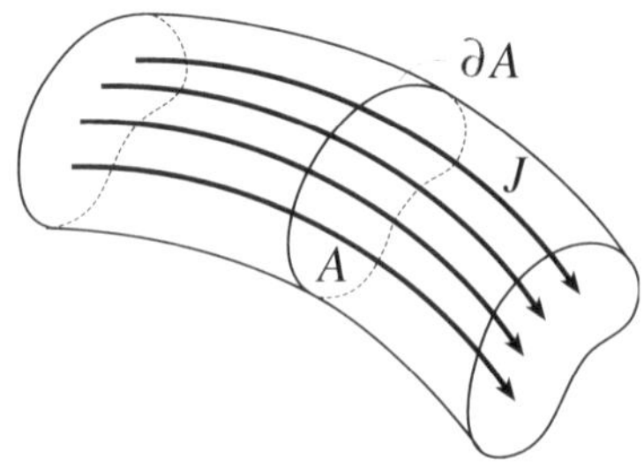

Current flow.

If the surface S is an arbitrarily oriented curved surface in three-dimensional space and the current density vector has the x-, y- and z-components J_x, J_y and J_z, the total current flowing through S is,

$$i = \int_S J_x dy \wedge dz + Jydz \wedge dx + J_z dx \wedge dy$$

The first term of the integrand concerns the integration of the x-component of the current density over the projection of the surface S on the yz-plane and so forth.

Let us introduce the current density two-form J

$$J = J_x dy \wedge dz + J_y dz \wedge dx + J_z dx \wedge dy$$

The total current i Flowing through S may be expressed in a compact notation as the integral of the differential form J

$$i = \int_S J$$

The dimension of the current density differential form J is A. The electric flux density two-form D has the dimension As, and the magnetic flux density two-form B has the dimension Vs. These differential forms represent the current or the flux through an infinitesimal area element.

Charge Density Three-form

The electric charge q is given by the volume integral of the electric charge density ρ. For the electric charge density we introduce the electric charge density three-form,

$$Q = \rho dx \wedge dy \wedge dz$$

Figure shown above the graphic visualization of a three-form by subdividing the volume into cells. The cell volume is inversely proportional to the charge density. The charge density form $\mathcal{Q}$ with the dimension As represents the charge in an infinitesimal volume element. We obtain the total charge q by performing the volume integral over the three-form $\mathcal{Q}$:

$$q = \int_V \mathcal{Q}$$

Electromagnetic Fields and Sources

Tables below summarize the p-forms that describe key electromagnetic Field and source quantities. The electric and magnetic fields are represented by both a one-form and a two-form. The one-form and two-form representations capture different physical properties of the field. The one-form represents the energy picture and shows the change in potential energy as a test charge moves across the one-form surfaces. The two-form shows the flux of the field that extends from positive charges to negative charges.

Table. Differential forms that represent electromagnetic fields and sources with units and the corresponding vector quantities.

Quantity	Form	Type	Units	Vector/Scalar
Electric Field Intensity	$\mathcal{E}$	one-form	V	E
Magnetic Field Intensity	$\mathcal{H}$	one-form	A	H
Electric Flux Density	$\mathcal{D}$	two-form	C	D
Magnetic Flux Density	$\mathcal{B}$	two-form	Wb	B
Electric Current Density	$\mathcal{J}$	two-form	A	J
Electric Charge Density	$\mathcal{Q}$	three-form	C	ρ

Table. Differential forms of electromagnetism expanded in components.

$$\mathcal{E} = E_x\,dx + E_y\,dy + E_z\,dz$$

$$\mathcal{H} = H_x\,dx + H_y\,dy + H_z\,dz$$

$$\mathcal{D} = D_x\,dy \wedge dz + D_y\,dz \wedge dx + D_z\,dx \wedge dy$$

$$\mathcal{B} = B_x\,dy \wedge dz + B_y\,dz \wedge dx + B_z\,dx \wedge dy$$

$$\mathcal{J} = J_x\,dy \wedge dz + J_y\,dz \wedge dx + J_z\,dx \wedge dy$$

$$\mathcal{S} = S_x\,dy \wedge dz + S_y\,dz \wedge dx + S_z\,dx \wedge dy$$

$$\mathcal{Q} = \rho dx \wedge dy \wedge dz$$

Hodge Star Operator and the Constitutive Relations

Since the electric field intensity $\mathcal{E}$ and the electric flux density $\mathcal{D}$ represent the same field, it is clear that there must be some relationship between $\mathcal{E}$ and $\mathcal{D}$. This relationship is expressed using the Hodge star operator. Simply stated, the star operator works by taking a differential form and converting it to a new form with the \missing" differentials. For one-forms and two-forms,

$$*dx = dydz, \; *dydz = dx$$
$$*dy = dzdx, \; *dzdx = dy$$
$$*dz = dxdy, \; *dxdy = dz$$

The one-form $\mathcal{A}$ and the two-form ? $*\mathcal{A}$ are both dual to the same vector. The star operator applied twice is the identity operator, so that $*\mathcal{A}\mathcal{A} = \mathcal{A}$ for any form $\mathcal{A}$.

The star operator has a simple graphical interpretation. Considering the fundamental one-form dx, the surfaces of dx are perpendicular to the x direction. Applying the star operator gives $dydz$, which has tubes in the x direction, so that the surfaces of dx are perpendicular to the tubes of ? dx. In general, the star operator acting on one-forms and two-forms converts surfaces into perpendicular tubes and tubes into perpendicular surfaces. This is illustrated in fgure below.

The relation between the electric field intensity one-form $\mathcal{E}$ and the electric flux density $\mathcal{D}$ depends on the material properties of the medium and the metric properties of space. In an isotropic medium, the electric field intensity vector and the flux density vector are proportional, and it is not always clear why the two different quantities are needed for the electric field. With differential forms notation,

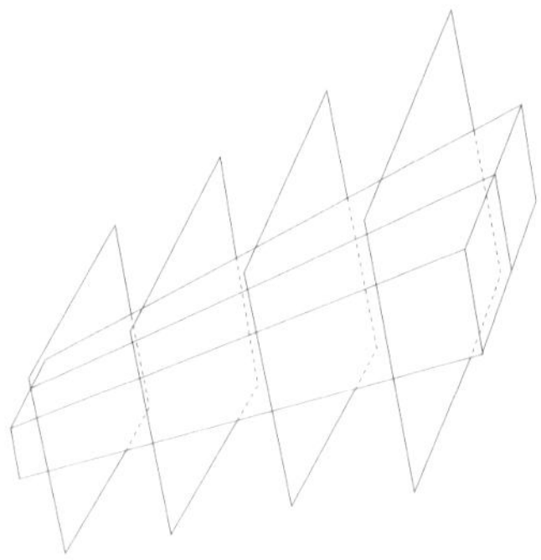

The star operator converts one-form surfaces into perpendicular two-form tubes.

field intensity and flux density are represented by different mathematical objects. The mathematical relationship between $\mathcal{E}$ and $\mathcal{D}$ is known as a constitutive relation. The constitutive relation is written using the Hodge star operator as

$$\mathcal{D} = \epsilon * \mathcal{E}$$

where $\in$ is the permittivity of a medium. The permittivity ϵ accounts for the material properties whereas the Hodge operator $*$ accounts for the metric properties of the space or the chosen coordinate system, respectively. This equation shows that surfaces of $\mathcal{E}$ yield perpendicular tubes of $\mathcal{D}$. The constant scales the sizes of the tubes without changing their direction. The magnetic field constitutive relation is

$$\mathcal{B} = \mu * \mathcal{H}$$

where μ is the permeability of the medium. The differential forms notation helps to elucidate why two different representations are used for one physical field. With vector notation, flux and field intensity are both vectors, but with differential forms, they become one-forms and two-forms, with different geometrical representations and physical interpretations.

Maxwell's Equations in Integral Form

The integral form of Maxwell's equations is given by:

$$\oint_{\partial s} \mathcal{H} = \frac{d}{dt}\int_s D + \int_S \mathcal{J} \qquad \text{Amp`ere's Law}$$

$$\oint_{\partial s} \varepsilon = -\frac{d}{dt}\int_S \mathcal{B}, \qquad \text{Faraday's Law}$$

$$\oint_{\partial V} \mathcal{B} = 0 \qquad \text{Magnetic Flux Continuity}$$

$$\oint_{\partial V} \mathcal{D} = \int_V \mathcal{Q}. \qquad \text{Gauss' Law}$$

Amp`ere's law relates current to magnetic field. It states that the sum of conduction and displacement currents through an area S equals the magnetic tension around the boundary ∂ S. Faraday's law relates the magnetic flux to the electric field. It says that the time derivative of the magnetic flux through an area S equals the electric tension around the boundary ∂ S. Both equations are tied together via the constitutive equations relating Flux densities to field intensities.

Point Charge

As a simple example of the application of Maxwell's laws in integral form, we consider the electric flux produced by a point charge. The tubes of flux from a point charge Q extend out radially from the

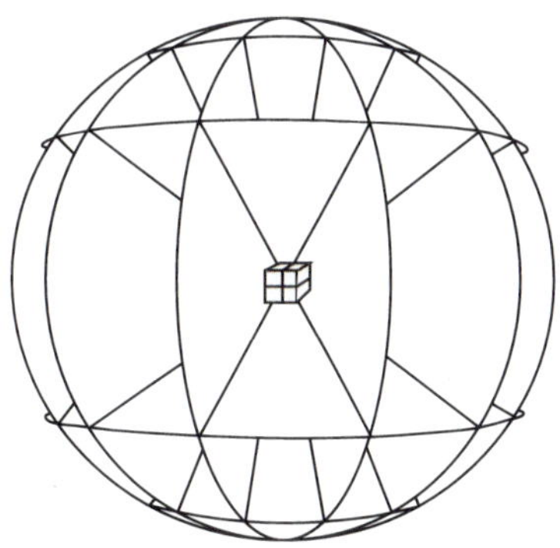

Electric flux density due to a point charge. Tubes of D extend out radially from the charge.

charge. Thus, the Flux density two-form $\mathcal{D}$ has to be a multiple of the differentials $d\theta d\phi$ in the spherical coordinate system. The tubes of $d\theta d\phi$ are denser near the poles $\theta = o$ and $\theta = \pi$ than at the equator. We need to include the correction factors $r^2 \sin è$ to have tubes with the same density everywhere in space. Thus, $\mathcal{D}$ has the form

$$\mathcal{D} = D_0 r^{2} \ sin \ \theta d\theta d\varphi$$

where $\mathcal{D}_o$ is a constant.

To apply Gauss law, we will choose the surface S to be a sphere around the charge. The right-hand side is

$$\int_V \mathcal{Q} = \mathcal{Q}$$

where V is the volume inside the sphere. The left-hand side is

$$\int_S D = \int_0^{2\pi} \int_0^{\pi} D_0 r^2 \sin\theta d\theta d\phi = 4\pi r^2 D_0$$

By Gauss's law, we know that $4\pi r^2 D_0 = Q$. Solving for D_0 and substituting into, we obtain

$$\mathcal{D} = \frac{Q}{4\pi} \sin\theta d\theta d\phi$$

for the electric flux density due to the point charge. Since 4π is the total amount of solid angle for a sphere and sin $\sin\theta d\theta d\phi$ is the differential element of solid angle, this expression implies that the amount of flux per solid angle is the same at any distance from the charge.

Potential Field Approach

Many times in the use and calculation of electric and magnetic fields, the approach used first computes an associated potential: the electric potential, φ, for the electric field, and the magnetic potential, **A**, for the magnetic field. The electric potential is a scalar field, while the magnetic potential is a vector field. This is why sometimes the electric potential is called the scalar potential and the magnetic potential is called the vector potential. These potentials can be used to find their associated fields as follows:

$$E = -\nabla\varphi - \frac{\partial A}{\partial t}$$

$$B = \nabla \times A$$

Maxwell's Equations in Potential Formulation

These relations can be substituted into Maxwell's equations to express the latter in terms of the potentials. Faraday's law and Gauss's law for magnetism reduce to identities (e.g., in the case of Gauss's Law for magnetism, 0 = 0). The other two of Maxwell's equations turn out less simply.

Maxwell's equations (*potential formulation*)

$$\nabla^2\varphi + \frac{\partial}{\partial t}(\nabla \cdot A) = -\frac{\rho}{\varepsilon_0}$$

$$\left(\nabla^2 A - \frac{1}{c^2}\frac{\partial^2 A}{\partial t^2}\right) - \nabla\left(\nabla \cdot A + \frac{1}{c^2}\frac{\partial\varphi}{\partial t}\right) = -\mu_0 J$$

These equations taken together are as powerful and complete as Maxwell's equations. Moreover, the problem has been reduced somewhat, as the electric and magnetic fields together had six components to solve for. In the potential formulation, there are only four components: the electric potential and the three components of the vector potential. However, the equations are messier than Maxwell's equations using the electric and magnetic fields.

Gauge Freedom

These equations can be simplified by taking advantage of the fact that only the electric and magnetic fields are physically meaningful quantities that can be measured; the potentials are not. There is a freedom to constrain the form of the potentials provided that this does not affect the resultant electric and magnetic fields, called gauge freedom. Specifically for these equations, for any choice of a twice-differentiable scalar function of position and time λ, if (φ, A) is a solution for a given system, then so is another potential (φ', A') given by:

$$\varphi' = \varphi - \frac{\partial \lambda}{\partial t}$$

$$A' = A + \nabla \lambda.$$

This freedom can be used to simplify the potential formulation. Either of two such scalar functions is typically chosen: the Coulomb gauge and the Lorenz gauge.

Coulomb Gauge

The Coulomb gauge is chosen in such a way that $\nabla \cdot A' = 0$, which corresponds to the case of magnetostatics. In terms of λ, this means that it must satisfy the equation,

$$\nabla^2 \lambda = -\nabla \cdot A.$$

This choice of function results in the following formulation of Maxwell's equations:

$$\nabla^2 \varphi' = -\frac{\rho}{\varepsilon_0}$$

$$\nabla^2 A' - \mu_0 \varepsilon_0 \frac{\partial^2 A'}{\partial t^2} = -\mu_0 J + \mu_0 \varepsilon_0 \nabla \left(\frac{\partial \varphi'}{\partial t} \right)$$

Several features about Maxwell's equations in the Coulomb gauge are as follows. Firstly, solving for the electric potential is very easy, as the equation is a version of Poisson's equation. Secondly, solving for the magnetic vector potential is particularly difficult. This is the big disadvantage of this gauge. The third thing to note, and something which is not immediately obvious, is that the electric potential changes instantly everywhere in response to a change in conditions in one locality.

For instance, if a charge is moved in New York at 1 pm local time, then a hypothetical observer in Australia who could measure the electric potential directly would measure a change in the potential at 1 pm New York time. This seemingly goes violates causality in special relativity, i.e. the impossibility of information, signals, or anything travelling faster than the speed of light. The resolution to this apparent problem lies in the fact that, as previously stated, no observers can measure the potentials; they measure the electric and magnetic fields. So, the combination of $\nabla\varphi$ and $\partial A/\partial t$ used in determining the electric field restores the speed limit imposed by special relativity for the electric field, making all observable quantities consistent with relativity.

Lorenz Gauge Condition

A gauge that is often used is the Lorenz gauge condition. In this, the scalar function λ is chosen such that,

$$\nabla \cdot A' = -\mu_0 \varepsilon_0 \frac{\partial \varphi'}{\partial t},$$

meaning that λ must satisfy the equation

$$\nabla^2 \lambda - \mu_0 \varepsilon_0 \frac{\partial^2 \lambda}{\partial t^2} = -\nabla \cdot A - \mu_0 \varepsilon_0 \frac{\partial \varphi}{\partial t}.$$

The Lorenz gauge results in the following form of Maxwell's equations:

$$\nabla^2 \varphi' - \mu_0 \varepsilon_0 \frac{\partial^2 \varphi'}{\partial t^2} = \Box^2 \varphi' = -\frac{\rho}{\varepsilon_0}$$

$$\nabla^2 \mathbf{A}' - \grave{\imath}_0 \mathring{a}_0 \frac{\partial^2 \mathbf{A}'}{\partial t^2} = \Box^2 \mathbf{A}' = -\grave{\imath}_0 \mathbf{J}$$

The operator $\Box^2$ is called the d'Alembertian (some authors denote this by only the square $\Box$). These equations are inhomogeneous versions of the wave equation, with the terms on the right side of the equation serving as the source functions for the wave. As with any wave equation, these equations lead to two types of solution: advanced potentials (which are related to the configuration of the sources at future points in time), and retarded potentials (which are related to the past configurations of the sources); the former are usually disregarded where the field is to analyzed from a causality perspective.

As pointed out above, the Lorenz gauge is no more valid than any other gauge since the potentials cannot be measured. Despite this, there are certain quantum mechanical phenomena in which potentials appear to affect particles in regions where the observable field vanishes throughout the region, for example as in the Aharonov–Bohm effect. However, these phenomena do not provide a means to directly measure the potentials nor to detect a difference between different but mutually gauge equivalent potentials. The Lorenz gauge has the further advantage of the equations being Lorentz invariant.

Extension to Quantum Electrodynamics

Canonical quantization of the electromagnetic fields proceeds by elevating the scalar and vector potentials; $\varphi(x)$, $A(x)$, from fields to field operators. Substituting $1/c^2 = \varepsilon_0 \mu_0$ into the previous Lorenz gauge equations gives:

$$\nabla^2 A - \frac{1}{c^2} \frac{\partial^2 A}{\partial t^2} = -\mu_0 J$$

$$\nabla^2 \varphi - \frac{1}{c^2} \frac{\partial^2 \varphi}{\partial t^2} = -\frac{\rho}{\varepsilon_0}$$

Here, J and ρ are the current and charge density of the *matter field*. If the matter field is taken so as to describe the interaction of electromagnetic fields with the Dirac electron given by the four-component Dirac spinor field ψ, the current and charge densities have form:

$$J = -e\psi^{\dagger}\alpha\psi \quad \rho = -e\psi^{\dagger}\psi,$$

where α are the first three Dirac matrices. Using this, we can re-write Maxwell's equations as:

Maxwell's equations (*QED*)

$$\nabla^2 A - \frac{1}{c^2}\frac{\partial^2 A}{\partial t^2} = \mu_0 e\psi^{\dagger}\acute{a}\psi$$

$$\nabla^2 \varphi - \frac{1}{c^2}\frac{\partial^2 \varphi}{\partial t^2} = \frac{1}{\varepsilon_0} e\psi^{\dagger}\psi$$

which is the form used in quantum electrodynamics.

Geometric Algebra Formulations

Analogous to the tensor formulation, two objects, one for the field and one for the current, are introduced. In geometric algebra (GA) these are multivectors. The field multivector, known as the Riemann–Silberstein vector, is

$$F = E + IcB = E^k\sigma_k + IcB^k\sigma_k$$

and the current multivector is,

$$c\rho - J = c\rho - J^k\sigma_k$$

where, in the algebra of physical space (APS) $C\ell_{3,0}(\mathbb{R})$ with the vector basis $\{\sigma_k\}$. The unit pseudoscalar is $I = \sigma_1\sigma_2\sigma_3$ (assuming an orthonormal basis). Orthonormal basis vectors share the algebra of the Pauli matrices, but are usually not equated with them. After defining the derivative

$$\nabla = \sigma^k \partial_k$$

Maxwell's equations are reduced to the single equation

Maxwell's equations *(APS formulation)*

$$\left(\frac{1}{c}\frac{\partial}{\partial t} + \nabla\right)F = \mu_0 c(c\rho - J).$$

In three dimensions, the derivative has a special structure allowing the introduction of a cross product:

$$\nabla F = \nabla \cdot F + \nabla \wedge F = \nabla \cdot F + I\nabla \times F$$

from which it is easily seen that Gauss's law is the scalar part, the Ampère–Maxwell law is the vector part, Faraday's law is the pseudovector part, and Gauss's law for magnetism is the pseudoscalar part of the equation. After expanding and rearranging, this can be written as,

$$\left(\nabla\cdot E-\frac{\rho}{\epsilon_0}\right)-c\left(\nabla\times B-\mu_0\epsilon_0\frac{\partial E}{\partial t}-\mu_0 J\right)+I\left(\nabla\times E+\frac{\partial B}{\partial t}\right)+Ic(\nabla\cdot B)=0$$

We can identify APS as a subalgebra of the spacetime algebra (STA) $C\ell_{1,3}(\mathbb{R})$, defining $\sigma_k=\gamma_k\gamma_0$ and $I=\gamma_0\gamma_1\gamma_2\gamma_3$. The γ_μ s have the same algebraic properties of the gamma matrices but their matrix representation is not needed. The derivative is now,

$$\nabla=\gamma^\mu\partial_\mu.$$

The Riemann–Silberstein becomes a bivector,

$$F=E+IcB=E^1\gamma_1\gamma_0+E^2\gamma_2\gamma_0+E^3\gamma_3\gamma_0-c(B^1\gamma_2\gamma_3+B^2\gamma_3\gamma_1+B^3\gamma_1\gamma_2),$$

and the charge and current density become a vector,

$$J=J^\mu\gamma_\mu=c\rho\gamma_0+J^k\gamma_k=\gamma_0(c\rho-J^k\sigma_k).$$

Owing to the identity,

$$\gamma_0\nabla=\gamma_0\gamma^0\partial_0+\gamma_0\gamma^k\partial_k=\partial_0+\sigma^k\partial_k=\frac{1}{c}\frac{\partial}{\partial t}+\nabla,$$

Maxwell's equations reduce to the single equation,

Maxwell's equations *(STA formulation)*

$$\nabla F=\mu_0 cJ.$$

The Exterior Derivative

In this section, we define the exterior derivative operator, which will allow us to express Maxwell's laws as differential equations. This operator has the symbol *d*, and acts on a p-form to produce a new form with degree p + 1. This (p + 1)-form characterizes the spatial variation of the p-form.

The exterior derivative operator can be written as,

$$d=\left(\frac{\partial}{\partial x}dx+\frac{\partial}{\partial x}dx+\frac{\partial}{\partial x}dz\right)\wedge$$

The d operator is similar to a one-form, except that the coefficients are partial derivative operators instead of functions. When this operator is applied to a differential form, the derivatives act on the coefficients of the form, and the differentials combine with those of the form according to the properties of the exterior product.

The sum and product rules for exterior differentiation are:

$$d\,(\mathcal{U}+\mathcal{V})=d\mathcal{U}+d\mathcal{V},$$

$$d\,(\mathcal{U}\wedge\mathcal{V})=d\mathcal{U}\wedge d\mathcal{V}+(-1)^{(\deg\mathcal{U})}\mathcal{U}\wedge d\mathcal{V}.$$

The exterior derivatives of *p*-forms are

zero – form :

$$\mathrm{d}f(\mathrm{x}) = \frac{\partial f}{\partial x}\mathrm{d}x + \frac{\partial f}{\partial y}\mathrm{d}y + \frac{\partial f}{\partial z}\mathrm{dz},$$

one – form :

$$\mathrm{d}\mathcal{U}(\mathrm{x}) = \left(\frac{\partial \mathrm{U_Z}}{\partial \mathrm{y}} - \frac{\partial \mathrm{U_y}}{\partial \mathrm{z}}\right)\mathrm{dy} \wedge \mathrm{dz} + \left(\frac{\partial \mathrm{U}_x}{\partial \mathrm{z}} - \frac{\partial \mathrm{U_Z}}{\partial x}\right)\mathrm{dz} \wedge \mathrm{d}x + \left(\frac{\partial \mathrm{U_y}}{\partial x} - \frac{\partial \mathrm{U}_x}{\partial \mathrm{y}}\right)\mathrm{d}x \wedge \mathrm{dy}$$

two – form :

$$\mathrm{d}\mathcal{V}(\mathrm{x}) = \left(\frac{\partial \mathrm{V}x}{\partial x} - \frac{\partial \mathrm{V_y}}{\partial_{\mathrm{Y}}} + \frac{\partial \mathrm{V_Z}}{\partial_{\mathrm{Z}}}\right)\mathrm{d}x \wedge \mathrm{dy} \wedge \mathrm{dz}$$

three – form :

$$\mathrm{d}\mathcal{Q}(\mathrm{x}) = 0.$$

Computing exterior derivatives is straightforward. One takes the partial derivative of a differential form by x and adds the differential $\mathrm{d}x$ from the left, repeats for y and z, and adds the three results. This process is very similar to implicit differentiation, except that one must combine new differentials with existing differentials using the exterior product. When taking the exterior derivative of a one-form or two-form, some terms may drop out due to repeated differentials. Table shows the correspondence between several vector derivative and exterior derivative operations.

Table: Differential operators.

Vector Differential Operator	*Exterior Differential Operator*
grad f	$\mathrm{d}f$
curl A	$\mathrm{d}\mathcal{A}$
div B	$\mathrm{d}\mathcal{B}$
curl grad f = o	$\mathrm{d}\,\mathrm{d}f = 0$
div curl A = o	$\mathrm{d}\,\mathrm{d}\mathcal{A} = 0$
div grad f	$\mathrm{d} * \mathrm{d}f$ or $* \mathrm{d} * \mathrm{d}f$
curl curl A	$\mathrm{d} * \mathrm{d}\mathcal{A}$ or $* \mathrm{d} * \mathrm{d}\mathcal{A}$

Poincare's Lemma

A form $\mathcal{V}$ for which $\mathrm{d}\mathcal{V} = 0$ is said to be closed, and a form $\mathcal{V}$ for which $\mathcal{V} = \mathrm{d}\mathcal{U}$ is said to be exact. For differential forms the statement $\mathcal{V} = \mathrm{d}\mathcal{U}$ implies $\mathrm{d}\mathcal{V} = 0$. The relation

$$\mathrm{dd}\mathcal{U} = 0$$

may be verified easily. In conventional vector notation this corresponds to curl grad = o and div curl = o. All exact forms are closed. However it may also be shown that all closed forms are exact. Poincar´e's lemma states that

$$\mathrm{d}\mathcal{V} = 0 \Leftrightarrow \mathcal{V} = \mathrm{d}\mathcal{U}.$$

If a differential form has zero exterior derivative, then it is itself the exterior derivative of another form.

Stokes' Theorem

In Ampere's Law and Faraday's Law line above, integrals over the boundary of the surface A are related to surface integrals over A. Figure below shows the relation between the orientation of the area A and the boundary ∂A. The line integral over the closed contour ∂A is called circulation. In Magnetic Flux Continuity and Gaus's Law ,the surface integrals are performed over the boundary ∂V of the volume V . Figure below shows the orientation of the boundary surface ∂V. The Stokes' theorem relates the integration of a p-form U over the closed p-dimensional boundary ∂V of a (p + 1)-dimensional volume V to the volume integral of $d\mathcal{U}$ over V via,

$$\oint_{\partial V} \mathcal{U} = \int_V d\mathcal{U}.$$

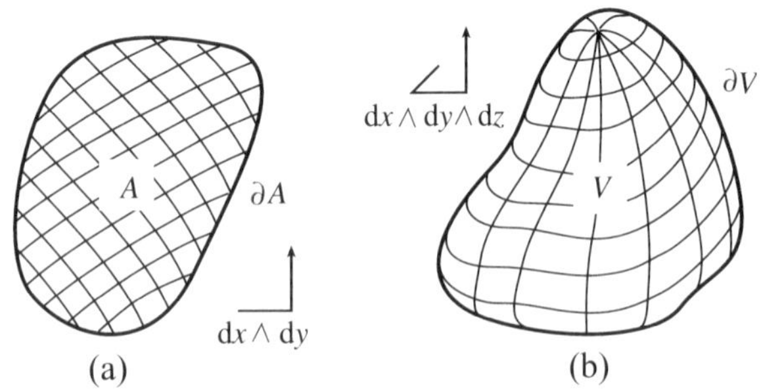

(a) Area A with boundary ∂A and (b) volume V with boundary ∂V

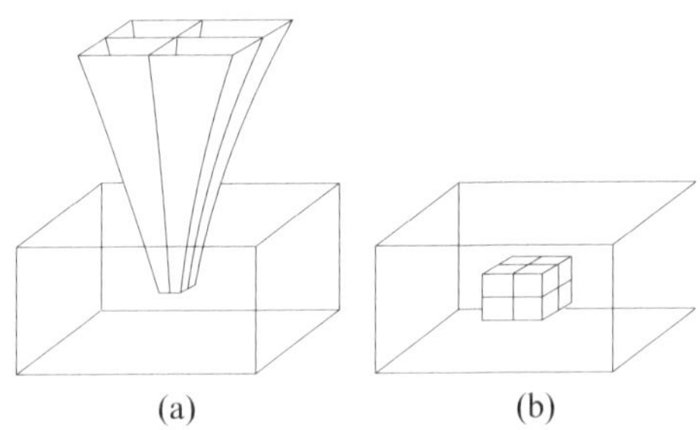

Stokes theorem for U a one-form. (a) The loop ∂V pierces three of the surfaces of U.
(b) Three tubes of dU pass through a surfaces V bounded by the loop ∂V .

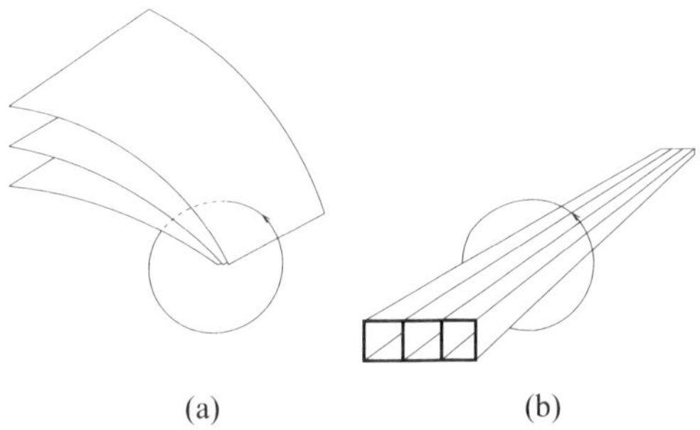

Stokes theorem for U a two-form.(a) Four tubes of U pass through the closed, rectangular surface ∂V .
(b) Four boxes of the three-form dU lie inside the surface.

This summarizes the Stokes' theorem and the Gauss' theorem of conventional vector notation.

The property of Stokes' theorem to relate the integration of a p-form over a closed boundary of a volume to the integration of the exterior derivative of the p-form over the volume enclosed by that boundary yields an interesting geometric interpretation of Poincar´e's lemma: since a boundary has no boundary the iterated application of the exterior derivative yields zero value.

If U is a zero-form, then Stokes' theorem states that $\int_a^b d\mathcal{U} = f(b) - f(a)$. This is the fundamental theorem of calculus. If $\mathcal{U}$ is a one-form, then ∂V has to be a closed path. V is a surface that has the path as its boundary. Graphically, Stokes' theorem says that the number of surfaces of $\mathcal{U}$ pierced by the path is equal to the number of tubes of the two-form $d\mathcal{U}$ that pass through the path.

If $\mathcal{U}$ is a two-form, then ∂V is a closed surface and V is the volume inside it. Stokes' theorem requires that the number of tubes of $\mathcal{U}$ that cross the surface is equal to the number of boxes of Du inside the surface, as shown in figure above.

Curvilinear Coordinates

In exterior calculus the field equations may be formulated without reference to a specific coordinate system. Depending on the problem the choice of a specific coordinate system may simplify the problem solution considerably. We introduce an orthogonal curvilinear coordinate system,

$$u = u(x, y, z), \quad v = v(x, y, z), \quad w = w(x, y, z)$$

The coordinate curves are obtained by setting two of the three coordinates u, v, w constant. Coordinate surfaces are defined by setting one of the three coordinates constant. In an orthogonal coordinate system in any point (except singular points) of the space the three coordinate curves are orthogonal. The sameholds for the three coordinate surfaces going through any point. The differentials dx, dy, dz by the differentials du, dv, dw are related to,

$$dx = \frac{\partial_x}{\partial u} du + \frac{\partial x}{\partial v} dv + \frac{\partial x}{\partial w} dw,$$

$$dy = \frac{\partial y}{\partial u} du + \frac{\partial y}{\partial v} dv + \frac{\partial y}{dw} dw,$$

$$dz = \frac{\partial z}{\partial u} du + \frac{\partial z}{\partial v} dv + \frac{\partial z}{\partial w} dw$$

The rules for transformation of the Cartesian basis two-forms $dx \wedge dy$, $dy \wedge dz$, $dz \wedge dx$ and the Cartesian basis three-form $dx \wedge dy \wedge dz$ follow directly from the above equations by applying the rules of the exterior product. Using the metric coefficients g_1, g_2 and g_3,

$$g_1^2 = \frac{\partial_x}{\partial u} \cdot \frac{\partial_x}{\partial u}, \quad g_2^2 = \frac{\partial_x}{\partial v} \cdot \frac{\partial_x}{\partial v}, \quad g_3^2 = \frac{\partial_x}{\partial w} \cdot \frac{\partial_x}{\partial w}$$

we introduce the unit one-forms,

$$s_1 = g_1 du, \quad s_2 = g_2 dv, \quad s_3 = g_3 dw$$

The integral of $s_1 = g_1 du$ along any path with v and w constant yields the length of the path. For the curvilinear unit differentials the Hodge star operator is,

$$* f = f s_1 \wedge s_2 \wedge s_3 \,,$$
$$*(A_u s_1 + A_v s_2 + A_w s_3) = A_u s_2 \wedge s_3 + A_v s_3 \wedge s_1 + A_w s_1 \wedge s_2,$$
$$*(A_u s_2 \wedge s_3 + A_v s_3 \wedge s_1 + A_w s_1 \wedge s_2) = A_u s_1 + A_v s_2 + A_w s_3,$$
$$*(f s_1 \wedge s_2 \wedge s_3) = f.$$

The application of the Hodge operator fully accounts for the metric properties of the used coordinate system. In contrast to this the exterior derivative is independent of the metric. This is a substantial advantage of exterior calculus.

Cylindrical Coordinates

In the cylindrical coordinate system, a point in space is specified by the radial distance of its x, y coordinates $\rho = \sqrt{x^2 + y^2}$ angle from the $+x$ axis in the $x - y$ plane ϕ, and height in the z direction. The differentials of the cylindrical coordinate system are $d\rho$, $d\phi$ and dz. To convert forms into unit vectors, the angular differential $d\phi$ must be made into a unit differential $\rho d\phi$. One-forms correspond to vectors by the rule,

$$d\rho \leftrightarrow \hat{\rho}$$
$$\rho d\phi \leftrightarrow \hat{\phi}$$
$$dz \leftrightarrow \hat{z}$$

Figure below shows the pictures of the differentials of the cylindrical coordinate system. The two-forms can be obtained by superimposing these surfaces. Tubes of $dz \wedge d\rho$, for example, are square donut-shaped and point in the ϕ direction.

In curvilinear coordinates, differentials must be converted to unit differentials before the star operator is applied. The star operator in cylindrical coordinates acts as follows:

$$*d\rho = \rho d\phi \wedge dz$$
$$*\rho d\phi = dz \wedge d\rho$$
$$*dz = d\rho \wedge \rho d\phi$$

so that $\tilde{n}$ must be included with $d\varphi$ before the star operator can be applied to make it a two-form with unit differentials. Also, $*1 = \rho d\rho d\phi dz$.. As with the rectangular coordinate system, $** = 1$, so that this same table can be used to convert two-forms to one-forms.

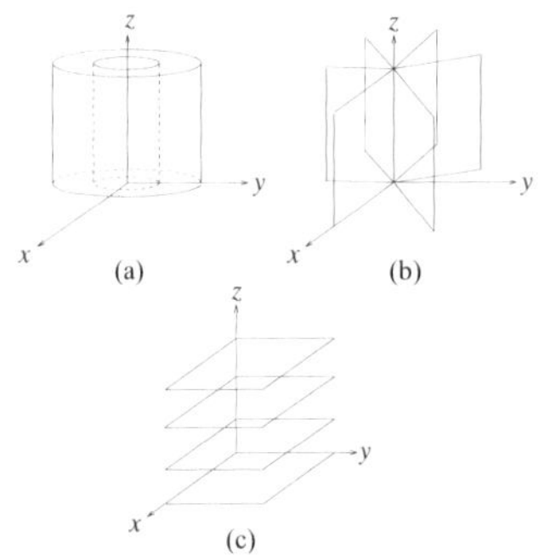

Surfaces of $d\rho$, $d\phi$ scaled by $3/\pi$ and dz.

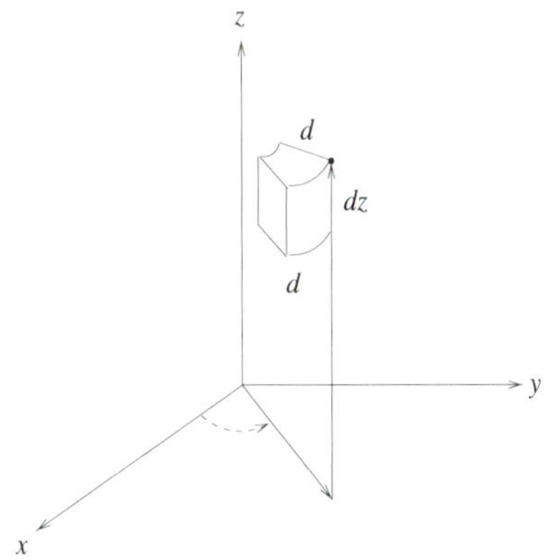

Unit differentials in cylindrical coordinates represented as faces of a differential volume.

Spherical Coordinates

In the spherical coordinate system, a point in space is specified by the radial distance from the origin angle $r = \sqrt{x^2 + y^2 + z^2}$ from the $+x$ axis in the $x - y$ plane ϕ, and angle from the z axis θ. The differentials of the spherical coordinate system are dr, $d\theta$ and $d\phi$. To convert forms into unit vectors, the angular differentials must be made into unit differentials $r\,d\theta$ and $r \sin \theta d\phi$. One-forms correspond to vectors by the rules,

$$
\begin{aligned}
dr &\leftrightarrow \hat{r} \\
rd\theta &\leftrightarrow \hat{\theta} \\
r \sin \theta d\phi &\leftrightarrow \hat{\phi}
\end{aligned}
$$

Figure below shows the pictures of the differentials of the spherical coordinate system.

As in the cylindrical coordinate system, differentials of the spherical coordinate system must be converted to length elements before the star operator is applied. The star operator acts on one-forms and two-forms as follows:

$$
\begin{aligned}
*dr &= rd\theta \wedge r \sin \theta d\phi \\
*rd\theta &= r \sin \theta d\phi \wedge dr \\
*r \sin \theta d\phi &= dr \wedge rd\theta
\end{aligned}
$$

Again, $** = 1$, so that this table can be used to convert two-forms to one-forms. The star operator applied to one is $r^2 \sin\theta dr d\theta d\phi$.

Exterior Derivative in Curvilinear Coordinates

Regardless of the particular coordinate system, the form of the exterior derivative operator remains the same. If the coordinates are (u, v, w), the exterior derivative operator is

$$
d = \left(\frac{\partial}{\partial u} du + \frac{\partial}{\partial v} d + \frac{\partial}{\partial w} dw \right) \wedge
$$

In cylindrical coordinates,

$$d = \left(\frac{\partial}{\partial \rho} d\rho + \frac{\partial}{\partial \phi} d\phi + \frac{\partial}{\partial z} dz \right) \wedge$$

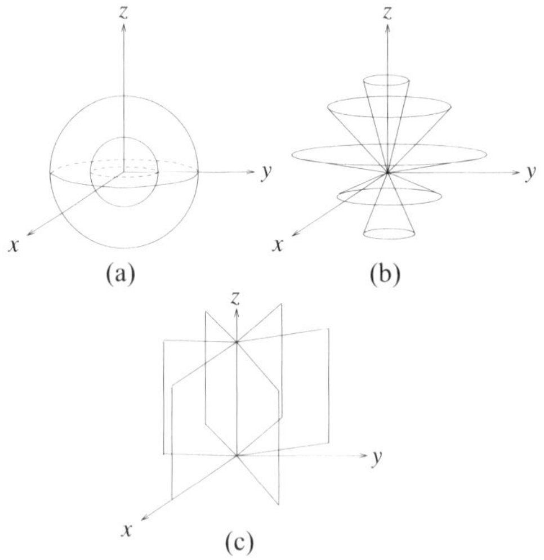

Surfaces of dr , $d\theta$ scaled by 10/π and $d\phi$ scaled by 3/π.

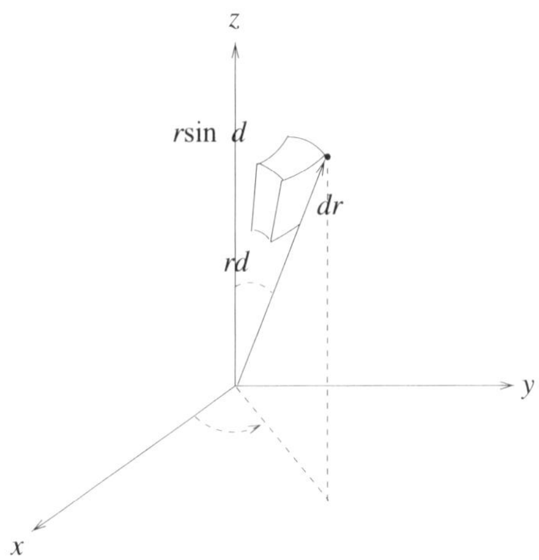

Unit differentials in spherical coordinates represented as faces of a differential volume.

which is the same as for rectangular coordinates but with the coordinates ρ, ϕ, z in the place of x, y, z. Note that the factor ρ associated with $d\phi$ must be present when converting to vectors or applying the star operator, but is not found in the exterior derivative operator.

The exterior derivative in spherical coordinates is,

$$\mathrm{d} = \left(\frac{\partial}{\partial r} dr + \frac{\partial}{\partial \theta} d\theta + \frac{\partial}{\partial \varphi} d\varphi \right) \wedge$$

From these expressions, we see that computing exterior derivatives in a curvilinear coordinate system is no different from computing in rectangular coordinates.

Maxwell's Equations In Local Form

Applying Stokes' theorem to the integral form of Maxwell's Equations we obtain the differential representation of Maxwell's equations:

$$d\mathcal{H} = \frac{\partial}{\partial t}\mathcal{D} + \mathcal{J}, \qquad \text{Ampere's Law}$$

$$d\mathcal{E} = -\frac{\partial}{\partial t}\mathcal{B}, \qquad \text{Faraday's Law}$$

$$d\mathcal{B} = 0, \qquad \text{Magnetic Flux Continuity}$$

$$d\mathcal{D} = \mathcal{Q}. \qquad \text{Gauss' Law}$$

Figure below shows the graphical representation of Maxwell's equations after Deschamps. We have two groups of equations, namely Faraday's law and magnetic flux continuity on the left-hand side of the diagram and Ampere's law and Gauss' law on the right-hand side. Within each group exterior differentiation is metric-independent and increases the order of the respective *p*-forms. Both groups are related via the constitutive relations depending on the material properties and the coordinate metrics and either incrementing or decrementing the order of the respective *p*-forms.

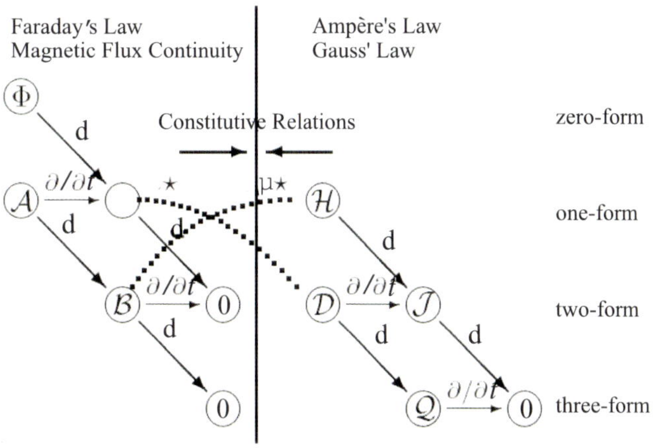

Graphical representation of Maxwells equations.

Curved Spacetime

TraditionaL FOrmulation

Matter and energy generate curvature of spacetime. This is the subject of general relativity. Curvature of spacetime affects electrodynamics. An electromagnetic field having energy and momentum also generates curvature in spacetime. Maxwell's equations in curved spacetime can be obtained by replacing the derivatives in the equations in flat spacetime with covariant derivatives. (Whether this is the appropriate generalization requires separate investigation.) The sourced and source-free equations become (cgs-Gaussian units):

$$\frac{4\pi}{c} j^{\beta} = \partial_{\alpha} F^{\alpha\beta} + \Gamma^{\alpha}{}_{\mu\alpha} F^{\mu\beta} + \Gamma^{\beta}{}_{\mu\alpha} F^{\alpha\mu} \stackrel{\text{def}}{=} \nabla_{\alpha} F^{\alpha\beta} \stackrel{\text{def}}{=} F^{\alpha\beta}{}_{;\alpha}$$

and

$$0 = \partial_\gamma F_{\alpha\beta} + \partial_\beta F_{\gamma\alpha} + \partial_\alpha F_{\beta\gamma} = \nabla_\gamma F_{\alpha\beta} + \nabla_\beta F_{\gamma\alpha} + \nabla_\alpha F_{\beta\gamma}.t$$

Here,

$$\Gamma^{\alpha}{}_{\mu\beta}$$

is a Christoffel symbol that characterizes the curvature of spacetime and ∇_α is the covariant derivative.

Formulation in Terms of Differential Forms

The formulation of the Maxwell equations in terms of differential forms can be used without change in general relativity. The equivalence of the more traditional general relativistic formulation using the covariant derivative with the differential form formulation can be seen as follows. Choose local coordinates x^α which gives a basis of 1-forms dx^α in every point of the open set where the coordinates are defined. Using this basis and cgs-Gaussian units we define,

- The antisymmetric field tensor $F_{\alpha\beta}$, corresponding to the field 2-form F

$$F = \frac{1}{2} F_{\alpha\beta}\, dx^\alpha \wedge dx^\beta.$$

- The current-vector infinitesimal 3-form J

$$J = \frac{4\pi}{c}\left(\frac{1}{6} j^\alpha \sqrt{-g}\, \epsilon_{\alpha\beta\gamma\delta} dx^\beta \wedge dx^\gamma \wedge dx^\delta.\right)$$

The epsilon tensor contracted with the differential 3-form produces 6 times the number of terms required.

Here g is as usual the determinant of the matrix representing the metric tensor, $g_{\alpha\beta}$. A small computation that uses the symmetry of the Christoffel symbols (i.e., the torsion-freeness of the Levi-Civita connection) and the covariant constantness of the Hodge star operator then shows that in this coordinate neighborhood we have:

- the Bianchi identity

$$dF = 2(\partial_\gamma F_{\alpha\beta} + \partial_\beta F_{\gamma\alpha} + \partial_\alpha F_{\beta\gamma}) dx^\alpha \wedge dx^\beta \wedge dx^\gamma = 0,$$

- the source equation

$$d \star F = \frac{1}{6} F^{\alpha\beta}{}_{;\alpha} \sqrt{-g}\, \epsilon_{\beta\gamma\delta\eta} dx^\gamma \wedge dx^\delta \wedge dx^\eta = J,$$

- the continuity equation

$$dJ = \frac{4\pi}{c} j^\alpha{}_{;\alpha} \sqrt{-g}\, \epsilon_{\alpha\beta\gamma\delta} dx^\alpha \wedge dx^\beta \wedge dx^\gamma \wedge dx^\delta = 0.$$

Classical Electrodynamics as the Curvature of a Line Bundle

An elegant and intuitive way to formulate Maxwell's equations is to use complex line bundles or principal bundles with fibre U(1). The connection ∇ on the line bundle has a curvature $F = \nabla^2$ which is a two-form that automatically satisfies $dF = 0$ and can be interpreted as a field-strength. If the line bundle is trivial with flat reference connection *d* we can write $\nabla = d + A$ and $F = dA$ with A the 1-form composed of the electric potential and the magnetic vector potential.

In quantum mechanics, the connection itself is used to define the dynamics of the system. This formulation allows a natural description of the Aharonov–Bohm effect. In this experiment, a static magnetic field runs through a long magnetic wire (e.g., an iron wire magnetized longitudinally). Outside of this wire the magnetic induction is zero, in contrast to the vector potential, which essentially depends on the magnetic flux through the cross-section of the wire and does not vanish outside. Since there is no electric field either, the Maxwell tensor $F = 0$ throughout the space-time region outside the tube, during the experiment. This means by definition that the connection ∇ is flat there.

However, as mentioned, the connection depends on the magnetic field through the tube since the holonomy along a non-contractible curve encircling the tube is the magnetic flux through the tube in the proper units. This can be detected quantum-mechanically with a double-slit electron diffraction experiment on an electron wave traveling around the tube. The holonomy corresponds to an extra phase shift, which leads to a shift in the diffraction pattern.

Effects of Electromagnetic Field and Radiation

Electromagnetic Radiation emitted from laptop computers, cell phones, and other electronic devices can be harmful to our bodies. Many studies have revealed a link between the use of these types of technological devices and various forms of illness, including tumors and infertility. Below are just some of the harmful effects of Electromagnetic Radiation.

Short-term Health Effects

- Headaches
- Tingling or burning sensations
- Aches and pains
- Decreased sperm motility
- Hands hurt

Long-term Health Effects

- Brain tumors
- Cancers – blood, breast and more
- Mutated Cells

- Fragmented DNA
- Toasted Skin Syndrome

Electrical Sensitivity

- Headaches
- Concentration or memory loss
- Cognitive impairment
- Tingling or burning sensations
- Sleeping problems
- Aches and pains
- Hand pain

A large body of literature exists on the response of tissues to electromagnetic fields, primarily in the extremely-low-frequency (ELF) and microwave-frequency ranges. In general, the reported effects of radiofrequency (RF) radiation on tissue and organ systems have been attributed to thermal interactions, although the existence of nonthermal effects at low field intensities is still a subject of active investigation. This chapter summarizes reported RF effects on major physiological systems and provides estimates of the threshold specific absorption rates (SARs) required to produce such effects. Organ and tissue responses to ELF fields and attempts to characterize field thresholds are also summarized. The relevance of these findings to the possible association of health effects with exposure to RF fields from GWEN antennas is assessed.

Nervous System

The effects of radiation on nervous tissues have been a subject of active investigation since changes in animal behavior and nerve electrical properties were first reported in the Soviet Union during the 1950s and 1960s.[1] RF radiation is reported to affect isolated nerve preparations, the central nervous system, brain chemistry and histology, and the blood-brain barrier.

In studies with in vitro nerve preparations, changes have been observed in the firing rates of Aplysia neurons and in the refractory period of isolated frog sciatic nerves exposed to 2.45-GHz microwaves at SAR values exceeding 5 W/kg. Those effects were very likely associated with heating of the nerve preparations, in that much higher SAR values have not been found to produce changes in the electrical properties of isolated nerves when the temperature was controlled. Studies on isolated heart preparations have provided evidence of bradycardia as a result of exposure to RF radiation at nonthermal power densities, although some of the reported effects might have been artifacts caused by currents induced in the recording electrodes or by nonphysiological conditions in the bathing medium. Several groups of investigators have reported that nonthermal levels of RF fields can alter Ca^{2+} binding to the surfaces of nerve cells in isolated brain hemispheres and neuroblastoma cells cultured in vitro. That phenomenon, however, is observed only when the RF field is amplitude-modulated at extremely low frequencies, the maximum effect occurs at a modulation frequency of 16 Hz. A similar effect has recently been reported in isolated frog hearts. The importance of changes in Ca^{2+} binding on the functional properties of nerve cells has not been established, and there is no clear evidence that the reported effect of low-intensity, amplitude-modulated RF fields poses a substantial health risk.

Results of in vivo studies of both pulsed and continuous-wave (CW) RF fields on brain electrical activity have indicated that transient effects can occur at SAR values exceeding 1 W/kg. Evidence has been presented that cholinergic activity of brain tissue is influenced by RF fields at SAR values as low as 0.45 W/kg. Exposure to nonthermal RF radiation has been reported to influence the electroencephalograms (EEGs) of cats when the field was amplitude-modulated at frequencies less than 25 Hz, which is the range of naturally occurring EEG frequencies. The rate of Ca^{2+} exchange from cat brain tissue in vivo was observed to change in response to similar irradiation conditions. Comparable effects on Ca^{2+} binding were not observed in rat cerebral tissue exposed to RF radiation, although the fields used were pulsed at EEG frequencies, rather than amplitude-modulated. As noted above, the physiological significance of small shifts in Ca^{2+} binding at nerve cell surfaces is unclear.

A wide variety of changes in brain chemistry and structure have been reported after exposure of animals to high-intensity RF fields. The changes include decreased concentrations of epinephrine, norepinephrine, dopamine, and 5-hydroxytryptamine; changes in axonal structure; a decreased number of Purkinje cells; and structural alterations in the hypothalamic region. Those effects have generally been associated with RF intensities that produced substantial local heating in the brain.

Extensive studies have been carried out to detect possible effects of RF radiation on the integrity of the blood-brain barrier. Although several reports have suggested that nonthermal RF radiation can influence the permeability of the blood-brain barrier, most of the experimental findings indicate that such effects result from local heating in the head in response to SAR values in excess of 2 W/kg. Changes in cerebral blood flow rate, rather than direct changes in permeability to tracer molecules, might also be incorrectly interpreted as changes in the properties of the blood-brain barrier.

Effects of pulsed and sinusoidal ELF fields on the electrical activity of the nervous system have also been studied extensively. In general, only high-intensity sinusoidal electric fields or rapidly pulsed magnetic fields induce sufficient current density in tissue (around 0.1-1.0 A/m^2 or higher) to alter neuronal excitability and synaptic transmission or to produce neuromuscular stimulation. Somewhat lower thresholds have been observed for the induction of visual phosphenes and for influencing the electrical activity of Aplysia pacemaker neurons when the frequency of the applied field matched the endogenous neuronal firing rate. Those effects, however, have been observed only with ELF frequencies and would not be expected to occur at the higher frequencies associated with GWEN transmitters. Recent studies with human volunteers exposed to 60-Hz electric and magnetic fields with intensities comparable with those of high-voltage power lines have shown no consistent effects on the EEG. Minor changes were observed in reaction time and heart rate, but the variations were within normal ranges.

Visual System

Cataract development as a result of exposure of the eye to high-intensity RF radiation has been studied for more than 30 years. Extensive experiments have been carried out with rabbits to determine the dependence of cataractogenesis on the frequency and intensity of RF fields and on exposure time. In general, the lowest thresholds for cataract induction have been observed with near-field exposure at 1-10 GHz, and a power density greater than 100 mW/cm² applied for at least an hour is required. Most of the evidence indicates that the mechanism of injury leading to lens opacity is thermal, and pulsed and CW microwave fields appear to have similar thresholds for producing cataracts.[19] Multiple subthreshold exposures do not lead to cataracts if the time between exposures is long enough to permit the tissue to return to its normal temperature.[1] At frequencies where the wavelength of the RF field is not well matched to the dimensions of the eye, cataracts are not produced even at extremely high power densities approaching the lethal levels. Although it is difficult to extrapolate results from laboratory animals to humans, the threshold power density required to produce cataracts is expected to be similar in rabbits and humans because of the structural similarities and comparable dimensions of the eyes in these species.

Results of recent studies with monkeys have indicated that vascular leakage can increase at relatively low power densities of pulsed 2.45-GHz radiation when the eye is pretreated with timolol maleate, which decreases intraocular pressure by reducing the production of aqueous humor. A power density as low as 1 mW/cm², corresponding to an intraocular SAR of 0.26 W/kg, has been observed to produce the effect. Those findings might have implications for RF ocular damage in humans being treated with timolol maleate for glaucoma. However, the threshold power density at 2.45 GHz is still well in excess of the intensity of the RF fields produced by GWEN antennae in areas accessible to the general public.

A visual phenomenon associated with exposure to ELF fields that has been studied for nearly a century is the induction of a flickering illumination known as phosphenes. Time-varying magnetic fields with either pulsed or sinusoidal waveforms and frequencies below 100 Hz have been shown to produce phosphenes when the time rate of change of the field exceeds 1.3 T/sec. The frequency most likely to produce magnetophosphenes with sinusoidal fields is 20 Hz; the threshold flux density for eliciting the visual effect is 8 mT. A similar frequency dependence has been observed for electrophosphenes produced by placing electrodes in contact with the forehead near the eyes. The locus of the effect is in the retina, and available evidence suggests that induced currents in the retina elicit visual responses similar to those resulting from photic stimulation. Changes in visually evoked potential (VEP) have also been reported in response to ELF magnetic fields with flux densities that are 5-10 times greater than those required to produce phosphenes Because phosphenes and VEP alterations are observed only with fields below 100 Hz, such phenomena are not expected to occur in response to the low-frequency fields associated with GWEN antenna.

Endocrine System

Many studies with rodents and monkeys have demonstrated that exposure to thermogenic levels of RF radiation produces endocrine alterations, the most consistent change being an increase in plasma corticosterone SAR values in excess of 3 W/kg produce an increase in plasma corticosterone in rats that depends on secretion of adrenocorticotropic hormones by the pituitary. Decreased thyroid hormone levels have also been observed in response to thermogenic levels of RF radiation, and this response has been associated with an inhibition of thyrotropin secretion by the pituitary In general, the alteration of hormone concentrations is reversible after termination of the RF exposure. Those findings indicate that RF heating alters the complex interactions of the hypothalamic, pituitary, adrenal, and thyroid systems that are important in the maintenance of homeostasis. It is noteworthy that a 2-yr exposure of rats to nonthermogenic pulsed 2.45-GHz microwaves did not produce detectable endocrine alterations.

Studies on the possible endocrine effects of ELF electric and magnetic fields have yielded inconsistent results. Increases, decreases, and no change in plasma steroid hormones have been reported. Results of studies with dogs and rats suggest that the threshold 60-Hz electric field required to produce changes in blood concentrations of corticosterone or testosterone is in excess of 10 kV/m. Results of experiments with monkeys exposed to 60-Hz electric and magnetic fields at intensities typical of those in the vicinity of high-voltage transmission lines indicated that a decrease in neurotransmitter concentrations occurs during chronic exposure. However, there were no other observations of behavioral or physiological changes in the exposed animals.

The most widely studied effect of ELF fields on the endocrine system is an apparent depression in the nocturnal rise of pineal melatonin Changes in pineal melatonin have reportedly occurred after 2-3 wk of exposure to electric fields with intensities exceeding 1.7 kV/m in air. The effect is reversible, with a return of nocturnal pineal melatonin to control values within 3 d of termination of exposure. A similar effect on pineal melatonin has been observed after exposure of rodents to a 0.05 mT static magnetic field that was continuously switched on and off in 5-min cycles for 1 h beginning 3.5 h after the onset of darkness. Interest in this phenomenon has centered around the effects of melatonin on cell proliferation and its possible carcinostatic effects. A major problem in the interpretation of the results of studies is lack of quantitative information on the threshold fields required to alter melatonin concentration. It is also unclear whether ELF fields directly alter pinealocyte functions or whether the reported alteration in pineal melatonin production is secondary to effects of the fields on the nervous system. Further studies are needed to assess the possible influence of field-induced changes in pineal melatonin on physiological regulation and the risk of endocrine-dependent cancers. It is not now possible to extrapolate the available information obtained with 60-Hz fields or intermittent DC magnetic fields to the possible effects of fields with higher frequencies.

Immune System

Effects of RF-field exposure on cellular components of the immune system have been reported with both in vitro and in vivo test systems Lymphoblast transformation and changes in responsiveness to mitogens have been reported, although the effects observed in different laboratories have been quite variable. It appears from available information that the threshold SAR for altering lymphocyte responses to mitogens exceeds 4 W/kg with both pulsed and CW microwaves. Thermogenic levels of exposure have been found to decrease natural killer cell activity and to activate macrophages. The changes observed in components of the immune system at RF power densities that produce tissue heating are consistent with the expected effects of increased release of steroid hormones into the circulation In one study involving a 2-yr exposure of rats to a nonthermal level of pulsed 2.45-GHz microwaves (SAR, 0.4 W/kg), no significant irreversible changes were found in the concentrations of lymphocytes or their responses to mitogen stimulation.

Numerous studies have been carried out to determine the effects of ELF electric and magnetic fields on components of the immune system. In general, sinusoidal ELF fields have been found to have no significant effects on immune competence after in vivo exposure of laboratory animals.

However, reduced mitogen responses and decreased target-cell toxicity have been reported for lymphocytes exposed in vitro to pulsed magnetic fields or 60-Hz amplitude-modulated RF fields. These effects might have resulted from the relatively high current densities induced in the cell suspensions. In a study involving 60-Hz electric and magnetic fields with a sinusoidal waveform and intensities comparable with those of the fields near high-voltage power lines, no effects were observed on the immunologic functions of peripheral human and canine lymphocytes obtained from donors that either were normal or had been challenged with specific antigens.

Hematologic and Cardiovascular Systems

Several studies have been performed to assess the effects of both thermogenic and nonthermogenic levels of RF radiation on blood chemistry and blood-cell counts.

Most were conducted with 2.45-GHz microwaves, and the results indicate that power densities producing an average whole-body SAR of less than 2.5 W/kg do not produce significant alterations in hematologic indexes. In one chronic-exposure study with rabbits subjected to 2.45-GHz microwaves 23 h/d for 180 d, a small decrease was observed in eosinophil count, serum albumin concentration, and calcium concentration. None of the other 38 blood characteristics measured in the study were found to change in response to chronic RF exposure.

At high levels of RF exposures of mice leading to rectal temperature increases of 2-40C, decreased lymphocyte counts and increased neutrophil counts were observed. It was suggested that the release of adrenal steroids into the blood as a result of heat stress could have produced the changes in blood-cell counts, Thermogenic levels of pulsed or CW microwaves might also affect the cellular composition and proliferative capacity of bone-marrow cells. Both bradycardia and tachycardia have been observed in laboratory animals exposed to thermogenic levels of RF radiation with SAR values in excess of 2.5 W/kg. In general, the effects on cardiac dynamics were transient and consistent with effects expected from body heating.

As reviewed by several authors, the threshold ELF fields for producing significant cardiovascular and hematologic alterations are high. For example, cardiovascular indexes were not affected by exposure to 60-Hz, 100-kV/m fields' Acute human exposures to ELF electric fields up to 200 kV/m and magnetic fields up to 5 mT have also failed to show any consistent hematologic or cardiovascular effects.

References

- Cloude, Shane (1995). An Introduction to Electromagnetic Wave Propagation and Antennas. Springer Science and Business Media. pp. 28–33. ISBN 038791501X
- Electromagnetic, radiation, glossary: abyss.uoregon.edu, Retrieved 31 March 2018
- Herschel, William (1 January 1800). "Experiments on the Refrangibility of the Invisible Rays of the Sun. By William Herschel, LL. D. F. R. S". Philosophical Transactions of the Royal Society of London. 90: 284–292. doi:10.1098/rstl.1800.0015. JSTOR 107057
- Electromagnetic-field, physics: byjus.com, Retrieved 15 March 2018
- "Spectroscopy". National Redshift Project. National Redshift Project. Archived from the original on 1 February 2017. Retrieved 19 January 2017

- Narayanan, DL; Saladi, RN; Fox, JL (September 2010). "Ultraviolet radiation and skin cancer". International Journal of Dermatology. 49 (9): 978–86. doi:10.1111/j.1365-4632.2010.04474.x. PMID 20883261
- Applications-of-electromagnetic-field-theory: winnerscience.com, Retrieved 25 May 2018
- Bettini, Alessandro (2016). A Course in Classical Physics, Vol. 4 - Waves and Light. Springer. pp. 95, 103. ISBN 3319483293
- Liebel, F.; Kaur, S.; Ruvolo, E.; Kollias, N.; Southall, M. D. (2012). "Irradiation of Skin with Visible Light Induces Reactive Oxygen Species and Matrix-Degrading Enzymes". Journal of Investigative Dermatology. 132 (7): 1901–1907. doi:10.1038/jid.2011.476. PMID 22318388
- What-are-sources-of-electromagnetic-fields: energyforliving.co.uk, Retrieved 30 June 2018
- Haneef, Deena T. Kochunni, Jazir. "7 Differences between Fluorescence and Phosphorescence". Archived from the original on 4 September 2017. Retrieved 4 September 2017
- Delgado, J. M.; Leal, J.; Monteagudo, J. L.; Gracia, M. G. (1982). "Embryological changes induced by weak, extremely low frequency electromagnetic fields". Journal of Anatomy. 134 (Pt 3): 533–551. PMC 1167891. PMID 7107514
- Harmful-effects-of-electromagnetic-radiation: defendershield.com, Retrieved 11 July 2018
- Tipler, Paul A. (1999). Physics for Scientists and Engeneers: Vol. 1: Mechanics, Oscillations and Waves, Thermodynamics. MacMillan. p. 454. ISBN 1572594918

3

Electrodynamics

Electrodynamics is a field of theoretical physics, which is concerned with the study of the interaction between electrical charges and currents. This chapter has been carefully written to provide an easy understanding of the varied aspects of Lorentz force, electromotive force, magnetomotive force and Maxwell's equations for an extensive understanding of electrodynamics.

Electrodynamics, study of phenomena associated with charged bodies in motion and varying electric and magnetic fields since a moving charge produces a magnetic field, electrodynamics is concerned with effects such as magnetism, electromagnetic radiation, and electromagnetic induction, including such practical applications as the electric generator and the electric motor. This area of electrodynamics, often known as classical electrodynamics, was first systematically explained by the physicist James Clerk Maxwell. Maxwell's equations, a set of differential equations, describe the phenomena of this area with great generality. A more recent development is quantum electrodynamics, which was formulated to explain the interaction of electromagnetic radiation with matter, to which the laws of the quantum theory apply. The physicists P. A. M. Dirac, W. Heisenberg, and W. Pauli were the pioneers in the formulation of quantum electrodynamics. When the velocities of the charged particles under consideration become comparable with the speed of light, corrections involving the theory of relativity must be made this branch of the theory is called relativistic electrodynamics. It is applied to phenomena involved with particle accelerators and with electron tubes that are subject to high voltages and carry heavy currents.

Electrodynamics is Everywhere

It is rarely the case that people know the principles behind the phenomena and effects they encounter in their everyday lives. This is sad since school should give us a general feeling for the mechanisms that are responsible for these phenomena. In this article we shall try to connect the course of classical electrodynamics to the phenomena the theory may explain. Some of these are lightning due to a huge electric discharge, the movement of a compass caused by the magnetic field of the earth, the occurence of rainbows caused by diffraction and reflection in water droplets and the function of blow-dryers based on an electric motor and ohmic heating.

It is clear that the connection between theory and phenomena can never be even remotely complete, so we may just try to explain the main topics of electrodynamics and give some intuitive examples that everyone should be familiar with. We hope that this connection may be a source of motivation for one or the other student.

Let us begin our journey towards an understanding of electrodynamics from a historical perspective: electrodynamics in terms of its forces. We will see that naturally humans were thinking in terms of electric and magnetic forces. Their combination and the introduction of the field concept later lead to the "Golden Age of Physics" that we will discuss afterwards.

Unification of Electric and Magnetic Forces

Thousands of years passed by before mankind finally understood the existance of electric charges, humans were fascinated by the possibility of amber to attract other objects. Now we know that if we rub this material, charges are transferred from the surface to the rubbing material causing an electric field between the piece of amber and the rubbing material, i.e. some fur. Very important concepts arise already in the simple experiment of rubbing amber: there are charges that may be transferred from one body to the other. These charged bodies may attract (or repell) each other. Furthermore, this force also depends on the distance of the bodies.

Scientists concluded that charges q cause an electric field $E_q(r)$. This electric field causes a force on another *charge* q', $F(r) = q'E_q(r,t)$ which gets weaker by the squared distance between q and q'. What the scientists deduced was the so-called Coulomb force,

$$F_{el}(r) = \frac{1}{4\pi\varepsilon_0} \frac{q'q}{|r'-r|^2} \cdot \frac{r'-r}{|r'-r|} .$$

The force is named after Charles Augustin de Coulomb who found it in 1785. Here, in SI-units, $\varepsilon_0 \approx 8.85 \times 10{-}12$As/Vm is the so-called permittivity of vacuum. Again: charges cause an electric field and electric fields in turn render a force on electric charges. In this way, electric charges interact via an electric field. But of course, electrons, the carrier of charges, have a certain mass. Then, if the electrons are free, the Coulomb force will set electrons in motion.

However, the Coulomb force may not be the only present force acting on electric charges. The whole situation changes if a magnetic field B(r) is present. Then, the actual motion of the charges has to be taken into account. Roughly speaking, moving charges can be seen as currents that cause a magnetic field and interact with it - just in the same manner as charges interact with the electric field. The magnetic force on a single charge is given by,

$$F_{ma}(r) = qv \times B(r) .$$

This force was found by Thomson and Heaviside around 1880.

If both electric and magnetic fields are present, the force on a charge must be combination of both forces. This combination is called the Lorentz force after Hendrik Lorentz,

$$F(r,t) = q(E(r,t) + v \times B(r,t)) ,$$

where we now also incorporate the time explicitly as a coordinate to emphasize possible time-dependencies in the electromagnetic field, i.e. caused by other moving charges. Historically it is not clear who the actual inventor of the force is - Maxwell, Heaviside or Lorentz. The important point is that it combines electrical and magnetic forces. In the present it seems almost trivial to consider both forces together: we just added them up. But there is a huge implication in this unification. We know that electric charges cause electric fields. However, if they are moved, they also cause magnetic fields. That means, electric fields and magnetic fields have to transform into each other if just the coordinate system is moved. Electric and magnetic forces, and hence electric and magnetic fields are then naturally just two different parts of the same medal. At the end of the 19th century, these thoughts were groundbreaking and the beginning of a “Golden Age of Physics” - it turned out that Maxwell's electrodynamics, naturally combining electric and magnetic fields, was just the beginning.

Electrodynamics and the Golden Age of Physics

At the beginning of the 20th century, it was common knowledge that there exists a medium called the ether in which light can propagate just as acoustic waves can propagate in air or water. But as we know now, this is not the case: light does not need a medium to propagate. This is a consequence of Maxwell's equations, which are, mathematically speaking, invariant under Lorentz transformations, a strong contradiction against the ether theory. Furthermore, Maxwell's equations incorporate an ultimate limit for the speed of electromagnetic wave propagation, a combination of the magnetic permeability of vacuum, $\mu_0 \approx 10.57 \times 10 - 7 Vs/Am$ and ε_0:

$$c \equiv \frac{1}{\sqrt{\mu_0 \varepsilon_0}} = 299{,}792{,}458\ m/s,$$

the speed of light. It was generally believed that the Lorentz transformation behaviour is wrong and scientists tried to incorporate the ether through modifications of Maxwell's equations. It was until 1905 when a 26-year-old physicist working in a patent office in Bern argumented in "Zur Elektrodynamik bewegter Körper", engl.: "On the Electrodynamics of Moving Bodies", that everything is alright with Maxwell's equations and what consequences arise, i.e. the principle of relativity. Albert Einstein's theory of relativity and his other invaluable contributions in the same year such as the quantum hypothesis kicked of a physics golden age which endured roughly until the 70's of the 20th century with the discovery of the renormalizability of Yang-Mills theories by scientists around Gerard 't Hooft and Martinus Veltman. One of the consequences was that all quantum field theories and general relativity are gauge field theories that have a geometrical interpretation. In such an interpretation, forces appear because of curvatures. The difference is that in general relativity space-time is curved, whereas in all quantum field theories, roughly speaking, so-called groups "attached" to space-time are curved.

Classical electrodynamics can be considered as the gauge field theory that is the easiest accessible. The reason is that the corresponding group is the group of rotations on a circle, the group *U(1)*. This group has the special property: two rotations on the circle can be performed in an arbitrary order and will always have the same result - *U(1)* is "abelian". During the study of electrodynamics students will learn why this correspond to the linearity of the Maxwell equations whereas, for example, the theory of the strong interaction, quantum chromodynamics, belongs to *SU(3)*, a group that can be thought of as closely related to rotations in three dimensions which are non-abelian. Thus, the field equations of quantum chromodynamics are inevetibly nonlinear and render the theory much more complicated than electrodynamics.

So even if electrodynamics and all of its more closely related fields are not in the focus of a student's interest, she/he is strongly advised to understand electrodynamics on a solid basis since it is the first gauge theory that will be introduced during the study and provides a natural feeling for the other theories in limiting cases. It is for example not a coincidence that Newton's theory of gravitation is, at least from the mathematical point of view, equivalent to electrostatics: the corresponding potentials follow Poisson's equation, $\ddot{A}\phi_{\text{grav,el}}(\mathbf{r}) \propto \rho_{\text{grav,el}}(\mathbf{r})$.

Now we went all the way from electrical forces towards a grand unified theory, the unification of all of the mentioned forces, the fundamental goal of physics in the 21st century. We have seen how much of an impact electrodynamics had on science in general as the first discovered gauge theory. In the following we will see how electrodynamics can be tought and what insights the main topics provide.

Classical Electrodynamics

There are different approaches to teach electrodynamics. We will refer to the standard way that is historically motivated and also used on this website. Let us motivate the different topics by the phenomena we can understand and why these topics are important during later courses.

Electrostatics

In the historic approach, students first discover how electric charges cause electric fields due to Gauss's law,

$$\nabla \cdot E(r) = \rho(r) / \varepsilon_0 \ .$$

Also, the concept of an electrostatic potential is introduced using $\nabla \times E(r) = 0$. This introduction can be seen as a bridge between highschool physics and more advanced theoretical approaches. The electrostatic potential is not only a very powerful concept, i.e. to introduce multipole moments, it also allows to build up very useful intuition for electrostatics. Soon after, boundary condition problems can be solved and the concept of eigenfunctions is introduced. Eigenfunctions will help to understand quantum mechanics in later courses. The notion of, say, "locally fixed dipoles" then leads to a description of dielectric media in terms of a polarization density $P(r)$ and the closely related understanding of a permittivity $\varepsilon(r)=\varepsilon_0\varepsilon_r(r)$ relating electric field E and the electric displacement field D linearly via $D(r)=\varepsilon(r)E(r)$. The generalization to a permittivity that varies not only in space, but also depends on the frequency will later be crucially important to understand optical phenomena such as dispersion.

Electrostatics provides a starting point for a thorough understanding of electric phenomena starting from the nanoscale, i.e. to understand interactions of molecules. But electrostatics is also useful on other length scales such as the attraction of hair to a rubbed balloon, the physics of capacitors or even charge transfer processes during lightning. Usually, electrostatics makes up for more than one third of the whole electrodynamics course.

Magnetostatics

A lot of the concepts from electrostatics can be directly applied to magnetostatics, at least in vacuum. In magnetostatics, we learn how currents cause magnetic fields and which tricks exist to calculate them using Ampère's law,

$$\nabla \times B(r) = \mu_0 j(r)$$

in integral and differential forms. Again, the introduction of a potential using $\nabla \cdot B(r)=0$ will be extremely helpful to understand magnetic phenomena in terms of approximations, for example couplings of magnetic fields to magnetic dipoles. Just as in electrostatics, locally attached dipoles will be used to introduce a so-called permeability $\mu(r)=\mu_0\mu_r(r)$ relating magnetic field $H(r)$ and magnetic induction $B(r)$ via $B(r)=\mu(r)H(r)$. This notion will allow to explain magnetic media.

Magnetostatics will be necessary to understand induction, superconductivity, but also large scale phenomena like the formation of galaxies. The theory is also used later to understand quantum bits and the Ising model, one of the working horses in quantum mechanics.

Circuit Theory

One of the main reasons for the success of electronics is that electronic circuits may be written in a modularized form, i.e. containing simple elements with certain, well-defined effects. During electrodynamics, a basic circuit theory is introduced based on slowly varying fields. That means, electric and magnetic fields may now be coupled via the Maxwell-Farady equation

$$\nabla \times E(r,t) = -\partial_t B(r,t) \ ,$$

or, simply, the law of induction. Usually, basic circuits consisting of resistors, capacitors and inductors are discussed. Even though nonlinear elements such as transistors are omitted, a lot of interesting phenomena can be explained like the oscillatory behaviour of RLC-circuits or reflection of waves in transmission lines.

It is needless to say that electrical engineering and its applications hugely rely on circuit theory concepts. Electrical engineering is a huge topic in itself involving power engineering, electronics itself, microelectronics, telecommunication and so on. However, a lot of other topics rely on the simplified concepts of circuit theory, for example a deeper understanding of processes in the life sciences.

Full Electrodynamics

Up to now, electrodynamics is tought using different approximations. In the last part of the course, this limitation is entirely lifted. This is accomplished by completing Ampère's law with Maxwell's correction to couple electric and magnetic fields symmetrically:

$$\nabla \times B(r,t) = \mu_0 j(r,t) + \frac{1}{c^2}\frac{\partial E(r,t)}{\partial t} \ .$$

The now complete set of Maxwell's equations forms the necessary starting point to understand: all radiation phenomena, including electromagnetic waves and multipolar radiation, electromagnetic modes to later understand their quantization in quantum electrodynamics, optics, waveguides, light-matter-interactions, relativistic electrodynamics and its generalization to a curved space-time, and, if you wish, microwave heating.

Lorentz force

Physics Concept

The Lorentz Force is the force on a charged particle due to electric and magnetic fields. A charged particle in an electric field will always feel a force due to this field. A charged particle in a magnetic field will only feel a force due to the magnetic field if it is moving relative to this field. These two effects are often studied separately at school but the sum of these two effects creates a force that we call the Lorentz force.

A charged particle moving through a magnetic field of strength B with a speed v will feel a Lorentz force with a magnitude of:

$$F = qvB \sin \theta$$

where θ is the angle between the velocity of the particle and the magnetic field (field lines are N to S), and q is the charge of the particle. This force acts at right angles to both the magnetic field and the velocity of the particle. Various mnemonics or memory aids are used in order to remember the direction of this force - a common one is known as the left-hand rule.

A current-carrying wire in a magnetic field will feel a Lorentz force in a direction given by Fleming's left hand rule, with a magnitude of:

$$F = BIl \sin \theta$$

where l is the length of the wire in the magnetic field, I is the current flowing through the wire and θ is the angle between the wire and the magnetic field.

Force on a Moving Change in a B Field

When a charged particle moves through a magnetic field, cutting through field lines, it experiences a Lorentz Force. This force acts at right angles to both the velocity of the particle v and the magnetic field B.

The direction of this force in various situations is shown in figure below- it depends on the direction of the velocity of the particle and the magnetic field, as well as the sign of the charge of the particle. Several ways of remembering the direction of this force are described in the tabs below.

Left Hand Rule - Thumb, First Finger and Centre Finger

A common mnemonic or memory aid for the direction of the force is the left hand rule using the Thumb, First finger and Centre fingers

These are held at right angles to each other and rotated so that:

- the First finger points in the direction of the magnetic Field
- the Centre finger points in the direction of the Current (remembering that the current due to an negatively charged particle is in the opposite direction to its velocity)
- the direction that the Thumb now points in is the direction of the Thrust or Force.

A more generally useful way of working out this direction is to write the Lorentz force using vector notation. This is useful because this form of notation, using the vector cross product, appears in many different branches of Physics, and so being able to use it is more useful than memorising a rule that only refers to one specific situation.

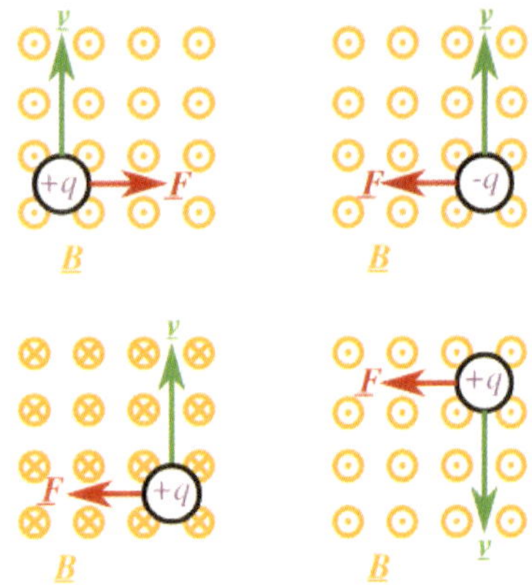

Direction of the Lorentz force in several different situations

The size of the Lorentz Force is given by:

$$F = qvB\sin\theta$$

where θ is the angle between the velocity of the particle and the magnetic field, and q is the charge of the particle.

If the particle is moving in the direction of the magnetic field, not cutting across any field lines, θ=0 and there is no Lorentz Force acting on the particle.

The Lorentz Force on particles with positive charge and those with negative charge acts in opposite directions, causing their paths to curve in opposite directions. This is how positrons (the anti-particle partner of electrons) were identified in cloud chambers.

Force on a current Carrying Wire in a Magnetic Field

A current is simply the movement of charged particles, so if a wire carrying current is inside a magnetic field then all of the charged particles flowing through it will experience a Lorentz force. This means that the overall force on the wire will be the sum of the forces on the moving charged particles, as shown in figure below.

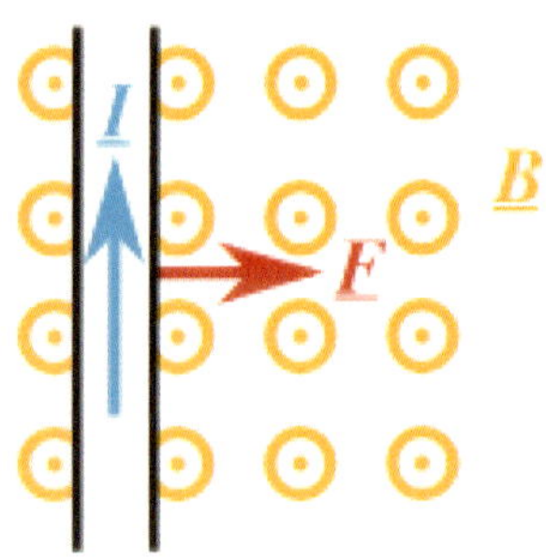

Lorentz force on a wire

For N charged particles, each with a charge q, moving at a speed v along a wire at an angle θ to a magnetic field of strength B, the total force on the wire would be:

$$F = NqvB\sin\theta$$

The magnitude of the current travelling down the wire I is given by:

$$I = nAve$$

where where n is the number density $\left(\frac{\text{number}}{\text{volume}}\right)$ of free electrons in the wire, A is the cross-sectional area of the wire, v is the speed of the electrons along the wire and e is the magnitude of the charge on an electron.

Re-arranging this to give v gives:

$$v = \frac{I}{nAe}$$

Substituting this into the equation for the magnitude of the force on the wire gives:

$$F = \frac{NqBIsin\theta}{nAe}$$

However for electrons q=−e and also by definition the number of electrons in the section of wire of length l and cross-section A is N=nAl. Using these, we get the equation for the magnitude of the force on the wire:

$$F = BIl \sin\theta$$

Electric Force

The Lorentz force is properly the total force on a charged particle moving through both a magnetic field and an electric field. The force on the particle due to the electric field is given by:

$$\underline{F} = q\underline{E}$$

as expected.

An electron exists at a point where the magnitude of the electric field is 1.0×105NC−11.0×105NC−1. What is the force on the electron due to this field?

Ans

1.6×10−14N in the direction opposite to the direction of the electric field at that point.

Lorentz Force Using Vector Notation

Using vector notation, the force on a moving charge q in a magnetic field B is given by:

$$\underline{F} = q\underline{v} \times \underline{B}$$

You will also see this written with the notation $\underline{F} = q\underline{v} \wedge \underline{B}$ in which the symbol for the cross product ∧ is now used instead of ×.

By the rules of vector notation, this means that F must be at right angles to both v and B, and in fact using the right-hand screw rule gives the correct direction for F. The magnitude of F is given by $F = qvBsin\theta$, where θ is the angle between v and B. This is the same as was used previously.

Similarly, the Lorentz force on a current-carrying wire in a magnetic field B can be written in vector notation as:

$$\underline{\boldsymbol{F}} = I\int^{\text{wire}} \mathrm{d}\underline{\boldsymbol{l}} \times \underline{\boldsymbol{B}}$$

where $d\underline{l}$ is an infinitesimal displacement along the wire, B is the magnetic field at that point, and the integral is taken over the entire length of the wire.

This is in effect $\underline{\boldsymbol{F}} = I\underline{\boldsymbol{l}} \times \underline{\boldsymbol{B}}$ but for a charged particle considered above, the force on it in a magnetic field is at a point, whereas the wire is an extended rigid body. So if different sections of the wire are in different strengths of magnetic field, then we must sum up the force on each section to obtain the resultant force on the wire.

By the rules of vector notation, this means that $\underline{\boldsymbol{F}}$ must be at right angles to both $\underline{v}$ and B, and in fact using the right-hand screw rule gives the correct direction for $\underline{\boldsymbol{F}}$.

Adding the force on a charged particle due to an electric field $(\underline{\boldsymbol{F}} = q\underline{E})$ gives a total Lorentz force on a particle of:

$$\underline{F} = q(\underline{E} + \underline{v} \times \underline{B})$$

Equation

Charged Particle

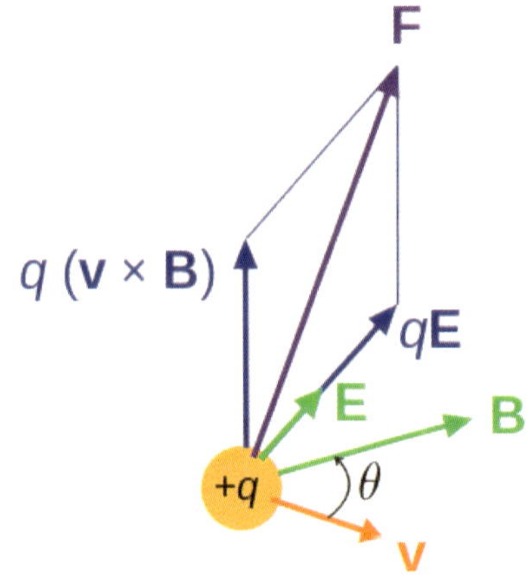

Lorentz force F on a charged particle (of charge q) in motion (instantaneous velocity **v**). The **E** field and **B** field vary in space and time.

The force F acting on a particle of electric charge q with instantaneous velocity v, due to an external electric field E and magnetic field B, is given by (in SI units):

$$\mathbf{F} = q(\mathbf{E} + \mathbf{v} \times \mathbf{B})$$

where × is the vector cross product (all boldface quantities are vectors). In terms of cartesian components, we have:

$$F_x = q(E_x + v_y B_z - v_z B_y),$$

$$F_y = q(E_y + v_z B_x - v_x B_z),$$

$$F_z = q(E_z + v_x B_y - v_y B_x).$$

In general, the electric and magnetic fields are functions of the position and time. Therefore, explicitly, the Lorentz force can be written as:

$$\mathbf{F}(\mathbf{r}, \dot{\mathbf{r}}, t, q) = q[\mathbf{E}(\mathbf{r}, t) + \dot{\mathbf{r}} \times \mathbf{B}(\mathbf{r}, t)]$$

in which r is the position vector of the charged particle, *t* is time, and the overdot is a time derivative.

A positively charged particle will be accelerated in the *same* linear orientation as the E field, but will curve perpendicularly to both the instantaneous velocity vector v and the B field according to the right-hand rule (in detail, if the fingers of the right hand are extended to point in the direction of v and are then curled to point in the direction of B, then the extended thumb will point in the direction of F).

The term qE is called the electric force, while the term qv × B is called the magnetic force. According to some definitions, the term "Lorentz force" refers specifically to the formula for the magnetic force, with the *total* electromagnetic force (including the electric force) given some other (nonstandard) name. In the text, the term "Lorentz force" will refer to the expression for the total force.

The magnetic force component of the Lorentz force manifests itself as the force that acts on a current-carrying wire in a magnetic field. In that context, it is also called the Laplace force.

The Lorentz force is a force exerted by the electromagnetic field on the charged particle, that is, it is the rate at which linear momentum is transferred from the electromagnetic field to the particle. Associated with it is the power which is the rate at which energy is transferred from the electromagnetic field to the particle. That power is,

$$\mathbf{v} \cdot \mathbf{F} = q\mathbf{v} \cdot \mathbf{E}\,.$$

Notice that the magnetic field does not contribute to the power because the magnetic force is always perpendicular to the velocity of the particle.

Continuous Charge Distribution

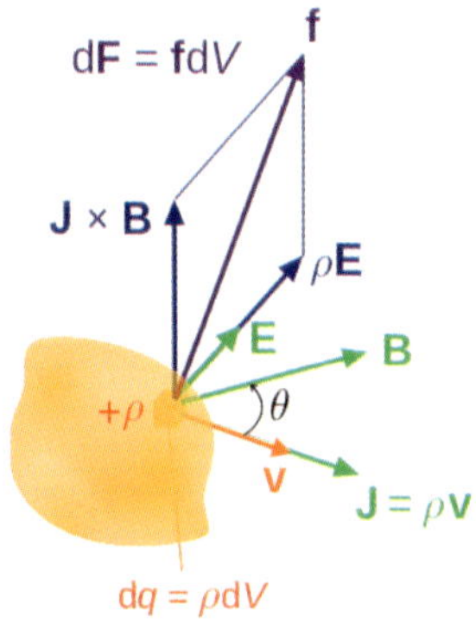

Lorentz force (per unit 3-volume) f on a continuous charge distribution (charge density ρ) in motion. The 3-current density J corresponds to the motion of the charge element dq in volume element dV and varies throughout the continuum.

For a continuous charge distribution in motion, the Lorentz force equation becomes:

$$d\mathbf{F} = dq(\mathbf{E} + \mathbf{v} \times \mathbf{B})$$

where $d\mathbf{F}$ is the force on a small piece of the charge distribution with charge dq. If both sides of this equation are divided by the volume of this small piece of the charge distribution dV, the result is:

$$\mathbf{f} = \rho(\mathbf{E} + \mathbf{v} \times \mathbf{B})$$

where f is the *force density* (force per unit volume) and ρ is the charge density (charge per unit volume). Next, the current density corresponding to the motion of the charge continuum is,

$$\mathbf{J} = \rho\mathbf{v}$$

so the continuous analogue to the equation is,

$$\mathbf{f} = \rho\mathbf{E} + \mathbf{J} \times \mathbf{B}$$

The total force is the volume integral over the charge distribution:

$$\mathbf{F} = \iiint(\rho\mathbf{E} + \mathbf{J} \times \mathbf{B})dV.$$

By eliminating ρ and J, using Maxwell's equations, and manipulating using the theorems of vector calculus, this form of the equation can be used to derive the Maxwell stress tensor σ, in turn this can be combined with the Poynting vector S to obtain the electromagnetic stress–energy tensor T used in general relativity.

In terms of σ and S, another way to write the Lorentz force (per unit volume) is

$$\mathrm{f} = \nabla \cdot \sigma - \frac{1}{c^2}\frac{\partial s}{\partial t}$$

where c is the speed of light and ∇· denotes the divergence of a tensor field. Rather than the amount of charge and its velocity in electric and magnetic fields, this equation relates the energy flux (flow of *energy* per unit time per unit distance) in the fields to the force exerted on a charge distribution.

The density of power associated with the Lorentz force in a material medium is,

$$\mathbf{J} \cdot \mathbf{E}.$$

If we separate the total charge and total current into their free and bound parts, we get that the density of the Lorentz force is,

$$\mathbf{f} = (\rho_f - \nabla \cdot \mathbf{P})\mathbf{E} + (\mathbf{J}_f + \nabla \times \mathbf{M} + \frac{\partial \mathbf{P}}{\partial t}) \times \mathbf{B}.$$

where: ρ_f is the density of free charge; P is the polarization density; J_f is the density of free current; and M is the magnetization density. In this way, the Lorentz force can explain the torque applied to a permanent magnet by the magnetic field. The density of the associated power is,

$$(\mathbf{J}_f + \nabla \times \mathbf{M} + \frac{\partial \mathbf{P}}{\partial t}) \cdot \mathbf{E}.$$

Equation in Cgs Units

The above-mentioned formulae use SI units which are the most common among experimentalists, technicians, and engineers. In cgs-Gaussian units, which are somewhat more common among theoretical physicists, one has instead,

$$\mathbf{F} = q_{\text{cgs}}\left(\mathbf{E}_{\text{cgs}} + \frac{\mathbf{v}}{c} \times \mathbf{B}_{\text{cgs}}\right).$$

where c is the speed of light. Although this equation looks slightly different, it is completely equivalent, since one has the following relations:

$$q_{\text{cgs}} = \frac{q_{\text{SI}}}{\sqrt{4\pi\epsilon_0}}, \quad \mathbf{E}_{\text{cgs}} = \sqrt{4\pi\epsilon_0}\,\mathbf{E}_{\text{SI}}, \quad \mathbf{B}_{\text{cgs}} = \sqrt{4\pi/\mu_0}\,\mathbf{B}_{\text{SI}}, \quad c = \frac{1}{\sqrt{\epsilon_0\mu_0}}.$$

where ε_0 is the vacuum permittivity and μ_0 the vacuum permeability. In practice, the subscripts "cgs" and "SI" are always omitted, and the unit system has to be assessed from context.

Trajectories of Particles Due to the Lorentz Force

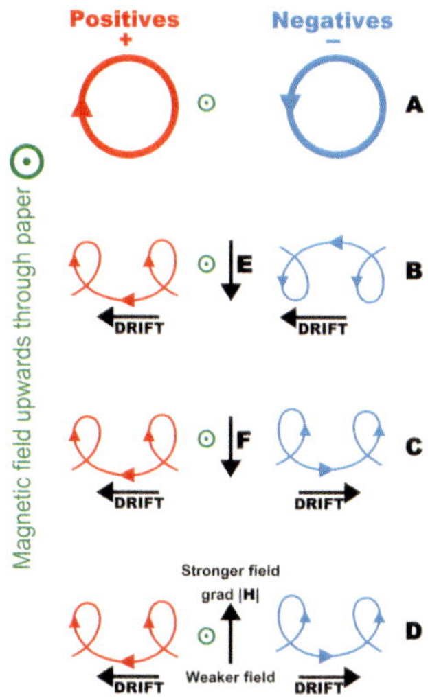

Charged particle drifts in a homogeneous magnetic field. (A) No disturbing force (B) With an electric field, E (C) With an independent force, F (e.g. gravity) (D) In an inhomogeneous magnetic field, grad H

In many cases of practical interest, the motion in a magnetic field of an electrically charged particle (such as an electron or ion in a plasma) can be treated as the superposition of a relatively fast circular motion around a point called the guiding center and a relatively slow drift of this point. The drift speeds may differ for various species depending on their charge states, masses, or temperatures, possibly resulting in electric currents or chemical separation.

Significance of the Lorentz Force

While the modern Maxwell's equations describe how electrically charged particles and currents or moving charged particles give rise to electric and magnetic fields, the Lorentz force law completes that picture by describing the force acting on a moving point charge q in the presence of electromagnetic fields. The Lorentz force law describes the effect of E and B upon a point charge, but such electromagnetic forces are not the entire picture. Charged particles are possibly coupled to other forces, notably gravity and nuclear forces. Thus, Maxwell's equations do not stand separate from other physical laws, but are coupled to them via the charge and current densities. The response of a point charge to the Lorentz law is one aspect; the generation of E and B by currents and charges is another.

In real materials the Lorentz force is inadequate to describe the collective behavior of charged particles, both in principle and as a matter of computation. The charged particles in a material medium not only respond to the E and B fields but also generate these fields. Complex transport equations must be solved to determine the time and spatial response of charges, for example, the Boltzmann equation or the Fokker–Planck equation or the Navier–Stokes equations. For example, see magnetohydrodynamics, fluid dynamics, electrohydrodynamics, superconductivity, stellar evolution. An entire physical apparatus for dealing with these matters has developed.

EMF

The magnetic force ($q\mathbf{v} \times$ B) component of the Lorentz force is responsible for *motional* electromotive force (or *motional EMF*), the phenomenon underlying many electrical generators. When a conductor is moved through a magnetic field, the magnetic field exerts opposite forces on electrons and nuclei in the wire, and this creates the EMF. The term "motional EMF" is applied to this phenomenon, since the EMF is due to the *motion* of the wire.

In other electrical generators, the magnets move, while the conductors do not. In this case, the EMF is due to the electric force (qE) term in the Lorentz Force equation. The electric field in question is created by the changing magnetic field, resulting in an *induced* EMF, as described by the Maxwell–Faraday equation (one of the four modern Maxwell's equations).

Both of these EMFs, despite their apparently distinct origins, are described by the same equation, namely, the EMF is the rate of change of magnetic flux through the wire. (This is Faraday's law of induction, see below.) Einstein's special theory of relativity was partially motivated by the desire to better understand this link between the two effects. In fact, the electric and magnetic fields are different facets of the same electromagnetic field, and in moving from one inertial frame to another, the solenoidal vector field portion of the E-field can change in whole or in part to a B-field or *vice versa*.

Lorentz Force and Faraday's Law of Induction

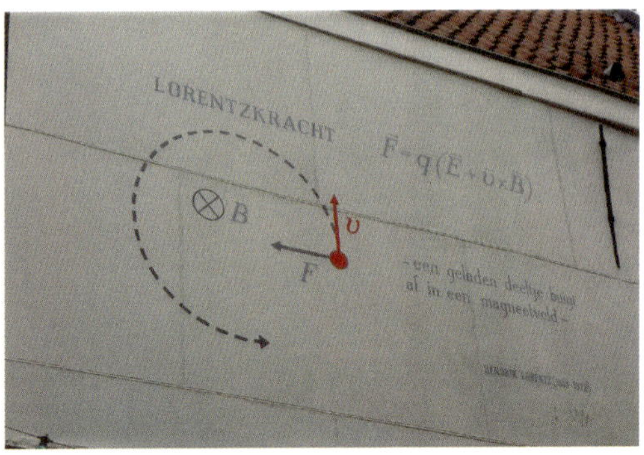

Lorentz force -image on a wall in Leiden

Given a loop of wire in a magnetic field, Faraday's law of induction states the induced electromotive force (EMF) in the wire is:

$$\mathcal{E} = -\frac{d\Phi_B}{dt}$$

where,

$$\Phi_B = \iint_{\Sigma(t)} d\mathbf{A} \cdot \mathbf{B}(\mathbf{r}, t)$$

is the magnetic flux through the loop, B is the magnetic field, $\Sigma(t)$ is a surface bounded by the closed contour $\partial\Sigma(t)$, at all at time t, dA is an infinitesimal vector area element of $\Sigma(t)$ (magnitude is the area of an infinitesimal patch of surface, direction is orthogonal to that surface patch).

The *sign* of the EMF is determined by Lenz's law. Note that this is valid for not only a *stationary* wire – but also for a *moving* wire.

From Faraday's law of induction (that is valid for a moving wire, for instance in a motor) and the Maxwell Equations, the Lorentz Force can be deduced. The reverse is also true, the Lorentz force and the Maxwell Equations can be used to derive the Faraday Law.

Let $\mathbf{\Sigma}(t)$ be the moving wire, moving together without rotation and with constant velocity $\mathbf{v}$ and $\Sigma(t)$ be the internal surface of the wire. The EMF around the closed path $\partial\Sigma(t)$ is given by:

$$\mathcal{E} = \oint_{\partial\Sigma(t)} d\ell \cdot \mathbf{F} / q$$

where,

$$\mathbf{E} = \mathbf{F} / q$$

is the electric field and $d\ell$ is an infinitesimal vector element of the contour $\partial\Sigma(t)$.

NB: Both $d\ell$ and dA have a sign ambiguity; to get the correct sign, the right-hand rule is used.

The above result can be compared with the version of Faraday's law of induction that appears in the modern Maxwell's equations, called here the *Maxwell–Faraday equation*:

$$\nabla \times \mathbf{E} = -\frac{\partial \mathbf{B}}{\partial t}.$$

The Maxwell–Faraday equation also can be written in an *integral form* using the Kelvin–Stokes theorem.

So we have, the Maxwell Faraday equation:

$$\oint_{\partial\Sigma(t)} d\ell \cdot \mathbf{E}(\mathbf{r}, t) = -\iint_{\Sigma(t)} d\mathbf{A} \cdot \frac{d\mathbf{B}(\mathbf{r}, t)}{dt}$$

and the Faraday Law,

$$\oint_{\partial\Sigma(t)} d\ell \cdot \mathbf{F} / q(\mathbf{r}, t) = -\frac{d}{dt} \iint_{\Sigma(t)} d\mathbf{A} \cdot \mathbf{B}(\mathbf{r}, t).$$

The two are equivalent if the wire is not moving. Using the Leibniz integral rule and that *div* B = 0, results in,

$$\oint_{\partial\Sigma(t)} d\ell \cdot \mathbf{F}/q(\mathbf{r},t) = -\iint_{\Sigma(t)} d\mathbf{A} \cdot \frac{\partial}{\partial t}\mathbf{B}(\mathbf{r},t) + \oint_{\partial\Sigma(t)} \mathbf{v} \times \mathbf{B} d\ell$$

and using the Maxwell Faraday equation,

$$\oint_{\partial\Sigma(t)} d\ell \cdot \mathbf{F}/q(\mathbf{r},t) = \oint_{\partial\Sigma(t)} d\ell \cdot \mathbf{E}(\mathbf{r},t) + \oint_{\partial\Sigma(t)} \mathbf{v} \times \mathbf{B}(\mathbf{r},t) d\ell$$

since this is valid for any wire position it implies that,

$$\mathbf{F} = q\mathbf{E}(\mathbf{r},t) + q\mathbf{v} \times \mathbf{B}(\mathbf{r},t).$$

Faraday's law of induction holds whether the loop of wire is rigid and stationary, or in motion or in process of deformation, and it holds whether the magnetic field is constant in time or changing. However, there are cases where Faraday's law is either inadequate or difficult to use, and application of the underlying Lorentz force law is necessary.

If the magnetic field is fixed in time and the conducting loop moves through the field, the magnetic flux Φ_B linking the loop can change in several ways. For example, if the B-field varies with position, and the loop moves to a location with different B-field, Φ_B will change. Alternatively, if the loop changes orientation with respect to the B-field, the B·dA differential element will change because of the different angle between B and dA, also changing Φ_B. As a third example, if a portion of the circuit is swept through a uniform, time-independent B-field, and another portion of the circuit is held stationary, the flux linking the entire closed circuit can change due to the shift in relative position of the circuit's component parts with time (surface $\partial\Sigma(t)$ time-dependent). In all three cases, Faraday's law of induction then predicts the EMF generated by the change in Φ_B.

Note that the Maxwell Faraday's equation implies that the Electric Field E is non conservative when the Magnetic Field **B** varies in time, and is not expressible as the gradient of a scalar field, and not subject to the gradient theorem since its rotational is not zero.

Lorentz Force in Terms of Potentials

The E and B fields can be replaced by the magnetic vector potential A and (scalar) electrostatic potential ϕ by,

$$\mathbf{E} = -\nabla\phi - \frac{\partial \mathbf{A}}{\partial t}$$

$$\mathbf{B} = \nabla \times \mathbf{A}$$

where ∇ is the gradient, ∇· is the divergence, ∇× is the curl.

The force becomes,

$$\mathbf{F} = q\left[-\nabla\phi - \frac{\partial \mathbf{A}}{\partial t} + \mathbf{v} \times (\nabla \times \mathbf{A})\right]$$

and using an identity for the triple product (with Feynman's subscript notation ∇_A to specify that ∇ operates only on **A**) simplifies to,

$$\mathbf{F} = q\left[-\nabla\phi - \frac{\partial \mathbf{A}}{\partial t} + \nabla_{\mathbf{A}}(\mathbf{v}\cdot\mathbf{A}) - (\mathbf{v}\cdot\nabla)\mathbf{A}\right].$$

Using the chain rule, the total derivative of **A** is:

$$\frac{\mathrm{d}\mathbf{A}}{\mathrm{d}t} = \frac{\partial \mathbf{A}}{\partial t} + (\mathbf{v}\cdot\nabla)\mathbf{A}$$

so the above expression can be rewritten as:

$$\mathbf{F} = q\left[-\nabla_{\phi,\mathbf{A}}(\phi - \mathbf{v}\cdot\mathbf{A}) - \frac{\mathrm{d}\mathbf{A}}{\mathrm{d}t}\right].$$

Dropping the Feynman notation now (with the understanding that ∇ still doesn't operate on $\mathbf{v} = \dot{\mathbf{x}}$), we can put the equation into the convenient Euler–Lagrange form,

$$\mathbf{F} = q\left[-\nabla_{\mathbf{x}}(\phi - \dot{\mathbf{x}}\cdot\mathbf{A}) + \frac{\mathrm{d}}{\mathrm{d}t}\nabla_{\dot{\mathbf{x}}}(\phi - \dot{\mathbf{x}}\cdot\mathbf{A})\right]$$

Where,

$$\nabla_{\mathbf{x}} = \hat{x}\frac{\partial}{\partial x} + \hat{y}\frac{\partial}{\partial y} + \hat{z}\frac{\partial}{\partial z}$$

and,

$$\nabla_{\dot{\mathbf{x}}} = \hat{x}\frac{\partial}{\partial \dot{x}} + \hat{y}\frac{\partial}{\partial \dot{y}} + \hat{z}\frac{\partial}{\partial \dot{z}}$$

Lorentz Force and Analytical Mechanics

The Lagrangian for a charged particle of mass m and charge q in an electromagnetic field equivalently describes the dynamics of the particle in terms of its *energy*, rather than the force exerted on it. The classical expression is given by:

$$L = \frac{m}{2}\dot{\mathbf{r}}\cdot\dot{\mathbf{r}} + q\mathbf{A}\cdot\dot{\mathbf{r}} - q\phi$$

where A and ϕ are the potential fields as above. Using Lagrange's equations, the equation for the Lorentz force can be obtained.

Derivation of Lorentz Force From Classical Lagrangian (SI units)

For an A field, a particle moving with velocity $v = \dot{\mathbf{r}}$ has potential momentum $q\mathbf{A}(\mathbf{r},t)$, so its potental energy is $q\mathbf{A}(\mathbf{r},t)\cdot\dot{\mathbf{r}}$. For a ϕ field, the particle's potential energy is $q\phi(\mathbf{r},t)$.

The total potential energy is then:

$$V = q\phi - q\mathbf{A}\cdot\dot{\mathbf{r}}$$

and the kinetic energy is:

$$T = \frac{m}{2}\dot{\mathbf{r}} \cdot \dot{\mathbf{r}}$$

hence the Lagrangian:

$$L = T - V = \frac{m}{2}\dot{\mathbf{r}} \cdot \dot{\mathbf{r}} + q\mathbf{A} \cdot \dot{\mathbf{r}} - q\phi$$

$$L = \frac{m}{2}(\dot{x}^2 + \dot{y}^2 + \dot{z}^2) + q(\dot{x}A_x + \dot{y}A_y + \dot{z}A_z) - q\phi$$

Lagrange's equations are

$$\frac{\mathrm{d}}{\mathrm{d}t}\frac{\partial L}{\partial \dot{x}} = \frac{\partial L}{\partial x}$$

(same for y and z). So calculating the partial derivatives:

$$\begin{aligned}
\frac{\mathrm{d}}{\mathrm{d}t}\frac{\partial L}{\partial \dot{x}} &= m\ddot{x} + q\frac{\mathrm{d}A_x}{\mathrm{d}t} \\
&= m\ddot{x} + \frac{q}{\mathrm{d}t}\left(\frac{\partial A_x}{\partial t}dt + \frac{\partial A_x}{\partial x}dx + \frac{\partial A_x}{\partial y}dy + \frac{\partial A_x}{\partial z}dz\right) \\
&= m\ddot{x} + q\left(\frac{\partial A_x}{\partial t} + \frac{\partial A_x}{\partial x}\dot{x} + \frac{\partial A_x}{\partial y}\dot{y} + \frac{\partial A_x}{\partial z}\dot{z}\right)
\end{aligned}$$

$$\frac{\partial L}{\partial x} = -q\frac{\partial \phi}{\partial x} + q\left(\frac{\partial A_x}{\partial x}\dot{x} + \frac{\partial A_y}{\partial x}\dot{y} + \frac{\partial A_z}{\partial x}\dot{z}\right)$$

equating and simplifying:

$$m\ddot{x} + q\left(\frac{\partial A_x}{\partial t} + \frac{\partial A_x}{\partial x}\dot{x} + \frac{\partial A_x}{\partial y}\dot{y} + \frac{\partial A_x}{\partial z}\dot{z}\right) = -q\frac{\partial \phi}{\partial x} + q\left(\frac{\partial A_x}{\partial x}\dot{x} + \frac{\partial A_y}{\partial x}\dot{y} + \frac{\partial A_z}{\partial x}\dot{z}\right)$$

$$\begin{aligned}
F_x &= -q\left(\frac{\partial \phi}{\partial x} + \frac{\partial A_x}{\partial t}\right) + q\left[\dot{y}\left(\frac{\partial A_y}{\partial x} - \frac{\partial A_x}{\partial y}\right) + \dot{z}\left(\frac{\partial A_z}{\partial x} - \frac{\partial A_x}{\partial z}\right)\right] \\
&= qE_x + q[\dot{y}(\nabla \times \mathbf{A})_z - \dot{z}(\nabla \times \mathbf{A})_y] \\
&= qE_x + q[\dot{\mathbf{r}} \times (\nabla \times \mathbf{A})]_x \\
&= qE_x + q(\dot{\mathbf{r}} \times \mathbf{B})_x
\end{aligned}$$

and similarly for the y and z directions. Hence the force equation is:

$$\mathbf{F} = q(\mathbf{E} + \dot{\mathbf{r}} \times \mathbf{B})$$

The potential energy depends on the velocity of the particle, so the force is velocity dependent, so it is not conservative.

The relativistic Lagrangian is,

$$L = -mc^2\sqrt{1-\left(\frac{\dot{\mathbf{r}}}{c}\right)^2} + q\mathbf{A}(\mathbf{r})\cdot\dot{\mathbf{r}} - q\phi(\mathbf{r})$$

The action is the relativistic arclength of the path of the particle in space time, minus the potential energy contribution, plus an extra contribution which quantum mechanically is an extra phase a charged particle gets when it is moving along a vector potential.

Derivation of Lorentz Force From Relativistic Lagrangian (SI units)

The equations of motion derived by extremizing the action:

$$\frac{d\mathbf{P}}{dt} = \frac{\partial L}{\partial \mathbf{r}} = q\frac{\partial \mathbf{A}}{\partial \mathbf{r}}\cdot\dot{\mathbf{r}} - q\frac{\partial \phi}{\partial \mathbf{r}}$$

$$\mathbf{P} - q\mathbf{A} = \frac{m\dot{\mathbf{r}}}{\sqrt{1-\left(\frac{\dot{\mathbf{r}}}{c}\right)^2}}$$

are the same as Hamilton's equations of motion:

$$\frac{d\mathbf{r}}{dt} = \frac{\partial}{\partial \mathbf{p}}\left(\sqrt{(\mathbf{P}-q\mathbf{A})^2 + (mc^2)^2} + q\phi\right)$$

$$\frac{d\mathbf{p}}{dt} = -\frac{\partial}{\partial \mathbf{r}}\left(\sqrt{(\mathbf{P}-q\mathbf{A})^2 + (mc^2)^2} + q\phi\right)$$

both are equivalent to the noncanonical form:

$$\frac{d}{dt}\left(\frac{m\dot{\mathbf{r}}}{\sqrt{1-\left(\frac{\dot{\mathbf{r}}}{c}\right)^2}}\right) = q\left(\mathbf{E} + \dot{\mathbf{r}}\times\mathbf{B}\right).$$

This formula is the Lorentz force, representing the rate at which the EM field adds relativistic momentum to the particle.

Relativistic Form of the Lorentz Force

Covariant Form of the Lorentz force

Using the metric signature $(1, -1, -1, -1)$, the Lorentz force for a charge q can be written in covariant form:

$$\frac{dp^{\alpha}}{d\tau} = qF^{\alpha\beta}U_{\beta}$$

where pα is the four-momentum, defined as,

$$p^{\alpha} = (p_0, p_1, p_2, p_3) = (\gamma mc, p_x, p_y, p_z),$$

τ the proper time of the particle, $F^{\alpha\beta}$ the contravariant electromagnetic tensor,

$$F^{\alpha\beta} = \begin{pmatrix} 0 & -E_x/c & -E_y/c & -E_z/c \\ E_x/c & 0 & -B_z & B_y \\ E_y/c & B_z & 0 & -B_x \\ E_z/c & -B_y & B_x & 0 \end{pmatrix}$$

and U is the covariant 4-velocity of the particle, defined as:

$$U_{\beta} = (U_0, U_1, U_2, U_3) = \gamma(c, -v_x, -v_y, -v_z),$$

in which,

$$\gamma(v) = \frac{1}{\sqrt{1-\frac{v^2}{c^2}}} = \frac{1}{\sqrt{1-\frac{v_x^2+v_y^2+v_z^2}{c^2}}}$$

is the Lorentz factor.

The fields are transformed to a frame moving with constant relative velocity by:

$$F'^{\mu\nu} = \Lambda^{\mu}{}_{\alpha}\Lambda^{\nu}{}_{\beta}F^{\alpha\beta},$$

where $\Lambda^{\mu}{}_{\alpha}$ is the Lorentz transformation tensor.

Translation to Vector Notation

The $\alpha = 1$ component (x-component) of the force is,

$$\frac{dp^1}{d\tau} = qU_{\beta}F^{1\beta} = q\left(U_0F^{10} + U_1F^{11} + U_2F^{12} + U_3F^{13}\right).$$

Substituting the components of the covariant electromagnetic tensor F yields

$$\frac{dp^1}{d\tau} = q\left[U_0\left(\frac{E_x}{c}\right) + U_2(-B_z) + U_3(B_y)\right].$$

Using the components of covariant four-velocity yields,

$$\begin{aligned}\frac{dp^1}{d\tau} &= q\gamma\left[c\left(\frac{E_x}{c}\right) + (-v_y)(-B_z) + (-v_z)(B_y)\right] \\ &= q\gamma\left(E_x + v_yB_z - v_zB_y\right) \\ &= q\gamma\left[E_x + (\mathbf{v}\times\mathbf{B})_x\right].\end{aligned}$$

The calculation for $\alpha = 2, 3$ (force components in the y and z directions) yields similar results, so collecting the 3 equations into one:

$$\frac{\mathrm{d}\mathbf{p}}{\mathrm{d}\tau} = q\gamma\left(\mathbf{E} + \mathbf{v} \times \mathbf{B}\right),$$

and since differentials in coordinate time dt and proper time $d\tau$ are related by the Lorentz factor,

$$dt = \gamma(v)d\tau,$$

so we arrive at,

$$\frac{\mathrm{d}\mathbf{p}}{\mathrm{d}t} = q\left(\mathbf{E} + \mathbf{v} \times \mathbf{B}\right).$$

This is precisely the Lorentz force law, however, it is important to note that **p** is the relativistic expression,

$$\mathbf{p} = \gamma(v)m_0\mathbf{v}.$$

Lorentz Force in Spacetime Algebra (STA)

The electric and magnetic fields are dependent on the velocity of an observer, so the relativistic form of the Lorentz force law can best be exhibited starting from a coordinate-independent expression for the electromagnetic and magnetic fields $\mathcal{F}$, and an arbitrary time-direction, γ_0. This can be settled through Space-Time Algebra (or the geometric algebra of space-time), a type of Clifford's Algebra defined on a pseudo-Euclidean space, as

$$\mathbf{E} = (\mathcal{F} \cdot \gamma_0)\gamma_0$$

and

$$i\mathbf{B} = (\mathcal{F} \wedge \gamma_0)\gamma_0$$

$\mathcal{F}$ is a space-time bivector (an oriented plane segment, just like a vector is an oriented line segment), which has six degrees of freedom corresponding to boosts (rotations in space-time planes) and rotations (rotations in space-space planes). The dot product with the vector γ_0 pulls a vector (in the space algebra) from the translational part, while the wedge-product creates a trivector (in the space algebra) who is dual to a vector which is the usual magnetic field vector. The relativistic velocity is given by the (time-like) changes in a time-position vector $v = \dot{x}$, where

$$1,$$

(which shows our choice for the metric) and the velocity is,

$$\mathbf{v} = cv \wedge \gamma_0 / (v \cdot \gamma_0).$$

The proper (invariant is an inadequate term because no transformation has been defined) form of the Lorentz force law is simply,

$$F = q\mathcal{F} \cdot v$$

Note that the order is important because between a bivector and a vector the dot product is anti-symmetric. Upon a space time split like one can obtain the velocity, and fields as above yielding the usual expression.

Lorentz Force in General Relativity

In the general theory of relativity the equation of motion for a particle with mass m and charge , moving in a space with metric tensor g_{ab} and electromagnetic field F_{ab}, is given as,

$$m\frac{du_c}{ds} - m\frac{1}{2}g_{ab,c}u^a u^b = eF_{cb}u^b,$$

where $u^a = dx^a / ds$ (dx^a is taken along the trajectory), $g_{ab,c} = \partial g_{ab} / \partial x^c$, and $ds^2 = g_{ab}dx^a dx^b$.

The equation can also be written as,

$$m\frac{du_c}{ds} - m\Gamma_{abc}u^a u^b = eF_{cb}u^b,$$

where Γ_{abc} is the Christoffel symbol (of the torsion-free metric connection in general relativity), or as,

$$m\frac{Du_c}{ds} = eF_{cb}u^b,$$

where D is the covariant differential in general relativity (metric, torsion-free).

Applications

The Lorentz force occurs in many devices, including:

- Cyclotrons and other circular path particle accelerators
- Mass spectrometers
- Velocity Filters
- Magnetrons
- Lorentz force velocimetry

In its manifestation as the Laplace force on an electric current in a conductor, this force occurs in many devices including:

- Electric motors
- Railguns
- Linear motors
- Loudspeakers
- Magnetoplasmadynamic thrusters
- Electrical generators
- Homopolar generators
- Linear alternators

Electromotive Force

Electromotive force, abbreviation E or emf, energy per unit electric charge that is imparted by an energy source, such as an electric generator or a battery. Energy is converted from one form to another in the generator or battery as the device does work on the electric charge being transferred within itself. One terminal of the device becomes positively charged, the other becomes negatively charged. The work done on a unit of electric charge, or the energy thereby gained per unit electric charge, is the electromotive force. Electromotive force is the characteristic of any energy source capable of driving electric charge around a circuit. It is abbreviated E in the international metric system but also, popularly, as emf.

Despite its name, electromotive force is not actually a force. It is commonly measured in units of volts, equivalent in the metre–kilogram–second system to one joule per coulomb of electric charge. In the electrostatic units of the centimetre–gram–second system, the unit of electromotive force is the statvolt, or one erg per electrostatic unit of charge.

Calculation of EMF

The EMF can be written in terms of the internal resistance of the battery (r) where: $\varepsilon = I(r + R)$

Which from Ohm's law, we can then rearrange this in terms of the terminal resistance: $\varepsilon = V + Ir$

The EMF of the cell can be determined by measuring the voltage across the cell using a voltmeter and the current in the circuit using an ammeter for various resistances. We can then set up a circuit to determine EMF as shown below.

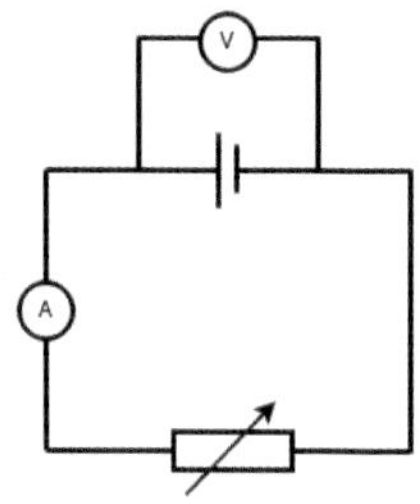

Motional Electromotive Force

An emf induced by the motion of the conductor across the magnetic field is a motional electromotive force. The equation is given by E = -vLB. This equation is true as long as the velocity, field, and length are mutually perpendicular. The minus sign associated with the Lenz's law.

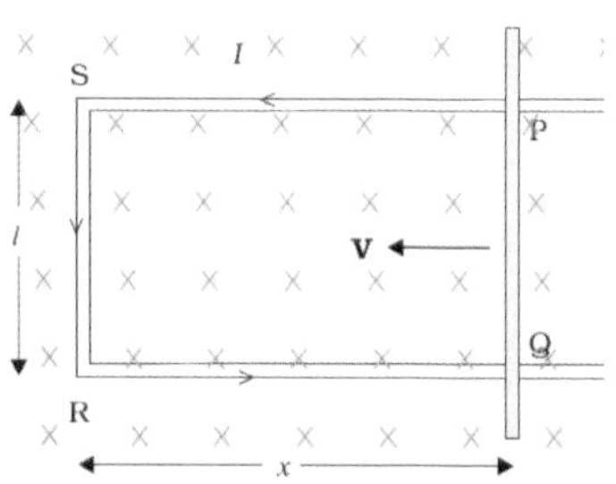

For us to understand the motional electromotive force, let us make a particular setup. Let us take a rectangular coil, a metal rod of length L, moving with velocity V, through a magnetic field B. There is a magnetic field at some location.

Length, velocity and magnetic field should always be at a right angle with each other. The direction of the magnetic field is going inside. Assume the metal rod is frictionless that means there is no loss of energy due to friction and we apply a uniform magnetic field. The conductor rod is moved with a constant velocity and placed in the magnetic field.

$$\Phi_B = Blx$$

But 'x' changes with time,

$$\mathrm{E} = -\frac{d\Phi_B}{dt} = -\frac{d}{dt}(\mathrm{Blx}) = -Bl\frac{dx}{dt}$$

$$\mathrm{E} = \mathbf{B}lv$$

The induced emf Blv is motion electromotive force. So we produce emf by moving a conductor inside the uniform magnetic field. The power required to move a conductor rod in a magnetic field is,

$$P = \frac{B^2 l^2 v^2}{R}$$

Where,

- B is the magnetic field,
- l is the length of the conductor
- v is the velocity of the conductor
- R is the resistance.

The magnetic flux associated with the coil is given by $\Phi = \mathrm{BA}\ \cos\theta$. We know that $\cos\theta = 0$, so $\Phi = \mathrm{BA}$. The motion of electromotive force can be further explained by Lorentz force which acts on free charge carriers. The Lorentz force on charge is:

$$\mathrm{F} = \mathrm{qVB}$$

Solved Questions

Q1. A coil having n turns and area A is initially placed with its plane normal to the magnetic field B. It is then rotated through 180° in 0.2 sec. The emf induced at the ends of the coils is:

A. 0.1 nAB

B. nAB

C. 5 nAB

D. 10 nAB

Answer: D. Total change in flux = $\Delta\Phi$ = 2 nAB

Total time of change = $\mathbf{\Delta t}$ = 0.2s

Emf induced = $\frac{\Delta\Phi}{\Delta t}$ = 10nAB

Q2. A straight line conductor of length 0. 4m is moved with a speed of $7ms^{1}$ perpendicular to a magnetic field of an intensity of $0.9wbm^{-2}$ The induced emf across the conductor is:

A. 25.2 V

B. 5.24 V

C. 2.52 V

D. 1.26 V

Answer: C. The induced emf across the conductor E= Blv

= 0.98 × 0.4 × 7 = 2.52V

Q.3 Two conducting rings of radii r and 2r move in opposite directions with velocities 2v and v respectively on a conducting surface S. There is a uniform magnetic field of magnitude B perpendicular to the plane of the rings. The potential difference between the highest points of the two rings is:

A. Zero

B. 2rvB

C. 4rvB

D. 8rvB

Answer: D. Replace the emf in the rings by the cells.

E_1= B2r(2V) = 4Brv

E_2 = B(4r)v = 4Brv

$V_2 - V_1$ = 8Brv

Internal Resistance

A 12-V truck battery is physically larger, contains more charge and energy, and can deliver a larger current than a 12-V motorcycle battery. Both are lead-acid batteries with identical emf, but, because of its size, the truck battery has a smaller internal resistance . Internal resistance is the inherent resistance to the flow of current within the source itself.

The emf (represented by a script E in the figure) and internal resistance are in series. The smaller the internal resistance for a given emf, the more current and the more power the source can supply.

Any voltage source (in this case, a carbon-zinc dry cell) has an emf related to its source of potential difference, and an internal resistance r related to its construction. (Note that the script E stands for emf.). Also shown are the output terminals across which the terminal voltage V is measured. Since $V = \text{emf} - Ir$, terminal voltage equals emf only if there is no current flowing.

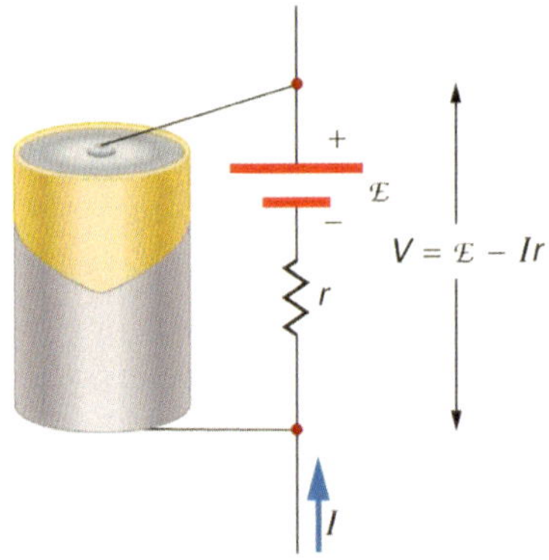

The internal resistance r can behave in complex ways. As noted, r increases as a battery is depleted. But internal resistance may also depend on the magnitude and direction of the current through a voltage source, its temperature, and even its history. The internal resistance of rechargeable nickel-cadmium cells, for example, depends on how many times and how deeply they have been depleted.

Things Great and Small: The Submicroscopic Origin of Battery Potential

Various types of batteries are available, with emfs determined by the combination of chemicals involved. We can view this as a molecular reaction (what much of chemistry is about) that separates charge.

The lead-acid battery used in cars and other vehicles is one of the most common types. The cathode (positive) terminal of the cell is connected to a lead oxide plate, while the anode (negative) terminal is connected to a lead plate. Both plates are immersed in sulfuric acid, the electrolyte for the system.

Artist's conception of a lead-acid cell. Chemical reactions in a lead-acid cell separate charge, sending negative charge to the anode, which is connected to the lead plates. The lead oxide plates are connected to the positive or cathode terminal of the cell. Sulfuric acid conducts the charge as well as participating in the chemical reaction.

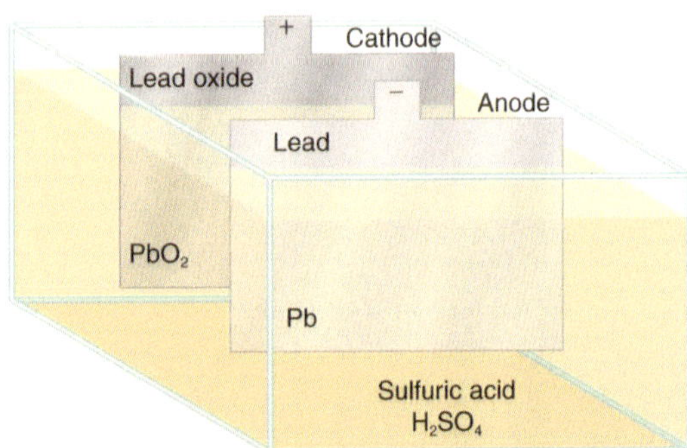

The details of the chemical reaction are left to the reader to pursue in a chemistry text, but their results at the molecular level help explain the potential created by the battery. Two electrons are placed on the anode, making it negative, provided that the cathode supplied two electrons. This leaves the cathode positively charged, because it has lost two electrons. In short, a separation of charge has been driven by a chemical reaction.

Note that the reaction will not take place unless there is a complete circuit to allow two electrons to be supplied to the cathode. Under many circumstances, these electrons come from the anode, flow through a resistance, and return to the cathode. Note also that since the chemical reactions involve substances with resistance, it is not possible to create the emf without an internal resistance.

Artist's conception of two electrons being forced onto the anode of a cell and two electrons being removed from the cathode of the cell. The chemical reaction in a lead-acid battery places two electrons on the anode and removes two from the cathode. It requires a closed circuit to proceed, since the two electrons must be supplied to the cathode.

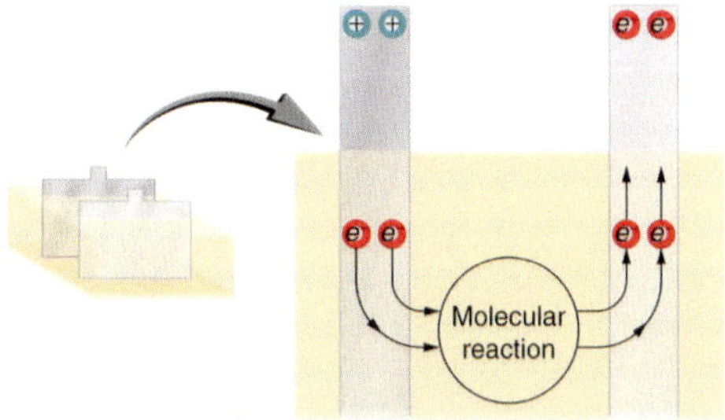

Why are the chemicals able to produce a unique potential difference? Quantum mechanical descriptions of molecules, which take into account the types of atoms and numbers of electrons in them, are able to predict the energy states they can have and the energies of reactions between them.

In the case of a lead-acid battery, an energy of 2 eV is given to each electron sent to the anode. Voltage is defined as the electrical potential energy divided by charge: $V = \frac{P_E}{q}$. An electron volt is the energy given to a single electron by a voltage of 1 V. So the voltage here is 2 V, since 2 eV is given to each electron. It is the energy produced in each molecular reaction that produces the voltage. A different reaction produces a different energy and, hence, a different voltage.

Terminal Voltage

The voltage output of a device is measured across its terminals and, thus, is called its terminal voltage V. Terminal voltage is given by,

$$V = \text{emf} - \text{Ir},$$

where r is the internal resistance and I is the current flowing at the time of the measurement.

You can see that the larger the current, the smaller the terminal voltage. And it is likewise true that the larger the internal resistance, the smaller the terminal voltage.

Since the resistances are in series, the total resistance in the circuit is $R_{\text{load}} + r$. Thus the current is given by Ohm's law to be,

$$I = \frac{\text{emf}}{R_{\text{load}} + r}.$$

Schematic of a voltage source and its load R_{load}. Since the internal resistance r is in series with the load, it can significantly affect the terminal voltage and current delivered to the load. (Note that the script E stands for emf.)

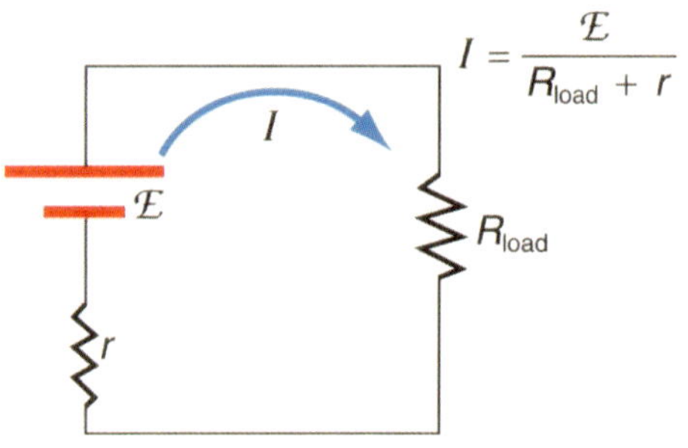

We see from this expression that the smaller the internal resistance r, the greater the current the voltage source supplies to its load R_{load}. As batteries are depleted, r increases. If r becomes a significant fraction of the load resistance, then the current is significantly reduced, as the following example illustrates.

Calculating Terminal Voltage, Power Dissipation, Current, and Resistance: Terminal Voltage and Load

A certain battery has a 12.0-V emf and an internal resistance of 0.100Ω. (a) Calculate its terminal voltage when connected to a $10.0\text{-}\Omega$ load. (b) What is the terminal voltage when connected to a $0.500\text{-}\Omega$ load? (c) What power does the $0.500\text{-}\Omega$ load dissipate? (d) If the internal resistance grows to $0.500\text{-}\Omega$, find the current, terminal voltage, and power dissipated by a $0.500\text{-}\Omega$ load.

Strategy

The analysis above gave an expression for current when internal resistance is taken into account. Once the current is found, the terminal voltage can be calculated using the equation $V = \text{emf} - \text{Ir}$. Once current is found, the power dissipated by a resistor can also be found.

Solution for (a)

Entering the given values for the emf, load resistance, and internal resistance into the expression above yields

$$I = \frac{\text{emf}}{R_{load} + r} = \frac{12.0\text{V}}{10.1\Omega} = 1.188\text{A}.$$

Enter the known values into the equation $V = \text{emf} - \text{Ir}$ **to get the terminal voltage:**

$$\begin{aligned} V &= \text{emf} - \text{Ir} = 12.0\text{ V} - (1.188\text{ A})(0.100\ ) \\ &= 11.9\text{ V}. \end{aligned}$$

Discussion for (a)

The terminal voltage here is only slightly lower than the emf, implying that 10.0Ω is a light load for this particular battery.

Solution for (b)

Similarly, with $R_{load} = 0.500\Omega$, the current is

$$I = \frac{\text{emf}}{R_{load} + r} = \frac{12.0\text{V}}{0.600\Omega} = 20.0\text{A}.$$

The terminal voltage is now

$$\begin{aligned} V &= \text{emf} - \text{Ir} = 12.0\ \text{V} - (20.0\ \text{A})(0.100\ ) \\ &= 10.0\ \text{V}. \end{aligned}$$

Discussion for (b)

This terminal voltage exhibits a more significant reduction compared with emf, implying $0.500\text{-}\Omega$ is a heavy load for this battery.

Solution for (c)

The power dissipated by the 0.500 - load can be found using the formula $P = I^2R$. Entering the known values gives

$$P_{load} = I^2R_{load} = (20.0\text{A})^2(0.500\ ) = 2.00\ 10^2\ \text{W}.$$

Discussion for (c)

Note that this power can also be obtained using the expressions $\frac{V^2}{R}$ or IV, where V is the terminal voltage (10.0 V in this case).

Solution for (d)

Here the internal resistance has increased, perhaps due to the depletion of the battery, to the point where it is as great as the load resistance. As before, we first find the current by entering the known values into the expression, yielding

$$I = \frac{\text{emf}}{R_{load} + r} = \frac{12.0\text{V}}{1.00\Omega} = 12.0\text{A}.$$

Now the terminal voltage is

$$\begin{aligned} V &= \text{emf} - \text{Ir} = 12.0\ \text{V} - (12.0\ \text{A})(0.500\ ) \\ &= 6.00\ \text{V}, \end{aligned}$$

and the power dissipated by the load is

$$P_{load} = I^2R_{load} = (12.0\ \text{A})^2(0.500\Omega) = 72.0\text{W}.$$

Discussion for (d)

We see that the increased internal resistance has significantly decreased terminal voltage, current, and power delivered to a load

Battery testers really test the internal resistance of the battery. If internal resistance is high, the battery is weak, as evidenced by its low terminal voltage.

These two battery testers measure terminal voltage under a load to determine the condition of a battery. The large device is being used by a U.S. Navy electronics technician to test large batteries aboard the aircraft carrier USS Nimitz and has a small resistance that can dissipate large amounts of power. The small device is used on small batteries and has a digital display to indicate the acceptability of their terminal voltage.

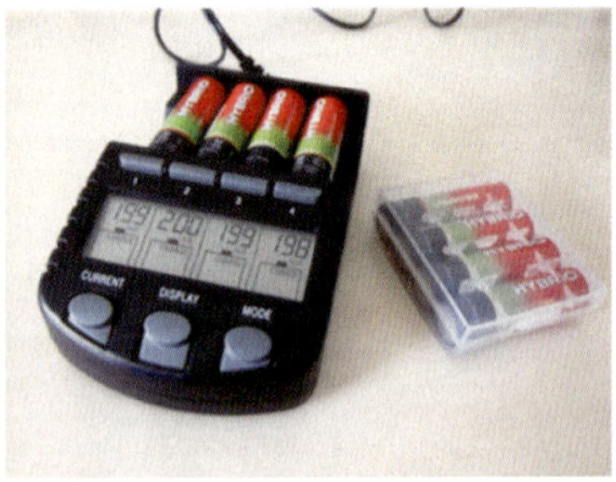

Some batteries can be recharged by passing a current through them in the direction opposite to the current they supply to a resistance. The voltage output of the battery charger must be greater than the emf of the battery to reverse current through it. This will cause the terminal voltage of the battery to be greater than the emf, since $V = \text{emf} - \text{Ir}$, and I is now negative.

A car battery charger reverses the normal direction of current through a battery, reversing its chemical reaction and replenishing its chemical potential.

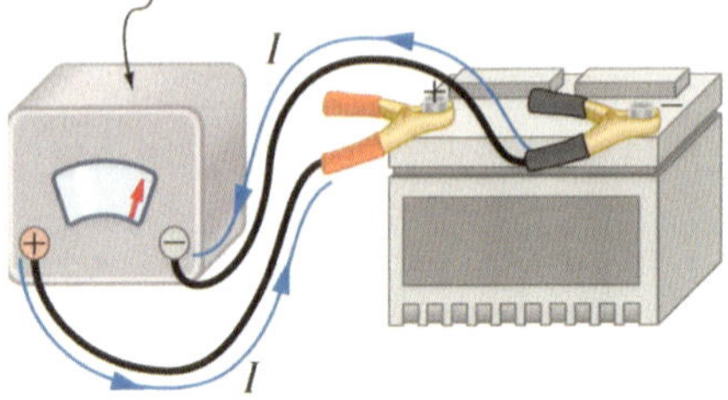

Multiple Voltage Sources

There are two voltage sources when a battery charger is used. Voltage sources connected in series are relatively simple. When voltage sources are in series, their internal resistances add and their emfs add algebraically. Series connections of voltage sources are common—for example, in flashlights, toys, and other appliances. Usually, the cells are in series in order to produce a larger total emf.

But if the cells oppose one another, such as when one is put into an appliance backward, the total emf is less, since it is the algebraic sum of the individual emfs.

The disadvantage of series connections of cells is that their internal resistances add. One of the authors once owned a 1957 MGA that had two 6-V batteries in series, rather than a single 12-V battery. This arrangement produced a large internal resistance that caused him many problems in starting the engine.

A series connection of two voltage sources. The emfs (each labeled with a script E) and internal resistances add, giving a total emf of $\mathrm{emf}_1 + \mathrm{emf}_2$ and a total internal resistance of $r_1 + r_2$.

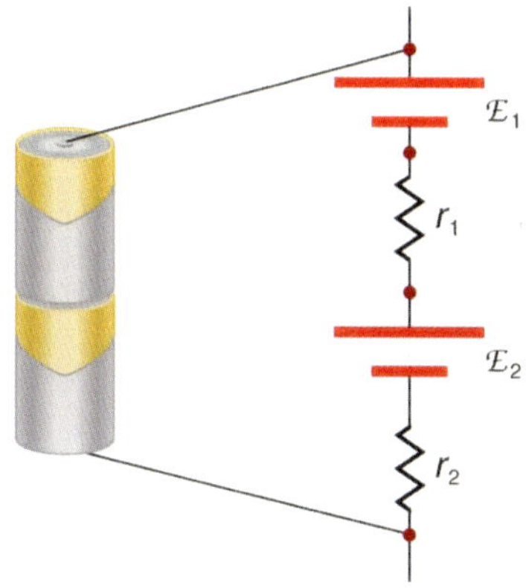

Batteries are multiple connections of individual cells, as shown in this modern rendition of an old print. Single cells, such as AA or C cells, are commonly called batteries, although this is technically incorrect.

If the *series* connection of two voltage sources is made into a complete circuit with the emfs in opposition, then a current of magnitude $I = \dfrac{(\mathrm{emf}_1 - \mathrm{emf}_2)}{r_1 + r_2}$ flows. For example, which shows a circuit exactly analogous to the battery charger discussed above. Then,

$$I = \frac{(\mathrm{emf}_1 + emf_2)}{r_1 + r_2 + R_{\mathrm{load}}} \text{flows.}$$

These two voltage sources are connected in series with their emfs in opposition. Current flows in the direction of the greater emf and is limited to $I = \dfrac{(\mathrm{emf}_1 - \mathrm{emf}_2)}{r_1 + r_2}$ by the sum of the internal resistances. (Note that each emf is represented by script E in the figure.) A battery charger connected to a battery is an example of such a connection. The charger must have a larger emf than the battery to reverse current through it.

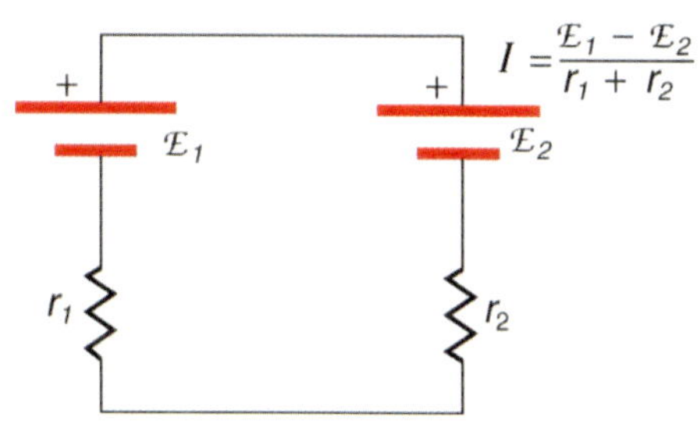

This schematic represents a flashlight with two cells (voltage sources) and a single bulb (load resistance) in series. The current that flows is $I = \dfrac{(\text{emf}_1 + \text{emf}_2)}{r_1 + r_2 + R_{\text{load}}}$. (Note that each emf is represented by script E in the figure.)

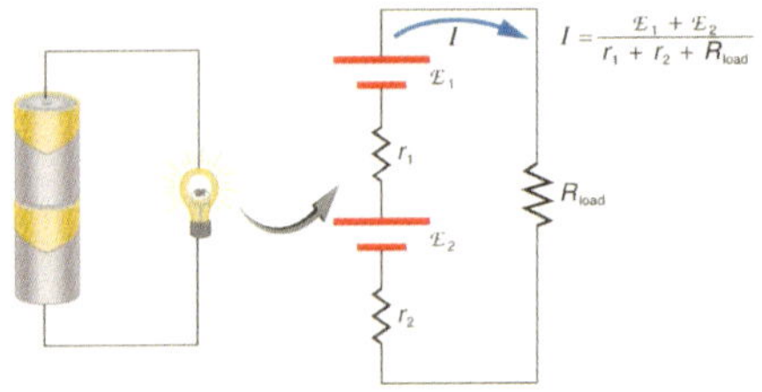

Take-home Experiment: Flashlight Batteries

Find a flashlight that uses several batteries and find new and old batteries. Based on the discussions in this module, predict the brightness of the flashlight when different combinations of batteries are used. Do your predictions match what you observe? Now place new batteries in the flashlight and leave the flashlight switched on for several hours. Is the flashlight still quite bright? Do the same with the old batteries. Is the flashlight as bright when left on for the same length of time with old and new batteries? What does this say for the case when you are limited in the number of available new batteries?

In this simple case, the total emf is the same as the individual emfs. But the total internal resistance is reduced, since the internal resistances are in parallel. The parallel connection thus can produce a larger current.

Here, $I = \dfrac{\text{emf}}{(r_{\text{tot}} + R_{\text{load}})}$ flows through the load, and r_{tot} is less than those of the individual batteries. For example, some diesel-powered cars use two 12-V batteries in parallel; they produce a total emf of 12 V but can deliver the larger current needed to start a diesel engine.

Two voltage sources with identical emfs (each labeled by script E) connected in parallel produce the same emf but have a smaller total internal resistance than the individual sources. Parallel combinations are often used to deliver more current. Here $I = \dfrac{\text{emf}}{(r_{\text{tot}} + R_{\text{load}})}$ flows through the load.

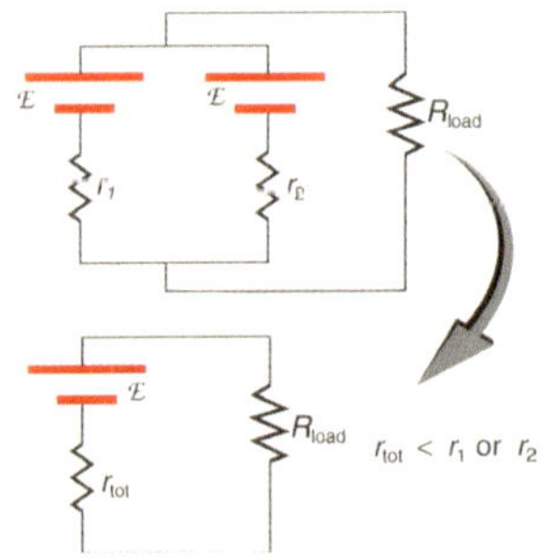

In Thermodynamics

When multiplied by an amount of charge dQ the emf E yields a thermodynamic work term EdQ that is used in the formalism for the change in Gibbs energy when charge is passed in a battery:

$$dG = -SdT + VdP + \mathcal{E}dQ\,,$$

where G is the Gibb's free energy, S is the entropy, V is the system volume, P is its pressure and T is its absolute temperature.

The combination (E, Q) is an example of a conjugate pair of variables. At constant pressure the above relationship produces a Maxwell relation that links the change in open cell voltage with temperature T (a measurable quantity) to the change in entropy S when charge is passed isothermally and isobarically. The latter is closely related to the reaction entropy of the electrochemical reaction that lends the battery its power. This Maxwell relation is:

$$\left(\frac{\partial \mathcal{E}}{\partial T}\right)_Q = -\left(\frac{\partial S}{\partial Q}\right)_T$$

If a mole of ions goes into solution (for example, in a Daniell cell, as discussed below) the charge through the external circuit is:

$$\Delta Q = -n_0 F_0 \,,$$

where n_0 is the number of electrons/ion, and F_0 is the Faraday constant and the minus sign indicates discharge of the cell. Assuming constant pressure and volume, the thermodynamic properties of the cell are related strictly to the behavior of its emf by:

$$\Delta H = -n_0 F_0 \left(\mathcal{E} - T\frac{d\mathcal{E}}{dT}\right),$$

where ΔH is the enthalpy of reaction. The quantities on the right are all directly measurable.

Generation

Chemical Sources

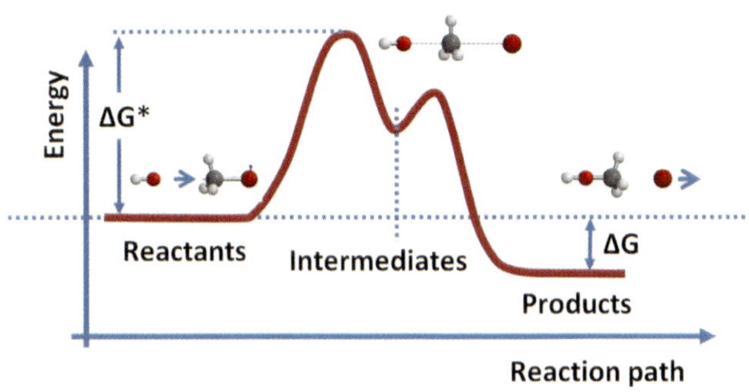

A typical reaction path requires the initial reactants to cross an energy barrier, enter an intermediate state and finally emerge in a lower energy configuration. If charge separation is involved, this energy difference can result in an emf.

The question of how batteries (galvanic cells) generate an emf is one that occupied scientists for most of the 19th century. The "seat of the electromotive force" was eventually determined by Walther Nernst to be primarily at the interfaces between the electrodes and the electrolyte.

Molecules are groups of atoms held together by chemical bonds, and these bonds consist of electrical forces between electrons (negative) and protons (positive). The molecule in isolation is a stable entity, but when different molecules are brought together, some types of molecules are able to steal electrons from others, resulting in charge separation. This redistribution of charge is accompanied by a change in energy of the system, and a reconfiguration of the atoms in the molecules. The gain of an electron is termed "reduction" and the loss of an electron is termed "oxidation". Reactions in which such electron exchange occurs (which are the basis for batteries) are called reduction-oxidation reactions or redox reactions. In a battery, one electrode is composed of material that gains electrons from the solute, and the other electrode loses electrons, because of these fundamental molecular attributes. The same behavior can be seen in atoms themselves, and their ability to steal electrons is referred to as their electronegativity.

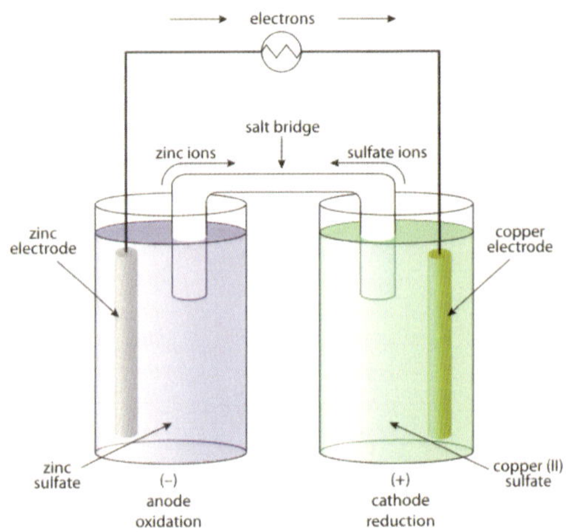

Galvanic cell using a salt bridge

As an example, a Daniell cell consists of a zinc anode (an electron collector), is oxidized as it dissolves into a zinc sulfate solution, the dissolving zinc leaving behind its electrons in the electrode according to the oxidation reaction (*s* = solid electrode; *aq* = aqueous solution):

$$Zn_{(s)} \rightarrow Zn^{2+}_{(aq)} + 2e^{-}$$

The zinc sulfate is the electrolyte in that half cell. It is a solution which contains zinc cations Zn^{2+}, and sulfate anions SO_4^{2-} with charges that balance to zero.

In the other half cell, the copper cations in a copper sulfate electrolyte are drawn to the copper cathode to which they attach themselves as they adopt electrons from the copper electrode by the reduction reaction:

$$Cu^{2+}_{(aq)} + 2e^{-} \rightarrow Cu_{(s)}$$

in effect leaving a deficit of electrons on the copper cathode. The difference of excess electrons on the anode and deficit of electrons on the cathode creates an electrical potential between the two electrodes.

If the cathode and anode are connected by an external conductor, electrons would pass through that external circuit (light bulb in figure), while the ions pass through the salt bridge to maintain charge balance until such a time as the anode and cathode reach electrical equilibrium of zero volts as chemical equilibrium is reached in the cell. In the process the zinc anode is dissolved while the copper electrode is plated with copper. The so-called "salt bridge" is not made of salt but could be made of material able to wick the cations and anions (salts) in the solutions, where the flow of positively charged cations along the "bridge" amounts to the same number of negative charges flowing in the opposite direction.

If the light bulb is removed (open circuit) the emf between the electrodes is opposed by the electric field due to charge separation, and the reactions stop.

For this particular cell chemistry, at 298 K (room temperature), the emf $E = 1.0934$ V, with a temperature coefficient of $dE/dT = -4.53\times10^{-4}$ V/K.

Voltaic Cells

Volta developed the voltaic cell about 1792, and presented his work March 20, 1800. Volta correctly identified the role of dissimilar electrodes in producing the voltage, but incorrectly dismissed any role for the electrolyte. Volta ordered the metals in a 'tension series', "that is to say in an order such that any one in the list becomes positive when in contact with any one that succeeds, but negative by contact with any one that precedes it." A typical symbolic convention in a schematic of this circuit (–||–) would have a long electrode 1 and a short electrode 2, to indicate that electrode 1 dominates. Volta's law about opposing electrode emfs implies that, given ten electrodes (for example, zinc and nine other materials), 45 unique combinations of voltaic cells ($10 \times 9/2$) can be created.

Of Cells

The electromotive force produced by primary (single-use) and secondary (rechargeable) cells is usually of the order of a few volts. The figures quoted below are nominal, because emf varies according to the size of the load and the state of exhaustion of the cell.

EMF	Cell chemistry			Common name
	Anode	Solvent, electrolyte	Cathode	
1.2 V	Cadmium	Water, potassium hydroxide	NiO(OH)	nickel-cadmium
1.2 V	Mischmetal (hydrogen absorbing)	Water, potassium hydroxide	Nickel	nickel–metal hydride
1.5 V	Zinc	Water, ammonium or zinc chloride	Carbon, manganese dioxide	Zinc carbon
2.1 V	Lead	Water, sulfuric acid	Lead dioxide	Lead–acid
3.6 V to 3.7 V	Graphite	Organic solvent, Li salts	$LiCoO_2$	Lithium-ion
1.35 V	Zinc	Water, sodium or potassium hydroxide	HgO	Mercury cell

Electromagnetic Induction

The principle of electromagnetic induction, noted above, states that a time-dependent magnetic field produces a circulating electric field. A time-dependent magnetic field can be produced either by motion of a magnet relative to a circuit, by motion of a circuit relative to another circuit (at least one of these must be carrying a current), or by changing the current in a fixed circuit. The effect on the circuit itself, of changing the current, is known as self-induction; the effect on another circuit is known as mutual induction.

For a given circuit, the electromagnetically induced emf is determined purely by the rate of change of the magnetic flux through the circuit according to Faraday's law of induction.

An emf is induced in a coil or conductor whenever there is change in the flux linkages. Depending on the way in which the changes are brought about, there are two types: When the conductor is moved in a stationary magnetic field to procure a change in the flux linkage, the emf is *statically induced.* The electromotive force generated by motion is often referred to as *motional emf.* When the change in flux linkage arises from a change in the magnetic field around the stationary conductor, the emf is *dynamically induced.* The electromotive force generated by a time-varying magnetic field is often referred to as *transformer emf.*

Contact Potentials

When solids of two different materials are in contact, thermodynamic equilibrium requires that one of the solids assume a higher electrical potential than the other. This is called the *contact potential.* Dissimilar metals in contact produce what is known also as a contact electromotive force or Galvani potential. The magnitude of this potential difference is often expressed as a difference in Fermi levels in the two solids when they are at charge neutrality, where the Fermi level (a name for the chemical potential of an electron system) describes the energy necessary to remove an electron from the body to some common point (such as ground). If there is an energy advantage in taking an electron from one body to the other, such a transfer will occur. The transfer causes a charge separation, with one body gaining electrons and the other losing electrons. This charge transfer causes a potential difference between the bodies, which partly cancels the potential originating from the contact, and eventually equilibrium is reached. At thermodynamic equilibrium, the Fermi levels are equal (the electron removal energy is identical) and there is now a built-in electrostatic potential between the bodies. The original difference in Fermi levels, before contact, is referred to as the emf.The contact potential cannot drive steady current through a load attached to its terminals because that current would involve a charge transfer. No mechanism exists to continue such transfer and, hence, maintain a current, once equilibrium is attained.

One might inquire why the contact potential does not appear in Kirchhoff's law of voltages as one contribution to the sum of potential drops. The customary answer is that any circuit involves not only a particular diode or junction, but also all the contact potentials due to wiring and so forth around the entire circuit. The sum of *all* the contact potentials is zero, and so they may be ignored in Kirchhoff's law.

Solar Cell

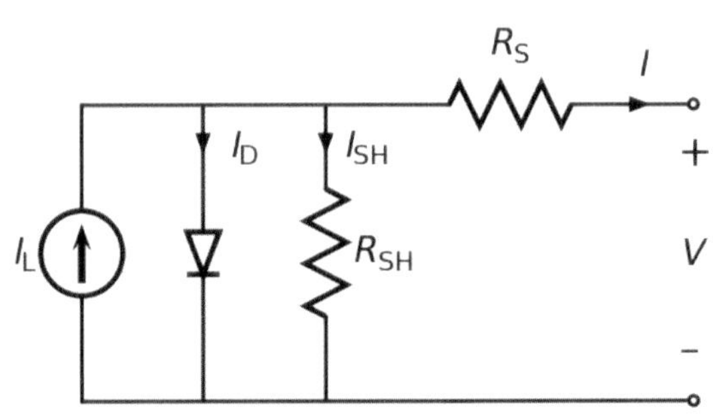

The equivalent circuit of a solar cell; parasitic resistances are ignored in the discussion of the text.

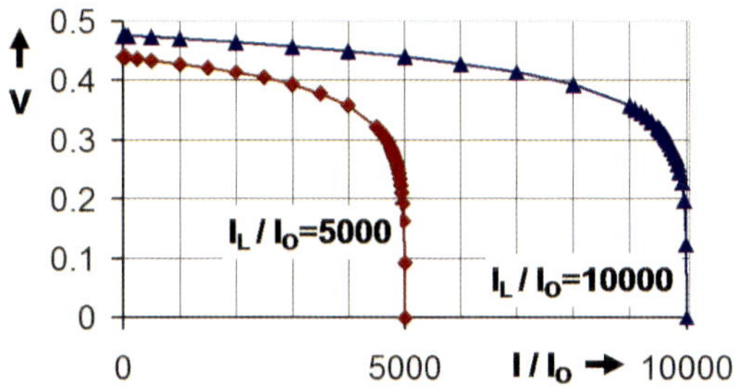

Solar cell voltage as a function of solar cell current delivered to a load for two light-induced currents I_L; currents as a ratio with reverse saturation current I_0.

Operation of a solar cell can be understood from the equivalent circuit at right. Light, of sufficient energy (greater than the bandgap of the material), creates mobile electron–hole pairs in a semiconductor. Charge separation occurs because of a pre-existing electric field associated with the p-n junction in thermal equilibrium (a contact potential creates the field). This charge separation between positive holes and negative electrons across a p-n junction (a diode) yields a *forward voltage*, the *photo voltage*, between the illuminated diode terminals. As has been noted earlier in the terminology section, the photo *voltage* is sometimes referred to as the photo *emf*, rather than distinguishing between the effect and the cause. The charge separation causes a photo voltage that drives current through any attached load.

The current available to the external circuit is limited by internal losses $I_0 = I_{SH} + I_D$:

$$I = I_L - I_0 = I_L - I_{SH} - I_D$$

Losses limit the current available to the external circuit. The light-induced charge separation eventually creates a current (called a forward current) I_{SH} through the cell's junction in the direction opposite that the light is driving the current. In addition, the induced voltage tends to forward bias the junction. At high enough levels, this forward bias of the junction will cause a forward current, I_D in the diode opposite that induced by the light. Consequently, the greatest current is obtained under short-circuit conditions, and is denoted as I_L (for light-induced current) in the equivalent circuit.Approximately, this same current is obtained for forward voltages up to the point where the diode conduction becomes significant.

The current delivered by the illuminated diode, to the external circuit is:

$$I = I_L - I_0 \left(e^{qV/(mkT)} - 1 \right),$$

where I_0 is the reverse saturation current. Where the two parameters that depend on the solar cell construction and to some degree upon the voltage itself are m, the ideality factor, and kT/q the thermal voltage (about 0.026 V at room temperature). This relation is plotted in the figure using a fixed value $m = 2$. Under open-circuit conditions (that is, as $I = 0$), the open-circuit voltage is the voltage at which forward bias of the junction is enough that the forward current completely balances the photocurrent. Solving the above for the voltage V and designating it the open-circuit voltage of the I–V equation as:

$$V_{oc} = m \frac{kT}{q} \ln\left(\frac{I_L}{I_0} + 1 \right),$$

which is useful in indicating a logarithmic dependence of V_{oc} upon the light-induced current. Typically, the open-circuit voltage is not more than about 0.5 V.

When driving a load, the photo voltage is variable. As shown in the figure, for a load resistance R_L, the cell develops a voltage that is between the short-circuit value $V = 0$, $I = I_L$ and the open-circuit value V_{oc}, $I = 0$, a value given by Ohm's law $V = I R_L$, where the current I is the difference between the short-circuit current and current due to forward bias of the junction, as indicated by the equivalent circuit (neglecting the parasitic resistances).

In contrast to the battery, at current levels delivered to the external circuit near I_L, the solar cell acts more like a *current source* rather than a voltage source(near vertical part of the two illustrated curves). The current drawn is nearly fixed over a range of load voltages, to one electron per converted photon. The quantum efficiency, or probability of getting an electron of photocurrent per incident photon, depends not only upon the solar cell itself, but upon the spectrum of the light.

The diode possesses a "built-in potential" due to the contact potential difference between the two different materials on either side of the junction. This built-in potential is established when the junction is manufactured and that voltage a by-product of thermodynamic equilibrium within the cell. Once established, this potential difference cannot drive a current, however, as connecting a load does not upset this equilibrium. In contrast, the accumulation of excess electrons in one region and of excess holes in another, due to illumination, results in a photo voltage that does drive a current when a load is attached to the illuminated diode. As noted above, this photo voltage also forward biases the junction, and so *reduces* the pre-existing field in the depletion region.

Faraday's Law Relate to EMF

Faraday's law states that any change in the magnetic field of a coil will act to induce an EMF in the coil (and hence also a current). It is proportional to minus the rate of change in magnetic flux (ϕ) (note N is the number of turns in the coil).

$$\varepsilon = -N\frac{\Delta\phi}{\Delta t}$$

Using Faraday's law, society has benefited from important technology such as transformers which are used in the transmission of electricity in the UK national grid which is now a necessity of our homes. Also it is used in the electric generators and motors such as hydroelectric dams, which produce the electricity that is now integral to our modern technological needs. A current research project at Birmingham, MAG-DRIVE, is finding ways to develop and improve permanent magnet materials that can be used in the next generation of electric vehicles. EMF is also generated from solar cells so it is important in renewable energy research topics.

Magnetomotive Force

The current flowing in an electric circuit is due to the existence of electromotive force similarly magnetomotive force (MMF) is required to drive the magnetic flux in the magnetic circuit. The magnetic pressure, which sets up the magnetic flux in a magnetic circuit is called Magnetomotive Force. The SI unit of MMF is Ampere-turn (AT), and their CGS unit is G (gilbert). The MMF for the inductive coil shown in the figure below is expressed as ,

$$F = NI$$

Where, N – numbers of turns of inductive coil

I – *current*

The strength of the MMF is equivalent to the product of the current around the turns and the number of turns of the coil. As per work law, The MMF is defined as the work done in moving the unit magnetic pole (1weber) once around the magnetic circuit.

The MMF is also known as the magnetic potential. It is the property of material to give rise the magnetic field. The magnetomotive force is the product of the magnetic flux and the magnetic reluctance. The reluctance is the opposition offers by the magnetic field to set up the magnetic flux on it. The MMF regarding reluctance and magnetic flux is given as,

$$F = \varnothing R$$

Where R – reluctance

Φ – *magnetic flux*

The magnetomotive force can measure regarding magnetic field intensity and the length of the substance. The magnetic field strength is the force act on the unit pole placed on the magnetic field. MMF regarding field intensity is expressed as,

$$F = Hl$$

Where H is the magnetic field strength, and l is the length of the substance.

Magnetomotive Force (mmf) Unit

Magnetomotive force (mmf), represented by $\Im$, is measured in ampere- turs (A-t) of the coil.

"One Ampere-turn is the amount of magnetic force or flux produced when one ampere flows through a single turn of an electrical conductor."

Magnetomotive force (MMF) Formula

The magnetomotive force (mmf) formula is expressed by the following relationship.

$$\mathfrak{F} = A - t$$

A coil having 10 turns and 1A of current produces the same magnetic flux or force as one having 5 turns and 2A current because the ampere-turns are the same.

Example

What is the magnetizing force of a coil having 4000 turns and 10 mA current flowing through it?

Solution

We have following formula to calculate ampere – turns (magnetizing force):

$$\mathfrak{F} = A - t$$

So, putting values of current and turns into formula,

$$\mathfrak{F} = 4000 * 0.01 = 40A - t$$

Thus, if a large amount of magnetizing force is needed and only a small amount of current is available, it is necessary to use a great number of turns on the coil.

Maxwell's Equations

Maxwell's Equations are a set of 4 complicated equations that describe the world of electromagnetics. These equations describe how electric and magnetic fields propagate, interact, and how they are influenced by objects.

James Clerk Maxwell was an Einstein/Newton-level genius who took a set of known experimental laws (Faraday's Law, Ampere's Law) and unified them into a symmetric coherent set of Equations known as Maxwell's Equations. Maxwell was one of the first to determine the speed of propagation of electromagnetic (EM) waves was the same as the speed of light - and hence to conclude that EM waves and visible light were really the same thing.

Maxwell's Equations are critical in understanding Antennas and Electromagnetics. They are formidable to look at - so complicated that most electrical engineers and physicists don't even really know what they mean. Shrouded in complex math (which is likely so "intellectual" people can feel superior in discussing them), true understanding of these equations is hard to come by.

This leads to the reason for this website - an intuitive tutorial of Maxwell's Equations. We will avoid if at all possible the mathematical difficulties that arise, and instead describe what the equations mean. And don't be afraid - the math is so complicated that those who do understand complex vector calculus still cannot apply Maxwell's Equations in anything but the simplest scenarios. For this reason, intuitive knowledge of Maxwell's Equations is far superior than mathematical manipulation-based knowledge. To understand the world, you must understand what equations mean, and not just know mathematical constructs. We believe the accepted methods of teaching electromagnetics and Maxwell's Equations do not produce understanding. And with that, let's say something about these equations.

Maxwell's Equations are laws - just like the law of gravity. These equations are rules the universe uses to govern the behavior of electric and magnetic fields. A flow of electric current will produce a magnetic field. If the current flow varies with time (as in any wave or periodic signal), the magnetic field will also give rise to an electric field. Maxwell's Equations shows that separated charge (positive and negative) gives rise to an electric field - and if this is varying in time as well will give rise to a propagating electric field, further giving rise to a propgating magnetic field.

Definitions of μ_0 and ε_0 Part

Ampere discovered that two parallel wires carrying electric currents in the same direction attract each other magnetically, the force in newtons per unit length being given by

$$F = 2\left(\frac{\mu_0}{4\pi}\right)\frac{I_1 I_2}{r},$$

for long wires a distance r apart. We are using the standard modern units (SI). The constant $\mu_0 / 4\pi$ that appears here is *exactly* 10^{-7}, this *defines* our *present* unit of current, the *ampere*. To repeat: $\mu_0 / 4\pi$ is not something to measure experimentally, it's just a funny way of writing the number 10^{-7}. That's not quite fair—it has dimensions to ensure that both sides of the above equation have the same dimensionality. (Of course, there's a historical reason for this strange convention). Anyway, if we bear in mind that dimensions have been taken care of, and *just* write the equation,

$$F = 2 \cdot 10^{-7} \cdot \frac{I_1 I_2}{r},$$

it's clear that this defines the unit current—one ampere—as that current in a long straight wire which exerts a magnetic force of 2×10^{-7} newtons per meter of wire on a parallel wire one meter away carrying the same current.

However, after we have established our unit of current—the ampere—*we have also thereby defined our unit of charge*, since current is a flow of charge, and the unit of charge must be the amount carried past a fixed point in unit time by unit current. Therefore, our unit of charge—the **coulomb**—is defined by stating that a *one amp* current in a wire carries *one coulomb per second* past a fixed point.

To be consistent, we must do electrostatics using this *same* unit of charge. Now, the electrostatic force between two charges is $(1/4\pi\varepsilon_0)q_1 q_2 / r^2$. The constant appearing here, now written $1/4\pi\varepsilon_0$, *must be experimentally measured*—its value turns out to be 9×10^9.

To find the value of $1/4\pi\varepsilon_0$, *two* experiments have to be performed. We must first establish the unit of charge from the unit of current by measuring the magnetic force between two current-carrying parallel wires. Second, we must find the electrostatic force between measured charges. (We could, alternatively, have defined some other unit of current from the start, then we would have had to find both μ_0 and ε_0 by experiments on magnetic and electrostatic attraction. In fact, the ampere *was* originally defined as the current that deposited a definite weight of silver per hour in an electrolytic cell).

Gauss's Law

Gauss's law describes the relationship between a static electric field and the electric charges that cause it: The static electric field points away from positive charges and towards negative charges, and the net outflow of the electric field through any closed surface is proportional to the charge enclosed by the surface. Picturing the electric field by its field lines, this means the field lines begin at positive electric charges and end at negative electric charges. ‹Counting' the number of field lines passing through a closed surface yields the total charge (including bound charge due to polarization of material) enclosed by that surface, divided by dielectricity of free space (the vacuum permittivity).

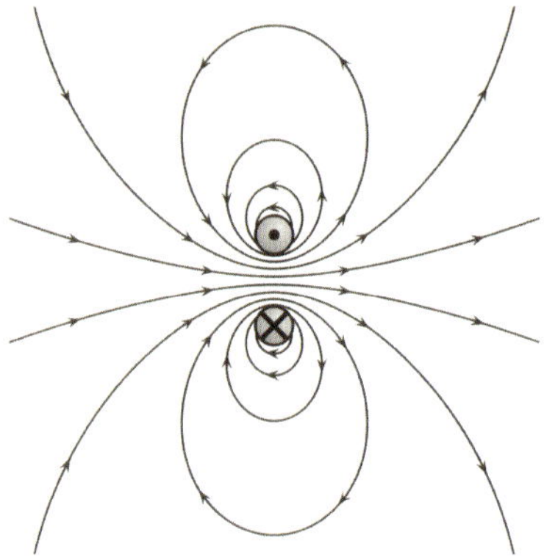

Gauss's law for magnetism: magnetic field lines never begin nor end but form loops or extend to infinity as shown here with the magnetic field due to a ring of current.

Gauss's Law for Magnetism

Gauss's law for magnetism states that there are no "magnetic charges" (also called magnetic monopoles), analogous to electric charges. Instead, the magnetic field due to materials is generated by a configuration called a dipole, and the net outflow of the magnetic field through any closed surface is zero. Magnetic dipoles are best represented as loops of current but resemble positive and negative 'magnetic charges', inseparably bound together, having no net 'magnetic charge'. In terms of field lines, this equation states that magnetic field lines neither begin nor end but make loops or extend to infinity and back. In other words, any magnetic field line that enters a given volume must somewhere exit that volume. Equivalent technical statements are that the sum total magnetic flux through any Gaussian surface is zero, or that the magnetic field is a solenoidal vector field.

Faraday's Law

In a geomagnetic storm, a surge in the flux of charged particles temporarily alters Earth's magnetic field, which induces electric fields in Earth's atmosphere, thus causing surges in electrical power grids.

The Maxwell–Faraday version of Faraday's law of induction describes how a time varying magnetic field creates ("induces") an electric field. In integral form, it states that the work per unit charge required to move a charge around a closed loop equals the rate of decrease of the magnetic flux through the enclosed surface.

The dynamically induced electric field has closed field lines similar to a magnetic field, unless superposed by a static (charge induced) electric field. This aspect of electromagnetic induction **is** the operating principle behind many electric generators: for example, a rotating bar magnet creates a changing magnetic field, which in turn generates an electric field in a nearby wire.

Ampère's Law with Maxwell's Addition

Magnetic core memory is an application of Ampère's law. Each corestores one bit of data.

Ampère's law with Maxwell's addition states that magnetic fields can be generated in two ways: by electric current (this was the original "Ampère's law") and by changing electric fields (this was "Maxwell's addition", which he called displacement current). In integral form, the magnetic field induced around any closed loop is proportional to the electric current plus displacement current (proportional to the rate of change of electric flux) through the enclosed surface.

Maxwell's addition to Ampère's law is particularly important: it makes the set of equations mathematically consistent for non static fields, without changing the laws of Ampere and Gauss for static fields. However, as a consequence, it predicts that a changing magnetic field induces an electric field and vice versa. Therefore, these equations allow self-sustaining "electromagnetic waves" to travel through empty space.

The speed calculated for electromagnetic waves, which could be predicted from experiments on charges and currents, exactly matches the speed of light; indeed, light *is* one form of electromagnetic radiation (as are X-rays, radio waves, and others). Maxwell understood the connection between electromagnetic waves and light in 1861, thereby unifying the theories of electromagnetism and optics.

Formulation in Terms of Electric and Magnetic Fields (Microscopic or in Vacuum Version)

In the electric and magnetic field formulation there are four equations that determine the fields for given charge and current distribution. A separate law of nature, the Lorentz force law, describes how, conversely, the electric and magnetic field act on charged particles and currents. A version of this law was included in the original equations by Maxwell but, by convention, is included no longer. The vector calculus formalism below, due to Oliver Heaviside, has become standard. It is manifestly rotation invariant, and therefore mathematically much more transparent than Maxwell's original 20 equations in x,y,z components. The relativistic formulations are even more symmetric and manifestly Lorentz invariant. For the same equations expressed using tensor calculus or differential forms.

The differential and integral equations formulations are mathematically equivalent and are both useful. The integral formulation relates fields within a region of space to fields on the boundary and can often be used to simplify and directly calculate fields from symmetric distributions of charges and currents. On the other hand, the differential equations are purely *local* and are a more natural starting point for calculating the fields in more complicated (less symmetric) situations, for example using finite element analysis.

Formulation in SI Units Convention

Name	Integral equations	Differential equations
Gauss's law	$\oiint_{\partial\Omega} \mathbf{E}\cdot d\mathbf{S} = \frac{1}{\varepsilon_0}\iiint_{\Omega}\rho dV$	$\nabla\cdot\mathbf{E} = \frac{\rho}{\varepsilon_0}$
Gauss's law for magnetism	$\oiint_{\partial\Omega} \mathbf{B}\cdot d\mathbf{S} = 0$	$\nabla\cdot\mathbf{B} = 0$
Maxwell–Faraday equation (Faraday's law of induction)	$\oint_{\partial\Sigma} \mathbf{E}\cdot d\mathbf{l} = -\frac{d}{dt}\iint_{\Sigma}\mathbf{B}\cdot d\mathbf{S}$	$\nabla\times\mathbf{E} = -\frac{\partial\mathbf{B}}{\partial t}$
Ampère's circuital law (with Maxwell's addition)	$\mu_0\left(\iint_{\Sigma}\mathbf{J}\cdot d\mathbf{S} + \varepsilon_0\frac{d}{dt}\iint_{\Sigma}\mathbf{E}\cdot d\mathbf{S}\right)$	$\nabla\times\mathbf{B} = \mu_0\left(\mathbf{J} + \varepsilon_0\frac{\partial\mathbf{E}}{\partial t}\right)$

Formulation in Gaussian Units Convention

The definitions of charge, electric field, and magnetic field can be altered to simplify theoretical calculation, by absorbing dimensioned factors of ε_0 and c into the units of calculation, by convention. With a corresponding change in convention for the Lorentz force law this yields the same physics, i.e. trajectories of charged particles, or work done by an electric motor. These definitions are often preferred in theoretical and high energy physics where it is natural to take the electric and magnetic field with the same units, to simplify the appearance of the electromagnetic tensor: the Lorentz covariant object unifying electric and magnetic field would then contain components with uniform unit and dimension. Such modified definitions are conventionally used with the Gaussian (CGS) units. Using these definitions and conventions, colloquially "in Gaussian units", the Maxwell equations become:

Name	Integral equations	Differential equations
Gauss's law	$\oiint_{\partial\Omega} \mathbf{E}\cdot d\mathbf{S} = \frac{1}{\varepsilon_0}\iiint_{\Omega}\rho dV$	$\nabla\cdot\mathbf{E} = \frac{\rho}{\varepsilon_0}$
Gauss's law for magnetism	$\oiint_{\partial\Omega} \mathbf{B}\cdot d\mathbf{S} = 0$	$\nabla\cdot\mathbf{B} = 0$
Maxwell–Faraday equation (Faraday's law of induction)	$\oint_{\partial\Sigma} \mathbf{E}\cdot d\ell = -\frac{1}{c}\frac{d}{dt}\iint_{\Sigma}\mathbf{B}\cdot d\mathbf{S}$	$\nabla\times\mathbf{E} = -\frac{1}{c}\frac{\partial\mathbf{B}}{\partial t}$
Ampère's circuital law (with Maxwell's addition)	$\oint_{\partial\Sigma}\mathbf{B}\cdot d\ell = \frac{1}{c}\left(4\pi\iint_{\Sigma}\mathbf{J}\cdot d\mathbf{S} + \frac{d}{dt}\iint_{\Sigma}\mathbf{E}\cdot d\mathbf{S}\right)$	$\nabla\times\mathbf{B} = \frac{1}{c}\left(4\pi\mathbf{J} + \frac{\partial\mathbf{E}}{\partial t}\right)$

Note that the equations are particularly readable when length and time are measured in compatible units like seconds and lightseconds i.e. in units such that c = 1 unit of length/unit of time. Ever since 1983, metres and seconds are compatible except for historical legacy since *by definition* c = 299 792 458 m/s (≈ 1.0 feet/nanosecond).

Further cosmetic changes, called rationalisations, are possible by absorbing factors of 4π depending on whether we want Coulomb's law or Gauss law to come out nicely, see Lorentz-Heaviside units (used mainly in particle physics). In theoretical physics it is often useful to choose units such that Plancks constant, the elementary charge, and even Newton's constant are 1.

Key to the Notation

Symbols in bold represent vector quantities, and symbols in *italics* represent scalar quantities, unless otherwise indicated.

The equations introduce the electric field, E, a vector field, and the magnetic field, B, a pseudovector field, each generally having a time and location dependence. The sources are

- the total electric charge density (total charge per unit volume), ρ, and
- the total electric current density (total current per unit area), J.

The universal constants appearing in the equations are

- the permittivity of free space, ε_0, and
- the permeability of free space, μ_0, and
- the speed of light, $c = \dfrac{1}{\sqrt{\varepsilon_0 \mu_0}}$

Differential Equations

In the differential equations,

- the nabla symbol, ∇, denotes the three-dimensional gradient operator, del,
- the $\nabla\cdot$ symbol (pronounced "del dot") denotes the divergence operator,
- the $\nabla\times$ symbol (pronounced "del cross") denotes the curl operator.

Integral Equations

In the integral equations,

- Ω is any fixed volume with closed boundary surface $\partial\Omega$, and
- Σ is any fixed surface with closed boundary curve $\partial\Sigma$,

Here a fixed volume or surface means that it does not change over time. The equations are correct, complete and a little easier to interpret with time-independent surfaces. For example, since the surface is time-independent, we can bring the differentiation under the integral sign in Faraday's law:

$$\frac{d}{dt}\iint_{\Sigma} \mathbf{B}\cdot d\mathbf{S} = \iint_{\Sigma} \frac{\partial \mathbf{B}}{\partial t}\cdot d\mathbf{S},$$

Maxwell's equations can be formulated with possibly time-dependent surfaces and volumes by using the differential version and using Gauss and Stokes formula appropriately.

- $\oiint_{\partial\Omega}$ is a surface integral over the boundary surface $\partial\Omega$, with the loop indicating the surface is closed
- $\iiint_{\Omega}$ is a volume integral over the volume Ω,

- $\oint_{\partial\Sigma}$ is a line integral around the boundary curve $\partial\Sigma$, with the loop indicating the curve is closed.
- $\iint_{\Sigma}$ is a surface integral over the surface Σ,
- The total electric charge Q enclosed in Ω is the volume integral over Ω of the charge density ρ.

  $$Q = \iiint_{\Omega} \rho \, \mathrm{d}V,$$

  where dV is the volume element.
- The net electric current I is the surface integral of the electric current density J passing through a fixed surface, Σ:

  $$I = \iint_{\Sigma} \mathbf{J} \cdot \mathrm{d}\mathbf{S},$$

where dS denotes the vector element of surface area *S*, normal to surface Σ. (Vector area is sometimes denoted by A rather than S, but this conflicts with the notation for magnetic potential).

Relationship Between Differential and Integral Formulations

The equivalence of the differential and integral formulations are a consequence of the Gauss divergence theorem and the Kelvin–Stokes theorem.

Flux and Divergence

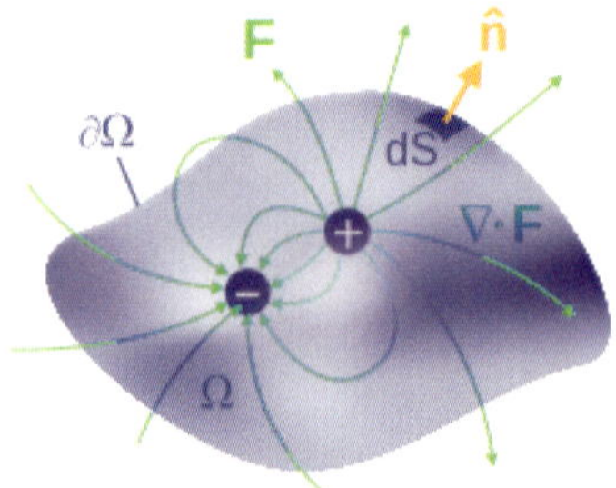

Volume Ω and its closed boundary $\partial\Omega$, containing (respectively enclosing) a source (+) and sink (−) of a vector field F. Here, F could be the E field with source electric charges, but *not*the B field, which has no magnetic charges as shown. The outward unit normal is n.

According to the (purely mathematical) Gauss divergence theorem the electric flux through the boundary surface $\partial\Omega$ can be rewritten as,

$$\oiint_{\partial\Omega} \mathbf{E} \cdot \mathrm{d}\mathbf{S} = \iiint_{\Omega} \nabla \cdot \mathbf{E} \mathrm{d}V$$

The integral version of Gauss's equation can thus be rewritten as,

$$\iiint_{\Omega} \left(\nabla \cdot \mathbf{E} - \frac{\rho}{\epsilon_0} \right) \mathrm{d}V = 0$$

Since Ω is arbitrary (e.g. an arbitrary small ball with arbitrary center), this is satisfied iff the integrand is zero. This is the differential equations formulation of Gauss equation up to a trivial rearrangement.

Similarly rewriting the magnetic flux in Gauss's law for magnetism in integral form gives,

$$\oiint_{\partial\Omega} \mathbf{B}\cdot \mathrm{d}\mathbf{S} = \iiint_{\Omega} \nabla\cdot\mathbf{B}\,\mathrm{d}V = 0$$

which is satisfied for all Ω iff $\nabla\cdot\mathbf{B}=0$.

Circulation and Curl

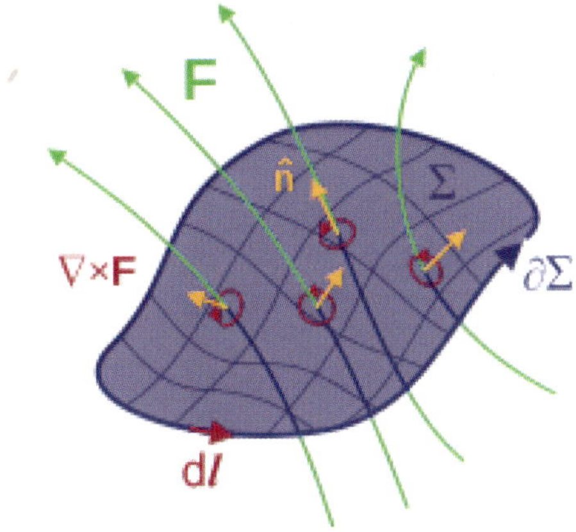

Surface Σ with closed boundary ∂Σ. F could be the E or B fields. Again, nis the unit normal. (The curl of a vector field doesn't literally look like the "circulations", this is a heuristic depiction.)

By the Kelvin–Stokes theorem we can rewrite the line integrals of the fields around the closed boundary curve ∂Σ to an integral of the "circulation of the fields" (i.e. their curls) over a surface it bounds, i.e.

$$\oint_{\partial\Sigma} B\cdot \mathrm{d}\ell = \iint_{\Sigma} (\nabla\times\mathbf{B})\cdot dS$$

Hence the modified Ampere law in integral form can be rewritten as,

$$\iint_{\Sigma}\left(\nabla\times\mathbf{B} - \mu_0\left(\mathbf{J} + \epsilon_0\frac{\partial\mathbf{E}}{\partial t}\right)\right)\cdot \mathrm{d}\mathbf{S} = 0$$

Since Σ can be chosen arbitrarily, e.g. as an arbitrary small, arbitrary oriented, and arbitrary centered disk, we conclude that the integrand is zero iff Ampere's modified law in differential equations form is satisfied. The equivalence of Faraday's law in differential and integral form follows likewise.

The line integrals and curls are analogous to quantities in classical fluid dynamics: the circulation of a fluid is the line integral of the fluid's flow velocity field around a closed loop, and the vorticity of the fluid is the curl of the velocity field.

Charge Conservation

The invariance of charge can be derived as a corollary of Maxwell's equations. The left hand side of the modified Ampere's Law has zero divergence by the div–curl identity. Combining the right hand side, Gauss's law, and interchange of derivatives gives:

$$0 = \nabla\cdot\nabla\times\mathbf{B} = \mu_0\left(\nabla\cdot\mathbf{J} + \varepsilon_0\frac{\partial}{\partial t}\nabla\cdot\mathbf{E}\right) = \mu_0\left(\nabla\cdot\mathbf{J} + \frac{\partial\rho}{\partial t}\right)$$

i.e.

$$\frac{\partial \rho}{\partial t} + \nabla \cdot \mathbf{J} = 0$$

By the Gauss Divergence Theorem, this means the rate of change of charge in a fixed volume equals the net current flowing through the boundary:

$$\frac{d}{dt} Q_{\Omega} = \frac{d}{dt} \iiint_{\Omega} \rho dV = -\oiint_{\partial\Omega} \mathbf{J} \cdot d\mathbf{S} = -I_{\partial\Omega}.$$

In particular, in an isolated system the total charge is conserved.

Vacuum Equations, Electromagnetic Waves and Speed of Light

In a region with no charges (ρ = 0) and no currents (J = 0), such as in a vacuum, Maxwell's equations reduce to:

$$\nabla \cdot \mathbf{E} = 0 \qquad \nabla \times \mathbf{E} = -\frac{\partial \mathbf{B}}{\partial t},$$

$$\nabla \cdot \mathbf{B} = 0 \qquad \nabla \times \mathbf{B} = \mu_0 \varepsilon_0 \frac{\partial \mathbf{E}}{\partial t}.$$

Taking the curl ($\nabla\times$) of the curl equations, and using the curl of the curl identity we obtain,

$$\mu_0 \varepsilon_0 \frac{\partial^2 \mathbf{E}}{\partial t^2} - \nabla^2 \mathbf{E} = 0$$

$$\mu_0 \varepsilon_0 \frac{\partial^2 \mathbf{B}}{\partial t^2} - \nabla^2 \mathbf{B} = 0$$

The quantity $\mu_0 \varepsilon_0$ has the dimension of (time/length)2. Defining $c = (\mu_0 \varepsilon_0)^{-1/2}$, the equations above have the form of the standard wave equations

$$\frac{1}{c^2} \frac{\partial^2 \mathbf{E}}{\partial t^2} - \nabla^2 \mathbf{E} = 0$$

$$\frac{1}{c^2} \frac{\partial^2 \mathbf{B}}{\partial t^2} - \nabla^2 \mathbf{B} = 0$$

Already during Maxwell's lifetime, it was found that the known values for ε_0 and μ_0 give $c \approx 2.998 \times 10^8$ m/s, then already known to be the speed of light in free space. This let him to propose that light and radio waves were propagating electromagnetic waves, since amply confirmed. In the old SI system of units, the values of $\mu_0 = 4\pi \cdot 10^{-7}$ and $c = 299792458$m/s are defined constants, (which means that by definition $\varepsilon_0 = 8.854...10^{-12}$F/m) that define the ampere and the metre. In the new SI system, only *c* keeps its defined value, and the electron charge gets a defined value.

In materials with relative permittivity, ε_r, and relative permeability, μ_r, the phase velocity of light becomes,

$$v_p = \frac{1}{\sqrt{\mu_0 \mu_r \varepsilon_0 \varepsilon_r}}$$

which is usually less than c.

In addition, E and B are perpendicular to each other and to the direction of wave propagation, and are in phase with each other. A sinusoidal plane wave is one special solution of these equations. Maxwell's equations explain how these waves can physically propagate through space. The changing magnetic field creates a changing electric field through Faraday's law. In turn, that electric field creates a changing magnetic field through Maxwell's addition to Ampère's law. This perpetual cycle allows these waves, now known as electromagnetic radiation, to move through space at velocity c.

Macroscopic Formulation

The above equations are the "microscopic" version of Maxwell's equations, expressing the electric and the magnetic fields in terms of the (possibly atomic-level) charges and currents present. This is sometimes called the "general" form, but the macroscopic version below is equally general, the difference being one of bookkeeping.

The microscopic version is sometimes called "Maxwell's equations in a vacuum": this refers to the fact that the material medium is not built into the structure of the equations, but appears only in the charge and current terms. The microscopic version was introduced by Lorentz, who tried to use it to derive the macroscopic properties of bulk matter from its microscopic constituents.

"Maxwell's macroscopic equations", also known as Maxwell's equations in matter, are more similar to those that Maxwell introduced himself.

Name	Integral equations (SI convention)	Differential equations (SI convention)	Differential equations (Gaussian convention)
Gauss's law	$\oiint_{\partial\Omega} \mathbf{D}\cdot d\mathbf{S} = \iiint_{\Omega} \rho_f \, dV$	$\nabla\cdot\mathbf{D} = \rho_f$	$\nabla\cdot\mathbf{D} = 4\pi\rho_f$
Gauss's law for magnetism	$\oiint_{\partial\Omega} \mathbf{B}\cdot d\mathbf{S} = 0$	$\nabla\cdot\mathbf{B} = 0$	$\nabla\cdot\mathbf{B} = 0$
Maxwell–Faraday equation (Faraday's law of induction)	$\oint_{\partial\Sigma} \mathbf{E}\cdot d\ell = -\frac{d}{dt}\iint_{\Sigma} \mathbf{B}\cdot d\mathbf{S}$	$\nabla\times\mathbf{E} = -\frac{\partial\mathbf{B}}{\partial t}$	$\nabla\times\mathbf{E} = -\frac{1}{c}\frac{\partial\mathbf{B}}{\partial t}$
Ampère's circuital law (with Maxwell's addition)	$\oint_{\partial\Sigma} \mathbf{H}\cdot d\ell = \iint_{\Sigma} \mathbf{J}_f \cdot dS + \frac{d}{dt}\iint_{\Sigma} D\cdot d\mathbf{S}$	$\nabla\times\mathbf{H} = \mathbf{J}_f + \frac{\partial\mathbf{D}}{\partial t}$	$\nabla\times\mathbf{H} = \frac{1}{c}\left(4\pi\mathbf{J}_f + \frac{\partial\mathbf{D}}{\partial t}\right)$

In the "macroscopic" equations, the influence of bound charge Q_b and bound current I_b is incorporated into the displacement field D and the magnetizing field H, while the equations depend only on the free charges Q_f and free currents I_f. This reflects a splitting of the total electric charge Q and current I (and their densities ρ and J) into free and bound parts:

$$Q = Q_f + Q_b = \iiint_{\Omega} (\rho_f + \rho_b) dV = \iiint_{\Omega} \rho dV$$

$$I = I_f + I_b = \iint_{\Sigma} (\mathbf{J}_f + \mathbf{J}_b)\cdot d\mathbf{S} = \iint_{\Sigma} \mathbf{J}\cdot d\mathbf{S}$$

The cost of this splitting is that the additional fields D and H need to be determined through phenomenological constituent equations relating these fields to the electric field E and the magnetic field B, together with the bound charge and current.

Bound Charge and Current

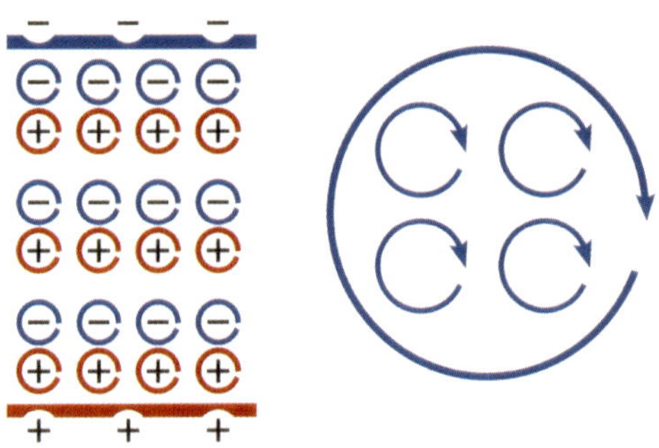

Left: A schematic view of how an assembly of microscopic dipoles produces opposite surface charges as shown at top and bottom. *Right:* How an assembly of microscopic current loops add together to produce a macroscopically circulating current loop. Inside the boundaries, the individual contributions tend to cancel, but at the boundaries no cancelation occurs.

When an electric field is applied to a dielectric material its molecules respond by forming microscopic electric dipoles – their atomic nuclei move a tiny distance in the direction of the field, while their electrons move a tiny distance in the opposite direction. This produces a *macroscopic bound charge* in the material even though all of the charges involved are bound to individual molecules. For example, if every molecule responds the same, similar to that shown in the figure, these tiny movements of charge combine to produce a layer of positive bound charge on one side of the material and a layer of negative charge on the other side. The bound charge is most conveniently described in terms of the polarization P of the material, its dipole moment per unit volume. If P is uniform, a macroscopic separation of charge is produced only at the surfaces where P enters and leaves the material. For non-uniform P, a charge is also produced in the bulk.

Somewhat similarly, in all materials the constituent atoms exhibit magnetic moments that are intrinsically linked to the angular momentum of the components of the atoms, most notably their electrons. The connection to angular momentum suggests the picture of an assembly of microscopic current loops. Outside the material, an assembly of such microscopic current loops is not different from a macroscopic current circulating around the material's surface, despite the fact that no individual charge is traveling a large distance. These *bound currents* can be described using the magnetization M.

The very complicated and granular bound charges and bound currents, therefore, can be represented on the macroscopic scale in terms of P and M, which average these charges and currents on a sufficiently large scale so as not to see the granularity of individual atoms, but also sufficiently small that they vary with location in the material. As such, *Maxwell's macroscopic equations* ignore many details on a fine scale that can be unimportant to understanding matters on a gross scale by calculating fields that are averaged over some suitable volume.

Auxiliary Fields, Polarization and Magnetization

The *definitions* (not constitutive relations) of the auxiliary fields are:

$$\mathbf{D}(\mathbf{r},t) = \varepsilon_0 \mathbf{E}(\mathbf{r},t) + \mathbf{P}(\mathbf{r},t)$$

$$\mathbf{H}(\mathbf{r},t) = \frac{1}{\mu_0}\mathbf{B}(\mathbf{r},t) - \mathbf{M}(\mathbf{r},t)$$

where P is the polarization field and M is the magnetization field, which are defined in terms of microscopic bound charges and bound currents respectively. The macroscopic bound charge density $\boldsymbol{\rho}_b$ and bound current density J_b in terms of polarization P and magnetization M are then defined as,

$$\rho_b = -\nabla \cdot \mathbf{P}$$

$$\mathbf{J}_b = \nabla \times \mathbf{M} + \frac{\partial \mathbf{P}}{\partial t}$$

If we define the total, bound, and free charge and current density by,

$$d\rho = \rho_b + \rho_f,$$

$$\mathbf{J} = \mathbf{J}_b + \mathbf{J}_f,$$

and use the defining relations above to eliminate D, and H, the "macroscopic" Maxwell's equations reproduce the "microscopic" equations.

Constitutive Relations

In order to apply 'Maxwell's macroscopic equations', it is necessary to specify the relations between displacement field D and the electric field E, as well as the magnetizing field H and the magnetic field B. Equivalently, we have to specify the dependence of the polarisation P (hence the bound charge) and the magnetisation M (hence the bound current) on the applied electric and magnetic field. The equations specifying this response are called constitutive relations. For real-world materials, the constitutive relations are rarely simple, except approximately, and usually determined by experiment.

For materials without polarization and magnetisation, the constitutive relations are (by definition),

$$\mathbf{D} = \varepsilon_0 \mathbf{E}, \quad \mathbf{H} = \frac{1}{\mu_0} \mathbf{B}$$

where $\boldsymbol{\varepsilon}_0$ is the permittivity of free space and $\boldsymbol{\mu}_0$ the permeability of free space. Since there is no bound charge, the total and the free charge and current are equal.

An alternative viewpoint on the microscopic equations is that they are the macroscopic equations *together* with the statement that vacuum behaves like a perfect linear "material" without additional polarisation and magnetisation. More generally, for linear materials the constitutive relations are,

$$\mathbf{D} = \varepsilon \mathbf{E}, \quad \mathbf{H} = \frac{1}{\mu} \mathbf{B}$$

where ε is the permittivity and μ the permeability of the material. For the displacement field D the linear approximation is usually excellent because for all but the most extreme electric fields or temperatures obtainable in the laboratory (high power pulsed lasers) the interatomic electric fields of materials of the order of 10^{11} V/m are much higher than the external field. For the magnetizing field **H**, however, the linear approximation can break down in common materials like iron leading to phenomena like hysteresis. Even the linear case can have various complications, however.

- For homogeneous materials, ε and μ are constant throughout the material, while for inhomogeneous materials they depend on location within the material (and perhaps time).
- For isotropic materials, ε and μ are scalars, while for anisotropic materials (e.g. due to crystal structure) they are tensors.
- Materials are generally dispersive, so ε and μ depend on the frequency of any incident EM waves.

Even more generally, in the case of non-linear materials, D and P are not necessarily proportional to E, similarly H or M is not necessarily proportional to B. In general Dand H depend on both E and B, on location and time, and possibly other physical quantities.

In applications one also has to describe how the free currents and charge density behave in terms of E and B possibly coupled to other physical quantities like pressure, and the mass, number density, and velocity of charge-carrying particles. E.g., the original equations given by Maxwell included Ohms law in the form,

$$\boldsymbol{J}_f = \sigma \boldsymbol{E}$$

Alternative Formulations

Following is a summary of some of the numerous other mathematical formalisms to write the microscopic Maxwell's equations, with the columns separating the two homogeneous Maxwell equations from the two inhomogeneous ones involving charge and current. Each formulation has versions directly in terms of the electric and magnetic fields, and indirectly in terms of the electrical potential φ and the vector potential A. Potentials were introduced as a convenient way to solve the homogeneous equations, but it was thought that all observable physics was contained in the electric and magnetic fields (or relativistically, the Faraday tensor). The potentials play a central role in quantum mechanics, however, and act quantum mechanically with observable consequences even when the electric and magnetic fields vanish (Aharonov–Bohm effect).

Each table describes one formalism.

Vector calculus		
Formulation	Homogeneous equations	Inhomogeneous equations
Fields 3D Euclidean space + time	$\nabla \cdot \mathbf{B} = 0$ $\nabla \times \mathbf{E} + \frac{\partial \mathbf{B}}{\partial t} = 0$	$\nabla \cdot \mathbf{E} = \frac{\rho}{\varepsilon_0}$ $\nabla \times \mathbf{B} - \frac{1}{c^2}\frac{\partial \mathbf{E}}{\partial t} = \mu_0 \mathbf{J}$
Potentials (any gauge) 3D Euclidean space + time	$\mathbf{B} = \nabla \times \mathbf{A}$ $\mathbf{E} = -\nabla\varphi - \frac{\partial \mathbf{A}}{\partial t}$	$-\nabla^2\varphi - \frac{\partial}{\partial t}(\nabla \cdot \mathbf{A}) = \frac{\rho}{\varepsilon_0}$ $\left(-\nabla^2 + \frac{1}{c^2}\frac{\partial^2}{\partial t^2}\right)\mathbf{A} + \nabla\left(\nabla \cdot \mathbf{A} + \frac{1}{c^2}\frac{\partial \varphi}{\partial t}\right) = \mu_0 \mathbf{J}$

Potentials (Lorenz gauge) 3D Euclidean space + time	$\mathbf{B} = \nabla \times \mathbf{A}$ $\mathbf{E} = -\nabla\varphi - \frac{\partial \mathbf{A}}{\partial t}\ \nabla \cdot \mathbf{A} = -\frac{1}{c^2}\frac{\partial \varphi}{\partial t}$	$\left(-\nabla^2 + \frac{1}{c^2}\frac{\partial^2}{\partial t^2}\right)\varphi = \frac{\rho}{\varepsilon_0}$ $\left(-\nabla^2 + \frac{1}{c^2}\frac{\partial^2}{\partial t^2}\right)\mathbf{A} = \mu_0 \mathbf{J}$
Tensor calculus		
Formulation	Homogeneous equations	Inhomogeneous equations
Fields space + time spatial metric independent of time	$\partial_{[i}B_{jk]} =$ $\nabla_{[i}B_{jk]} = 0$ $\partial_{[i}E_{j]} + \frac{\partial B_{ij}}{\partial t} =$ $\nabla_{[i}E_{j]} + \frac{\partial B_{ij}}{\partial t} = 0$	$\frac{1}{\sqrt{h}}\partial_i \sqrt{h}E^i =$ $\nabla_i E^i = \frac{\rho}{\epsilon_0}$ $-\frac{1}{\sqrt{h}}\partial_i \sqrt{h}B^{ij} - \frac{1}{c^2}\frac{\partial}{\partial t}E^j =$ $-\nabla_i B^{ij} - \frac{1}{c^2}\frac{\partial E^j}{\partial t} = \mu_0 J^j$
Potentials space (with topological restrictions) + time spatial metric independent of time	$B_{ij} = \partial_{[i}A_{j]}$ $= \nabla_{[i}A_{j]}$ $E_i = -\frac{\partial A_i}{\partial t} - \partial_i \phi$ $= -\frac{\partial A_i}{\partial t} - \nabla_i \phi$	$-\frac{1}{\sqrt{h}}\partial_i \sqrt{h}\left(\partial^i \phi + \frac{\partial A^i}{\partial t}\right) =$ $-\nabla_i \nabla^i \phi - \frac{\partial}{\partial t}\nabla_i A^i = \frac{\rho}{\epsilon_0}$ $-\frac{1}{\sqrt{h}}\partial_i\left(\sqrt{h}h^{im}h^{jn}\partial_{[m}A_{n]}\right) + \frac{1}{c^2}\frac{\partial}{\partial t}\left(\frac{\partial A^j}{\partial t} + \partial^j \phi\right) =$ $-\nabla_i \nabla^i A^j + \frac{1}{c^2}\frac{\partial^2 A^j}{\partial t^2} + R_i^j A^i + \nabla^j\left(\nabla_i A^i + \frac{1}{c^2}\frac{\partial \phi}{\partial t}\right) = \mu_0 J^j$
Potentials (Lorenz gauge) space (with topological restrictions) + time spatial metric independent of time	$B_{ij} = \partial_{[i}A_{j]}$ $= \nabla_{[i}A_{j]}$ $E_i = -\frac{\partial A_i}{\partial t} - \partial_i \phi$ $= -\frac{\partial A_i}{\partial t} - \nabla_i \phi$ $\nabla_i A^i = -\frac{1}{c^2}\frac{\partial \phi}{\partial t}$	$-\nabla_i \nabla^i \phi + \frac{1}{c^2}\frac{\partial^2 \phi}{\partial t^2} = \frac{\rho}{\epsilon_0}$ $-\nabla_i \nabla^i A^j + \frac{1}{c^2}\frac{\partial^2 A^j}{\partial t^2} + R_i^j A^i = \mu_0 J^j$

Differential forms		
Formulation	Homogeneous equations	Inhomogeneous equations
Fields Any space + time	$dB = 0$ $dE + \frac{\partial B}{\partial t} = 0$	$d * E = \frac{\rho}{\epsilon_0}$ $d * B - \frac{1}{c^2}\frac{\partial * E}{\partial t} = \mu_0 J$
Potentials (any gauge) Any space (with topological restrictions) + time	$dB = dA$ $E = -d\phi - \frac{\partial A}{\partial t}$	$-d * \left(d\phi + \frac{\partial A}{\partial t}\right) = \frac{\rho}{\epsilon_0}$ $d * dA + \frac{1}{c^2}\frac{\partial}{\partial t} * \left(d\phi + \frac{\partial A}{\partial t}\right) = \mu_0 J$
Potential (Lorenz Gauge) Any space (with topological restrictions) + time spatial metric independent of time	$dB = dA$ $E = -d\phi - \frac{\partial A}{\partial t}$ $d * A = - * \frac{1}{c^2}\frac{\partial \phi}{\partial t}$	$* \left(-\Delta\phi + \frac{1}{c^2}\frac{\partial^2}{\partial t^2}\phi\right) = \frac{\rho}{\epsilon_0}$ $* \left(-\Delta A + \frac{1}{c^2}\frac{\partial^2 A}{\partial^2 t}\right) = \mu_0 J$

Relativistic Formulations

The Maxwell equations can also be formulated on a spacetime-like Minkowski space where space and time are treated on equal footing. The direct spacetime formulations make manifest that the

Maxwell equations are relativistically invariant. Because of this symmetry electric and magnetic field are treated on equal footing and are recognised as components of the Faraday tensor. This reduces the four Maxwell equations to two, which simplifies the equations, although we can no longer use the familiar vector formulation. In fact the Maxwell equations in the space + time formulation are not Galileo invariantand ha ve Lorentz invariance as a hidden symmetry. This was a major source of inspiration for the development of relativity theory. To repeat: the space + time formulation is not a non-relativistic approximation and it describes the same physics by simply renaming variables. For this reason the relativistic invariant equations are usually called the Maxwell equations as well.

Each table describes one formalism.

	Tensor calculus	
Formulation	Homogeneous equations	Inhomogeneous equations
Fields Minkowski space	$\partial_{[\alpha} F_{\beta\gamma]} = 0$	$\partial_\alpha F^{\alpha\beta} = \mu_0 J^\beta$
Potentials (any gauge) Minkowski space	$F_{\alpha\beta} = 2\partial_{[\alpha} A_{\beta]}$	$2\partial_\alpha \partial^{[\alpha} A^{\beta]} = \mu_0 J^\beta$
Potentials (Lorenz gauge) Minkowski space	$F_{\alpha\beta} = 2\partial_{[\alpha} A_{\beta]}$ $\partial_\alpha A^\alpha = 0$	$\partial_\alpha \partial^\alpha A^\beta = \mu_0 J^\beta$
Fields Any spacetime	$\partial_{[\alpha} F_{\beta\gamma]} =$ $\nabla_{[\alpha} F_{\beta\gamma]} = 0$	$\frac{1}{\sqrt{-g}} \partial_\alpha (\sqrt{-g} F^{\alpha\beta}) =$ $\nabla_\alpha F^{\alpha\beta} = \mu_0 J^\beta$
Potentials (any gauge) Any spacetime (with topological restrictions)	$F_{\alpha\beta} = 2\partial_{[\alpha} A_{\beta]}$ $= 2\nabla_{[\alpha} A_{\beta]}$	$\frac{2}{\sqrt{-g}} \partial_\alpha (\sqrt{-g} g^{\alpha\mu} g^{\beta\nu} \partial_{[\mu} A_{\nu]}) =$ $2\nabla_\alpha (\nabla^{[\alpha} A^{\beta]}) = \mu_0 J^\beta$
Potentials (Lorenz gauge) Any spacetime (with topological restrictions)	$F_{\alpha\beta} = 2\partial_{[\alpha} A_{\beta]}$ $= 2\nabla_{[\alpha} A_{\beta]}$ $\nabla_\alpha A^\alpha = 0$	$\nabla_\alpha \nabla^\alpha A^\beta - R^\beta{}_\alpha A^\alpha = \mu_0 J^\beta$

	Differential forms	
Formulation	Homogeneous equations	Inhomogeneous equations
Fields Any spacetime	$\mathrm{d}F = 0$	$\mathrm{d} \star F = \mu_0 J$
Potentials (any gauge) Any spacetime (with topological restrictions)	$F = \mathrm{d}A$	$\mathrm{d} \star \mathrm{d}A = \mu_0 J$
Potentials (Lorenz gauge) Any spacetime (with topological restrictions)	$F = \mathrm{d}A$ $\mathrm{d} \star A = 0$	$\star \sqcap A = \mu_0 J$

- In the tensor calculus formulation, the electromagnetic tensor $F_{\alpha\beta}$ is an antisymmetric covariant rank 2 tensor; the four-potential, A_α, is a covariant vector; the current, J^α, is a vector; the square brackets, [ ], denote antisymmetrization of indices; ∂_α is the derivative with respect to the coordinate, x^α . In Minkowski space coordinates are chosen with respect to an inertial frame; $(x^\alpha) = (ct,x,y,z)$, so that the metric tensor used to raise and lower indices is $\eta_{\alpha\beta} = diag(1,-1,-1,-1)$. The d'Alembert operator on Minkowski space is $\square = \partial_\alpha \partial^\alpha$ as in the vector formulation. In general spacetimes, the coordinate system x^α is arbitrary, the covariant derivative ∇_α, the Ricci tensor, $R_{\alpha\beta}$ and raising and lowering of indices are defined by the Lorentzian metric, $g_{\alpha\beta}$ and the d'Alembert operator is defined as $\square = \nabla_\alpha \nabla^\alpha$. . The topological restriction is that the second real cohomology group of the space vanishes (see the differential form formulation for an explanation). Note that this is violated for Minkowski space with a line removed, which can model a (flat) spacetime with a point-like monopole on the complement of the line.

- In the differential form formulation on arbitrary space times, $F = F_{\alpha\beta} dx^\alpha \wedge dx^\beta$ is the electromagnetic tensor considered as a 2-form, $A = A_\alpha dx^\alpha$ is the potential 1-form, J is the current 3-form, d is the exterior derivative, and $\star$ is the Hodge star on forms defined (up to its an orientation, i.e. its sign) by the Lorentzian metric of spacetime. Note that in the special case of 2-forms such as F, the Hodge star $\star$ depends on the metric tensor only for its local scale. This means that, as formulated, the differential form field equations are conformally invariant, but the Lorenz gauge condition breaks conformal invariance. The operator $\Box = (-\star \mathrm{d} \star \mathrm{d} - \mathrm{d} \star \mathrm{d} \star)$ is the d'Alembert–Laplace–Beltrami operator on 1-forms on an arbitrary Lorentzian spacetime. The topological condition is again that the second real cohomology group is trivial. By the isomorphism with the second de Rham cohomology this condition means that every closed 2-form is exact.

Other formalisms include the geometric algebra formulation and a matrix representation of Maxwell's equations. Historically, a quaternionic formulation was used.

Solutions

Maxwell's equations are partial differential equations that relate the electric and magnetic fields to each other and to the electric charges and currents. Often, the charges and currents are themselves dependent on the electric and magnetic fields via the Lorentz force equation and the constitutive relations. These all form a set of coupled partial differential equations which are often very difficult to solve: the solutions encompass all the diverse phenomena of classical electromagnetism. Some general remarks follow.

As for any differential equation, boundary conditions and initial conditions are necessary for a unique solution. For example, even with no charges and no currents anywhere in spacetime, there are the obvious solutions for which E and B are zero or constant, but there are also non-trivial solutions corresponding to electromagnetic waves. In some cases, Maxwell's equations are solved over the whole of space, and boundary conditions are given as asymptotic limits at infinity. In other cases, Maxwell's equations are solved in a finite region of space, with appropriate conditions on the boundary of that region, for example an artificial absorbing boundary representing the rest of the universe, or periodic boundary conditions, or walls that isolate a small region from the outside world (as with a waveguide or cavity resonator).

Jefimenko's equations (or the closely related Liénard–Wiechert potentials) are the explicit solution to Maxwell's equations for the electric and magnetic fields created by any given distribution of charges and currents. It assumes specific initial conditions to obtain the so-called "retarded solution", where the only fields present are the ones created by the charges. However, Jefimenko's equations are unhelpful in situations when the charges and currents are themselves affected by the fields they create.

Numerical methods for differential equations can be used to compute approximate solutions of Maxwell's equations when exact solutions are impossible. These include the finite element method and finite-difference time-domain method.

Overdetermination of Maxwell's Equations

Maxwell's equations *seem* overdetermined, in that they involve six unknowns (the three components of E and B) but eight equations (one for each of the two Gauss's laws, three vector components each for Faraday's and Ampere's laws). (The currents and charges are not unknowns, being freely specifiable subject to charge conservation.) This is related to a certain limited kind of redundancy in Maxwell's equations: It can be proven that any system satisfying Faraday's law and Ampere's law *automatically* also satisfies the two Gauss's laws, as long as the system's initial condition does. This explanation was first introduced by Julius Adams Stratton in 1941. Although it is possible to simply ignore the two Gauss's laws in a numerical algorithm (apart from the initial conditions), the imperfect precision of the calculations can lead to ever-increasing violations of those laws. By introducing dummy variables characterizing these violations, the four equations become not overdetermined after all. The resulting formulation can lead to more accurate algorithms that take all four laws into account.

Both identities $\nabla\cdot\nabla\times\mathbf{B}\equiv 0, \nabla\cdot\nabla\times\mathbf{E}\equiv 0$, which reduce eight equations to six independent ones, are the true reason of overdetermination.

Maxwell Equations as the Classical Limit of QED

Maxwell's equations and the Lorentz force law (along with the rest of classical electromagnetism) are extraordinarily successful at explaining and predicting a variety of phenomena; however they are not exact, but a classical limit of quantum electrodynamics (QED).

Some observed electromagnetic phenomena are incompatible with Maxwell's equations. These include photon–photon scattering and many other phenomena related to photons or virtual photons, "nonclassical light" and quantum entanglement of electromagnetic fields. E.g. quantum cryptography cannot be described by Maxwell theory, not even approximately. The approximate nature of Maxwell's equations becomes more and more apparent when going into the extremely strong field regime or to extremely small distances.

Finally, Maxwell's equations cannot explain any phenomenon involving individual photons interacting with quantum matter, such as the photoelectric effect, Planck's law, the Duane–Hunt law, and single-photon light detectors. However, many such phenomena may be approximated using a halfway theory of quantum matter coupled to a classical electromagnetic field, either as external field or with the expected value of the charge current and density on the right hand side of Maxwell's equations.

Variations

Popular variations on the Maxwell equations as a classical theory of electromagnetic fields are relatively scarce because the standard equations have stood the test of time remarkably well.

Magnetic Monopoles

Maxwell's equations posit that there is electric charge, but no magnetic charge (also called magnetic monopoles), in the universe. Indeed, magnetic charge has never been observed (despite extensive searches) and may not exist. If they did exist, both Gauss's law for magnetism and Faraday's law would need to be modified, and the resulting four equations would be fully symmetric under the interchange of electric and magnetic fields.

Ampere's Law Go Wrong

A simple example to see that *something* must be wrong with Ampere's Law in the general case is given by Feynman in his *Lectures in Physics*. Suppose we use a hypodermic needle to insert a spherically symmetric blob of charge in the middle of a large vat of solidified jello (which we assume conducts electricity). Because of electrostatic repulsion, the charge will dissipate, currents will flow outwards in a spherically symmetric way. *Question*: does this outward-flowing current distribution generate a magnetic field? The answer must be *no* , because since we have a completely spherically symmetric situation, it could only generate a spherically symmetric magnetic field. But the only possible such fields are one pointing outwards everywhere and one pointing inwards everywhere, both corresponding to non-existent monopoles. So, there can be no magnetic field.

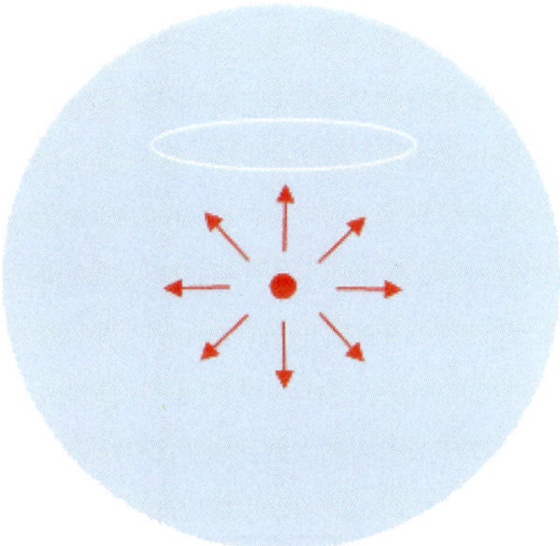

However, imagine we now consider checking Ampere's law by taking as a path a horizontal circle with its center above the point where we injected the charge (think of a halo above someone's head.) Obviously, the left hand side of Ampere's equation is zero, since there can be no magnetic field. (It would have to be spherically symmetric, meaning radial.) On the other hand, the right hand side is most definitely not zero, since some of the outward flowing current is going to go through our circle. So the equation must be wrong.

Ampere's law was established as the result of large numbers of careful experiments on all kinds of current distributions. So how could it be that something of the kind we describe above was overlooked? The reason is really similar to why electromagnetic induction was missed for so long. No-one thought about looking at *changing* fields, all the experiments were done on steady situations. With our ball of charge spreading outward in the jello, there is obviously a changing electric field. Imagine yourself in the jello near where the charge was injected: at first, you would feel a strong field from the nearby concentrated charge, but as the charge spreads out spherically, some of it going past you, the field will decrease with time.

Maxwell's Example

Maxwell himself gave a more practical example: consider Ampere's law for the usual infinitely long wire carrying a steady current I, but now break the wire at some point and put in two large circular metal plates, a capacitor, maintaining the steady current I in the wire everywhere else, so that charge is simply piling up on one of the plates and draining off the other.

Looking now at the wire some distance away from the plates, the situation appears normal, and if we put the usual circular path around the wire, application of Ampere's law tells us that the magnetic field at distance r, from,

$$\oint \vec{B} \cdot d\vec{\ell} = \mu_0 I$$

is just,

$$B = \mu_0 I / 2\pi r.$$

Recall, however, that we defined the current threading the path in terms of current punching through a soap film spanning the path, and said this was independent of whether the soap film was flat, bulging out on one side, or whatever. With a single infinite wire, there was no escape— no contortions of this covering surface could wriggle free of the wire going through it (actually, if you distort the surface enough, the wire could penetrate it several times, but you have to count the net flow across the surface, and the new penetrations would come in pairs with the current crossing the surface in opposite directions, so they would cancel).

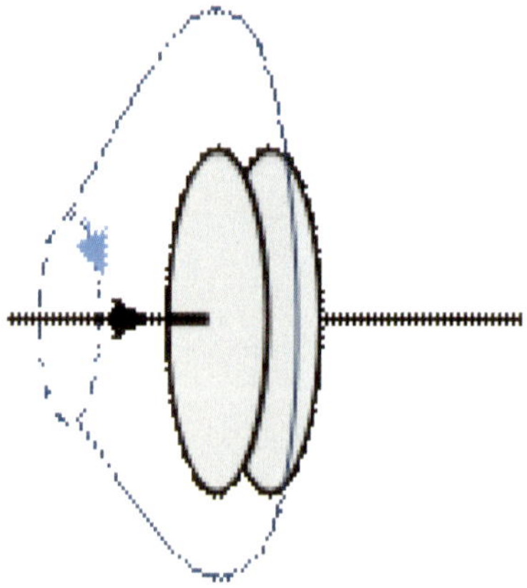

Once we bring in Maxwell's parallel plate capacitor, however, there *is* a way to distort the surface so that no current penetrates it at all: we can run it between the plates.

The question then arises: can we rescue Ampere's law by adding another term just as the electrostatic version of the third equation was rescued by adding Faraday's induction term? The answer is of course yes: although there is no *current* crossing the surface if we put it between the capacitor plates, there is certainly a *changing electric field* , because the capacitor is charging up as the current I flows in. Assuming the plates are close together, we can take all the electric field lines from the charge q on one plate to flow across to the other plate, so the total electric flux across the surface between the plates,

$$\int \vec{E} \cdot d\vec{A} \rightarrow = q / \varepsilon_0 .$$

Now, the current in the wire , I, is just the rate of change of charge on the plate,

$$I = dq / dt.$$

Putting the above two equations together, we see that

$$I = \frac{d}{dt}\left(\varepsilon_0 \int \vec{E} \cdot d\vec{A}\right).$$

Ampere's law can now be written in a way that is correct no matter where we put the surface spanning the path we integrate the magnetic field around:

$$\oint \vec{B} \cdot d\vec{\ell} = \mu_0 \left(I + \frac{d}{dt}\left(\varepsilon_0 \int \vec{E} \cdot d\vec{A}\right)\right).$$

This is *Maxwell's fourth equation.*

Notice that in the case of the wire, *either* the current in the wire, *or* the increasing electric field, contribute on the right hand side, depending on whether we have the surface simply cutting through the wire, or positioned between the plates. (Actually, more complicated situations are possible—we could imaging the surface partly between the plates, then cutting through the plates to get out! In this case, we would have to figure out the current actually in the plate to get the right hand side, but the equation would still apply).

Displacement Current

Displacement current, in electromagnetism, a phenomenon analogous to an ordinary electric current, posited to explain magnetic fields that are produced by changing electric fields. Ordinary electric currents, called conduction currents, whether steady or varying, produce an accompanying magnetic field in the vicinity of the current. The British physicist James Clerk Maxwell in the 19th century predicted that a magnetic field also must be associated with a changing electric field even in the absence of a conduction current, a theory that was subsequently verified experimentally. As magnetic fields had long been associated with currents, the predicted magnetic field also was thought of as stemming from another kind of current. Maxwell gave it the name displacement current, which was proportional to the rate of change of the electric field that kept cropping up naturally in his theoretical formulations.

As electric charges do not flow through the insulation from one plate of a capacitor to the other, there is no conduction current; instead, a displacement current is said to be present to account for the continuity of the magnetic effects. In fact, the calculated size of the displacement current between the plates of a capacitor being charged and discharged in an alternating-current circuit is equal to the size of the conduction current in the wires leading to and from the capacitor. Displacement currents play a central role in the propagation of electromagnetic radiation, such as light and radio waves, through empty space. A traveling, varying magnetic field is everywhere associated with a periodically changing electric field that may be conceived in terms of a displacement current. Maxwell's insight on displacement current, therefore, made it possible to understand electromagnetic waves as being propagated through space completely detached from electric currents in conductors.

The calculation of the magnetic field of a current distribution can, in principle, be carried out using Ampere's law which relates the path integral of the magnetic field around a closed path to the current intercepted by an arbitrary surface that spans this path:

$$\int_{path} \overline{\mathrm{B}} \cdot d\overline{\mathrm{L}} = \mu_0 \, \mathrm{I}$$

Ampere's law is independent of the shape of the surface chosen as long as the current flows along a continuous, unbroken circuit. However, consider the case in which the current wire is broken and connected to a parallel-plate capacitor (see Figure). A current will flow through the wire during the charging process of the capacitor. This current will generate a magnetic field and if we are far away from the capacitor, this field should be very similar to the magnetic field produced by an infinitely long, continuous, wire. However, the current intercepted by an arbitrary surface now depends on the surface chosen. For example, the surface shown in figure below does not intercept any current. Clearly, Ampere's law can not be applied in this case to find the magnetic field generated by the current.

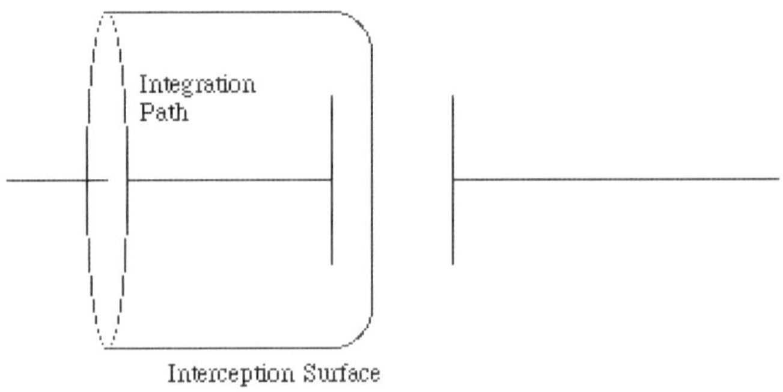

Ampere's law in a capacitor circuit.

Although the surface shown in figure does not intercept any current, it intercepts electric flux. Suppose the capacitor is an ideal capacitor, with a homogeneous electric field E between the plates and no electric field outside the plates. At a certain time t the charge on the capacitor plates is Q. If the plates have a surface area A then the electric field between the plates is equal to,

$$E = \frac{Q}{\varepsilon_0 A}$$

The electric field outside the capacitor is equal to zero. The electric flux, $[Phi]_E$, intercepted by the surface shown in figure is equal to,

$$\Phi_E = EA = \frac{Q}{\varepsilon_0}$$

If a current I is flowing through the wire, then the charge on the capacitor plates will be time dependent. The electric flux will therefore also be time dependent, and the rate of change of electric flux is equal to,

$$\frac{d\Phi_E}{dt} = \frac{1}{\varepsilon_0}\frac{dQ}{dt} = \frac{I}{\varepsilon_0}$$

The magnetic field around the wire can now be found by modifying Ampere's law,

$$\int_{path} \bar{B} \cdot d\bar{L} = \mu_0 \varepsilon_0 \frac{d\Phi_E}{dt}$$

where $[Phi]_E$ is the electric flux through the surface indicated in Figure In the most general case, the surface spanned by the integration path of the magnetic field can intercept current and electric flux. In such a case, the effects of the electric flux and the electric current must be combined, and Ampere's law becomes,

$$\int_{path} \bar{B} \cdot d\bar{L} = \mu_0 I + \mu_0 \varepsilon_0 \frac{d\Phi_E}{dt}$$

The current I is the current intercepted by whatever surface is used in the calculation, and is not necessarily the same as the current in the wires. Equation is frequently written as,

$$\int_{path} \bar{B} \cdot d\bar{L} = \mu_0 (I + I_d)$$

where I_d is called the displacement current and is defined as,

$$I_d = \varepsilon_0 \frac{d\Phi_E}{dt}$$

Example

Problem

A parallel-plate capacitor has circular plates of area A separated by a distance d. A thin straight wire of length d lies along the axis of the capacitor and connects the two plates. This wire has a resistance R. The exterior terminals of the plates are connected to a source of alternating emf with a voltage $V = V_0$ sin([omega] t).

a) What is the current in the thin wire ?

b) What is the displacement current through the capacitor ?

c) What is the current arriving at the outside terminals of the capacitor ?

d) What is the magnetic field between the capacitor plates at a distance r from the axis ? Assume that r is less than the radius of the plates.

a) The setup can be regarded as a parallel circuit of a resistor with resistance R and a capacitor with capacitance C. The current in the thin wire can be obtained using Ohm's law,

$$I_R(t) = \frac{\varepsilon(t)}{R} = \frac{V_0 \sin(wt)}{R}$$

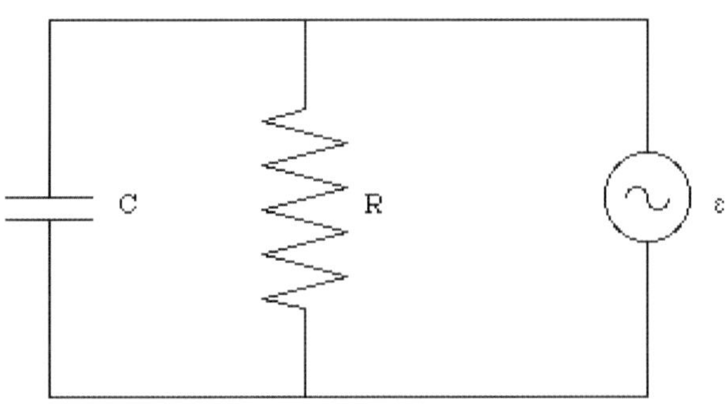

Circuit Problem

b) The voltage across the capacitor is equal to the external emf. The electric field between the capacitor plates is therefore equal to,

$$E(t) = \frac{\varepsilon(t)}{d} = \frac{V_0 \sin(wt)}{d}$$

The electric flux through the capacitor is therefore equal to,

$$\Phi_E(t) = A\,E(t) = \frac{A\,V_0 \sin(wt)}{d}$$

The displacement current I_d can be obtained by substituting eq $\Phi_E(t) = A\,E(t) = \frac{A\,V_0 \sin(wt)}{d}$ into eq $I_d = \varepsilon_0 \frac{d\,\Phi_E}{dt}$,

$$I_d = \varepsilon_0 \frac{d\,\Phi_E(t)}{dt} = \frac{\varepsilon_0\,A\,V_0\,w}{d}\cos(wt)$$

c) The current at the outside terminals of the capacitor is the sum of the current used to charge the capacitor and the current through the resistor. The charge on the capacitor is equal to,

$$Q(t) = C\,V(t)\ \frac{\varepsilon_0\,A\,V_0}{d}\sin(wt)$$

The charging current is thus equal to,

$$I_c(t) = \frac{dQ(t)}{dt} = \frac{\varepsilon_0\,A\,V_0\,w}{d}\cos(wt)$$

The total current is therefore equal to,

$$I_{to}(t) = I_R(t) + I_c(t) = V_0\left\{\frac{1}{R}\sin(wt) + \frac{\varepsilon_0\,A\,w}{d}\cos(wt)\right\}$$

d) The magnetic field lines inside the capacitor will form concentric circles, centered around the resistor . The path integral of the magnetic field around a circle of radius r is equal to,

$$\int_{path} \overline{B}\cdot d\overline{L} = 2\,\pi r\,B(r)$$

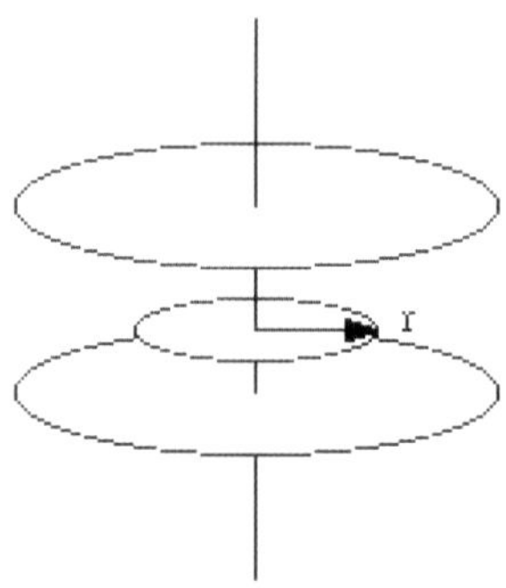

Amperian loop used to determine the magnetic field inside a capacitor

The surface to be used to determine the current and electric flux intercepted is the disk of radius r shown figure above. The electric flux through this disk is equal to,

$$\Phi_E(t) = \pi r^2 \, E(t) = \frac{\pi r^2 \, V_0}{d} \sin(wt)$$

The displacement current intercepted by this surface is equal to,

$$I_d(t) = \varepsilon_0 \frac{d\Phi_E(t)}{dt} = \frac{\varepsilon_0 \, \pi \, r^2 \, V_0 \, w}{d} \cos(wt)$$

The current intercepted by the surface is equal to the current through the resistor (eq $I_R\,(t) = \frac{\varepsilon(t)}{R} = \frac{V_0 \sin(wt)}{R}$). Ampere's law thus requires,

$$\int_{path} \bar{B} \cdot d\bar{L} = 2\,\pi r \, B(r) =$$

$$= \mu_0 \left(I_R + I_d \right) =$$

$$= \mu_0 \, V_0 \left(\frac{1}{R} \sin(wt) + \frac{\varepsilon_0 \, \pi r^2 w}{d} \cos(wt) \right)$$

The strength of the magnetic field is thus equal to,

$$B(r) = \frac{\mu_0}{2\pi} V_0 \left(\frac{1}{rR} \sin(wt) + \frac{\varepsilon_0 \, \pi r w}{d} \cos(wt) \right)$$

References

- Griffiths, David J. (1999). Introduction to electrodynamics (3rd ed.). Upper Saddle River, [NJ.]: Prentice-Hall. ISBN 0-13-805326-X
- Everyone-should-study-electrodynamics: photonics101.com, Retrieved 17 March 2018
- J Rosen. "Redundancy and superfluity for electromagnetic fields and potentials". American Journal of Physics. 48 (12): 1071. Bibcode:1980AmJPh..48.1071R. doi:10.1119/1.12289
- Electromotive-force, science: britannica.com, Retrieved 27 May 2018
- Littlejohn, Robert (Fall 2007). "Gaussian, SI and Other Systems of Units in Electromagnetic Theory" (PDF). Physics 221A, University of California, Berkeley lecture notes. Retrieved 2008-05-06
- Motional-electromotive-force, electromagnetic-induction, physics: toppr.com, Retrieved 07 July 2018
- B Jiang & J Wu & L. A. Povinelli (1996). "The Origin of Spurious Solutions in Computational Electromagnetics". Journal of Computational Physics. 125 (1): 104. Bibcode:1996JCoPh.125..104J. doi:10.1006/jcph.1996.0082
- What-is-magnetomotive-force: circuitglobe.com, Retrieved 29 May 2018
- Jackson, John David (1999). Classical electrodynamics (3rd ed.). New York, [NY.]: Wiley. ISBN 0-471-30932-X
- Magnetomotive-force-mmf-definition-unit-formula, electromagnetism: electricalacademia.com, Retrieved 30 April 2018

4

Electrostatics

Electrostatics is the study of electric charges that are at rest. Coulomb's law describes the forces of attraction and repulsion exerted by charges on each other. All the diverse concepts central to the field of electrostatics, such as electric field, electric potential, electric flux, Gauss's law, etc. have been dealt within this chapter.

Electrostatics is a branch of science that involves the investigation of phenomena related to what appear to be stationary electric charges. Such phenomena are known as electrostatic phenomena.

It has been known since ancient times that the rubbing of some materials gives them the ability to attract small, light objects. Such materials were found to have acquired an electric charge. Examples of materials that can acquire a significant charge when rubbed together include amber rubbed with wool, glass rubbed with silk, and hard rubber rubbed with fur. This property, first recorded by Thales of Miletus, suggested the word "electricity" from the Greek word for amber.

Although electrostatically induced forces seem to be rather weak, the electrostatic force between an electron (which is negatively charged) and a proton (which is positively charged), together forming a hydrogen atom, is about 40 orders of magnitude stronger than the gravitational force acting between them.

The principles of electrostatics play an essential role in a number of technologies, including batteries, fuel cells, electroplating, thermocouples, radio detector diodes, photocells, light-emitting diodes (LEDs), thermoelectric cells, air filters, and xerography. However, undesirable charge accumulation has led to the damage of electronic components during manufacture, and the apparently spontaneous explosion of grain silos.

The natural phenomenon of static electricity has been known since antiquity. It was recorded by Thales of Miletus in the sixth century B.C.E. Scientific research into the subject began when machines were built to create static electricity artificially, such as the friction generator developed by Otto von Guericke in the seventeenth century. The connection between static electricity and storm clouds was famously demonstrated by Benjamin Franklin in 1750.

In the late eighteenth century, scientists developed sensitive instruments for detecting "electrification," otherwise known as electrostatic charge imbalance. The phenomenon of electrification by contact, also known as contact electrification or contact tension, was quickly discovered. When two objects were touched together, sometimes the objects became spontaneously charged. One object developed a net negative charge, while the other developed an equal and opposite positive charge.

Based on the observation of contact electrification, scientists attempted to explain how electricity is generated in an electric battery, or the Voltaic pile, as it was then called. This theoretical view was eventually replaced by the current theory of electrochemistry, namely, that electricity is generated by the action of chemistry and the exchange of electrons between atoms making up the battery. An important fact leading to rejection of the theory of contact tension was the observation that corrosion (chemical degradation of the battery) seemed unavoidable with its use, and that the more electricity was drawn from the battery, the faster the corrosion proceeded.

The Volta effect does indeed correspond to a weak electric potential developed by the contact of different metals. This effect was first discovered by Alessandro Volta, and can be measured using a capacitance electroscope comprised of different metals. However, the actual effect is not sufficiently strong to account for the action of electric batteries.

A number of high voltage dry piles were invented between the early 1800s and the 1830s, in an attempt to explain static electricity, and specifically to support Volta's hypothesis of contact tension. The Oxford Electric Bell is one example.

Until 1832, physicists thought that static electricity was somehow different from other forms of electricity. Then in 1832, Michael Faraday published the results of an experiment demonstrating that the electricity induced by a magnet, voltaic electricity produced by a battery, and static electricity were all the same. Since the time of Faraday's result, the history of static electricity merged with the study of electricity in general.

Generation of Static Electricity

Lightning over Oradea in Romania.

Static electricity is the buildup of electric charge on the surface of objects when they come in contact with each other and are then pulled apart. One surface acquires a positive charge, the other, a negative charge. Friction (rubbing of the surfaces) is not required, although in many situations it greatly enhances the phenomenon. Electrostatics involves study of the properties associated with such charge buildup.

Although charge exchange happens whenever any two surfaces come in contact and separate, the effects of charge exchange are usually noticed only when at least one of the surfaces has a high resistance to electrical flow. This is because the charges that transfer to or from the highly resistive surface are more or less trapped there for a long enough time for their effects to be observed. These charges then remain on the object until they either bleed off to ground or are quickly neutralized by a discharge.

For example, electric charge is built up when wool is rubbed against plastic or the soles of shoes rub on carpet. The mild shock one receives when touching a grounded object after walking on carpet is an example of excess electrical charge accumulating in one's body from frictional charging between one's shoes and the carpet. The resultant charge buildup within the body can generate a strong electrical discharge. Lightning is an example of giant electrical discharge in the atmosphere.

The process of how charges are built up on surfaces that come in contact with each other is not well understood. It may be caused by electron-stealing via quantum tunneling, or by transfer of surface ions. When negatively charged material touches a positively charged conductor, a spark is produced or a static shock is felt. A similar phenomenon occurs when a material that is positively charged touches a negatively charged conductor.

Static Electricity and Chemical Industry

Although experimenting with static electricity may be fun, sparks produced during electric discharge can create severe hazards in chemical industries that deal with flammable substances. In such cases, a small electrical spark may ignite explosive mixtures with devastating consequences.

A similar charging mechanism can occur within low-conductivity fluids flowing through pipelines—a process called *flow electrification*. Fluids that have low electrical conductivity (below 50 pico siemens/cm, where pico siemens/cm is a measure of electrical conductivity), are called *accumulators*. Fluids having conductivities above 50 pico siemens/cm are called *non-accumulators*. In non-accumulators, charges recombine as fast as they are separated and hence electrostatic charge generation is not significant. In the petrochemical industry, 50 pico siemens/cm is the recommended minimum value of electrical conductivity for adequate removal of charge from a fluid.

An important concept for insulating fluids is the static relaxation time. This is similar to the time constant (tau) within an RC circuit. For insulating materials, it is the ratio of the static dielectric constant divided by the electrical conductivity of the material. For hydrocarbon fluids, this is sometimes approximated by dividing the number 18 by the electrical conductivity of the fluid. Thus a fluid that has an electrical conductivity of one pico siemens /cm will have an estimated relaxation time of about 18 seconds. The excess charge within a fluid will be almost completely dissipated after four to five times the relaxation time, or 90 seconds for the fluid in the above example.

Charge generation increases at higher fluid velocities and larger pipe diameters, becoming quite significant in pipes 8 inches (200 mm) or larger. Static charge generation in these systems is best controlled by limiting fluid velocity.

Bonding and earthing are the usual ways by which charge buildup can be prevented. For fluids with electrical conductivity below ten pico siemens/cm, bonding and earthing are not adequate for charge dissipation, and anti-static additives may be required.

Electrostatic Generators

Usually insulators, that is, substances that do not conduct electricity, are good at both generating and holding a surface charge. Some examples of these substances are rubber, plastic, glass, and pith. Conductive objects only rarely generate charge imbalance except, for example, when a metal surface is impacted by solid or liquid nonconductors. The charge that is transferred during contact electrification is stored on the surface of each object. Static electric generators, devices which produce very high voltage at very low current and used for classroom physics demonstrations, rely on this effect.

Charge Neutralization

Natural electrostatic phenomena are most familiar as an occasional annoyance in seasons of low humidity, but they can be destructive and harmful in some situations (e.g. electronics manufacturing). When working in direct contact with integrated circuit electronics (especially delicate MOSFETs), or in the presence of flammable gas, care must be taken to avoid accumulating and suddenly discharging a static charge. To protect sensitive electronic components, manufacturers may use a number of antistatic devices.

Charge Induction

Charge induction occurs when a negatively charged object repels electrons from the surface of a second object. This creates a region in the second object that is more positively charged. An attractive force is then exerted between the objects. For example, when a balloon is rubbed, the balloon will stick to the wall as an attractive force is exerted by two oppositely charged surfaces (the surface of the wall gains an electric charge due to charge induction, as the free electrons at the surface of the wall are repelled by the negative balloon, creating a positive wall surface, which is subsequently attracted to the surface of the balloon). One can explore the effect with a simulation of the balloon and static electricity.

Applications

Studies in the field of electrostatics have led to the development of a variety of modern electrical and electronic technologies such as batteries, fuel cells, electroplating, thermocouples, and semiconductor junction devices including radio detector diodes, photocells, light-emitting diodes (LEDs), and thermoelectric cells. In addition, the principles of electrostatics are commonly applied in xerography, air filters, and some automotive paints.

Mathematical Foundations

Coulomb's Law

The fundamental equation of electrostatics is Coulomb's law, which describes the force between two point charges. The magnitude of the electrostatic force between two point electric charges is directly proportional to the product of the magnitudes of each charge and inversely proportional to the square of the distance between the charges, Q_1 and Q_2:

$$F = \frac{Q_1 Q_2}{4\pi\varepsilon_0 r^2},$$

where ε_o is the electric constant, a defined value:

$$\mathring{a}_o \stackrel{\text{def}}{=} \frac{1}{\mu_o c_o^{\ 2}} = 8.854\ 187\ 817 \times 10^{-12} \quad \text{in } A^2 s^4\ kg^{-1} m^{-3} \text{ or } C^2 N^{-1} m^{-2} \text{ or } F\, m^{-1}.$$

The Electric Field

The electric field (in units of volts per meter) is defined as the force (in newtons) per unit charge (in coulombs). From this definition and Coulomb's law, it follows that the magnitude of the electric field E created by a single point charge Q is

$$E = \frac{Q}{4\pi\varepsilon_o r^2}.$$

Gauss's Law

Gauss' law states that "the total electric flux through a closed surface is proportional to the total electric charge enclosed within the surface." The constant of proportionality is the permittivity of free space.

Mathematically, Gauss's law takes the form of an integral equation:

$$\oint_S \varepsilon_o \vec{E} \cdot d\vec{A} = \int_V \rho \cdot dV.$$

Alternatively, in differential form, the equation becomes

$$\vec{\nabla} \cdot \varepsilon_o \vec{E} = \rho.$$

The Electrostatic Approximation

The validity of the electrostatic approximation rests on the assumption that the electric field is irrotational:

$$\vec{\nabla} \times \vec{E} = 0.$$

From Faraday's law, this assumption implies the absence or near-absence of time-varying magnetic fields:

$$\frac{\partial \vec{B}}{\partial t} = 0.$$

In other words, electrostatics does not require the absence of magnetic fields or electric currents. Rather, if magnetic fields or electric currents *do* exist, they must not change with time, or in the worst-case, they must change with time only *very slowly*. In some problems, both electrostatics and magnetostatics may be required for accurate predictions, but the coupling between the two can still be ignored.

Electrostatic Potential (Voltage)

Because the electric field is irrotational, it is possible to express the electric field as the gradient of a scalar function, called the electrostatic potential (also known as the voltage). An electric field, E, points from regions of high potential, φ, to regions of low potential, expressed mathematically as

$$\vec{E} = -\vec{\nabla}\phi.$$

Poisson's equation

The definition of electrostatic potential, combined with the differential form of Gauss's law (above), provides a relationship between the potential φ and the charge density ρ:

$$\nabla^2\phi = -\frac{\rho}{\varepsilon_0}.$$

This relationship is a form of Poisson's equation. Where ε_0 is Vacuum permittivity.

Laplace's Equation

In the absence of unpaired electric charge, the equation becomes

$$\nabla^2\phi = 0,$$

which is Laplace's equation.

Electrostatics Examples

There are many examples of electrostatic phenomena:

- The attraction of the plastic wrap to your hand after you remove it from a package.
- The attraction of paper to a charged scale.
- The apparently spontaneous explosion of grain silos
- The damage of electronic components during manufacturing
- Photocopier & laser printer operation.

Electric Charge

In physics, charge, also known as electric charge, electrical charge, or electrostatic charge and symbolized q, is a characteristic of a unit of matter that expresses the extent to which it has more or fewer electrons than protons. In atoms, the electron carries a negative elementary or unit charge; the proton carries a positive charge. The two types of charge are equal and opposite.

In an atom of matter, an electrical charge occurs whenever the number of protons in the nucleus differs from the number of electrons surrounding that nucleus. If there are more electrons than protons, the atom has a negative charge. If there are fewer electrons than protons, the atom has a positive charge. The amount of charge carried by an atom is always a multiple of the elementary charge, that is, the charge carried by a single electron or a single proton. A particle, atom, or object with negative charge is said to have negative electric polarity; a particle, atom, or object with positive charge is said to have positive electric polarity.

In an object comprised of many atoms, the net charge is equal to the arithmetic sum, taking polarity into account, of the charges of all the atoms taken together. In a massive sample, this can amount to a considerable quantity of elementary charges. The unit of electrical charge in the International System of Units is the coulomb (symbolized C), where 1 C is equal to approximately 6.24×10^{18} elementary charges. It is not unusual for real-world objects to hold charges of many coulombs.

An electric field, also called an electrical field or an electrostatic field, surrounds any object that has charge. The electric field strength at any given distance from an object is directly proportional to the amount of charge on the object. Near any object having a fixed electric charge, the electric field strength diminishes in proportion to the square of the distance from the object (that is, it obeys the inverse square law).

When two objects having electric charge are brought into each other's vicinity, an electrostatic force is manifested between them. If the electric charges are of the same polarity, the electrostatic force is repulsive. If the electric charges are of opposite polarity, the electrostatic force is attractive. In free space (a vacuum), if the charges on the two nearby objects in coulombs are q_1 and q_2 and the centers of the objects are separated by a distance r in meters, the net force F between the objects, in newtons, is given by the following formula:

$$F = (q_1 q_2)/(4\pi\varepsilon_0 r^2)$$

where ε_0 is the permittivity of free space, a physical constant, and π is the ratio of a circle's circumference to its diameter, a dimensionless mathematical constant. A positive net force is repulsive, and a negative net force is attractive. This relation is known as Coulomb's law.

Types of Electric Charges

There are two types of electric charges namely positive and negative. Electrons are called negatively charged particles and protons are called positively charged particles.

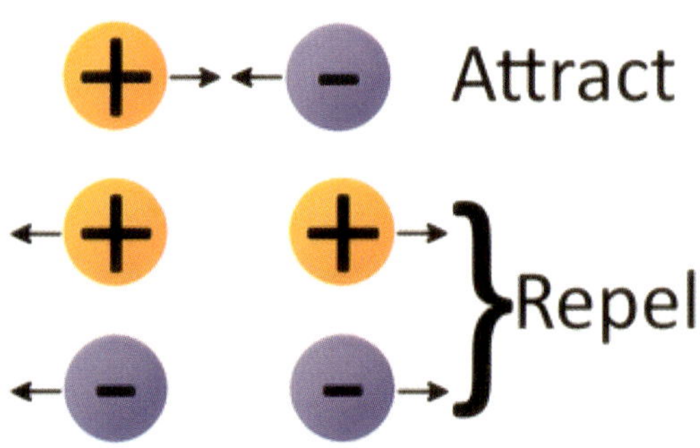

If two charged particles of similar charge are placed close to each other, they get repelled because the force present between them is repulsive. On the other hand, if two charged particles of opposite charge are placed close to each other, they get attracted because the force present between them is attractive. We can also simply define it as opposite charges attract and like charges repel.

An object said to be negatively charged if it has excess number of electrons than protons. On the other hand, if the object has excess number of protons than electrons then the object is said to be positively charged. When the number of electrons and protons are equal then the object is said to be neutral. The SI unit of electric charge is coulomb (C). The coulomb is defined as the quantity of charge that is passed through the electrical conductor carrying one ampere in one second. The symbol Q is used to denote the charge.

Electro-meter is used to measure the quantity of electric charge. It can also measured indirectly by using ballistic galvanometer.

Around 600 BC, ancient Greek philosopher "Thales of Miletus" proposed that charge or electricity could be gathered by rubbing fur (hair of non-human animals) on substances like amber.

Properties of Electric Charge

The various properties of electric charge include:

- Additivity of charges
- Charge is conserved
- Quantization of charge

Additivity of Charges

If a system contains two point charges q_1 and q_2, then the total charge of the system is obtained by simply adding q_1 and q_2, i.e., charges add up like real numbers.

If a system contains n number of charges $q_1, q_2, q_3, q_4, ------, q_n$, then the total charge of the system is $q_1 + q_2 + q_3 + q_4 + --------- + q_n$.

Charge is a scalar quantity; it has magnitude but no direction, similar to mass. However, there is one difference between charge and mass. Mass of a body is always positive whereas charge can be either positive or negative.

Let us take for example, the system containing four charges $q_1 = +2C$, $q_2 = +3C$, $q_3 = -3C$, $q_4 = +4C$, then the total charge of the system is,

$$q = q_1 + q_2 + q_3 + q_4$$

$$= (+2) + (+3) + (-3) + (+4)$$

$$= +6C$$

Therefore, the total charge of the system is +6C and it is positively charged.

Charge is Conserved

The law of conservation of charge states that charge cannot be created or destroyed. However, a charge can be transferred from one object to other.

Let us consider two objects, object A and object B. Object A has equal number of electrons and protons. So, it is electrically neutral. Similarly, object B has equal number of electrons and protons. So, it is also electrically neutral.

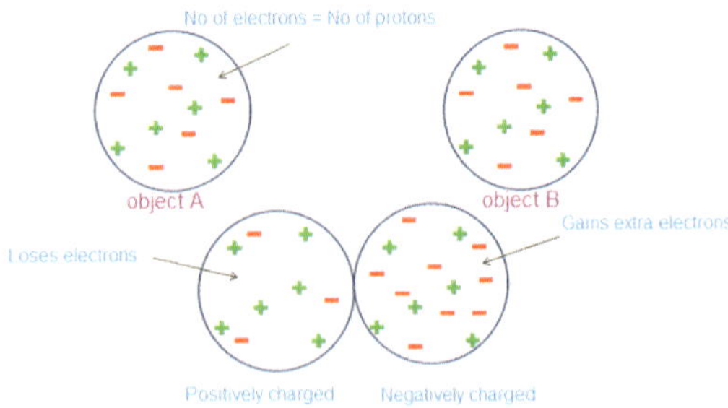

When object A and object B are rubbed with each other, negative charges from object A can be transferred to object B. Hence, object B has more number of electrons than protons due to gaining of extra electrons. Similarly, object A has lesser number of electrons than protons due to losing of some electrons.

Therefore, object A becomes positively charged and object B becomes negatively charged. However, the total charge of an isolated system remains constant.

Quantization of charge

The charge of any object is equal to integer multiples of the elementary charge. This is known as quantization of charge. It is given by

$q = ne$ (or) $n(-e)$

Where q = electric charge of any object or body

n = any integer positive or negative

$-e$ = elementary charge = charge carried by single electron.

e = elementary charge = charge carried by single proton.

The charge on an electron is written as -e and charge on a proton is written as +e. The quantization of charge was first suggested by the experimental laws of electrolysis discovered by Faraday. It was experimentally proved by Millikan. The total charge on a object is equal to the algebraic sum of individual charges present within the object.

If an object contains n1 electrons and n2 protons, then the total charge on the object is $n1 \times (-e) + n2 \times e$. For example, if the object contains 150 electrons and 200 protons, then the total charge on the object is $-150e + 200e = 50e$. Hence, the object is positively charged. The object charge can be exactly 0e or 1e, 2e - - - or -1e, -2e - - - but not 1/2, 1/4 etc.

Electric Field

Electric field, an electric property associated with each point in space when charge is present in any form. The magnitude and direction of the electric field are expressed by the value of E, called electric field strength or electric field intensity or simply the electric field. Knowledge of the value of the electric field at a point, without any specific knowledge of what produced the field, is all that is needed to determine what will happen to electric charges close to that particular point.

Instead of considering the electric force as a direct interaction of two electric charges at a distance from each other, one charge is considered the source of an electric field that extends outward into the surrounding space, and the force exerted on a second charge in this space is considered as a direct interaction between the electric field and the second charge. The strength of an electric field Eat any point may be defined as the electric, or Coulomb, force F exerted per unit positive electric charge q at that point, or simply $E = F/q$. If the second, or test, charge is twice as great, the resultant force is doubled; but their quotient, the measure of the electric field E, remains the same at any given point. The strength of the electric field depends on the source charge, not on the test charge. Strictly speaking, the introduction of a small test charge, which itself has an electric field, slightly modifies the existing field. The electric field may be thought of as the force per unit positive charge that would be exerted before the field is disturbed by the presence of the test charge.

The direction of the force that is exerted on a negative charge is opposite that which is exerted on a positive charge. Because an electric field has both magnitude and direction, the direction of the force on a positive charge is chosen arbitrarily as the direction of the electric field. Because positive charges repel each other, the electric field around an isolated positive charge is oriented radially outward. When they are represented by lines of force, or field lines, electric fields are depicted as starting on positive charges and terminating on negative charges. A line tangent to a field line indicates the direction of the electric field at that point. Where the field lines are close together, the electric field is stronger than where they are farther apart. The magnitude of the electric field around an electric charge, considered as source of the electric field, depends on how the charge is distributed in space. For a charge concentrated nearly at a point, the electric field is directly proportional to the amount of charge; it is inversely proportional to the square of the distance radially away from the centre of the source charge and depends also upon the nature of the medium. The presence of a material medium always diminishes the electric field below the value it has in a vacuum.

At times the electric field itself may become detached from the source charge and form closed loops, as in the case of charges accelerating up and down the transmitting antenna of a television station. The electric field with an accompanying magnetic field is propagated through space as a radiated wave at the same speed as that of light. Such electromagnetic waves indicate that electric fields are generated not only from electric charges but also from changing magnetic fields.

What is the Direction of Electric Field?

Electrostatic-force between two electrical charges is either repulsive or attractive depending upon the nature of charges, hence it is a vector quantity. So, an electric field is a vector quantity, since an electric field is a force per unit positive charge, so the direction of the electrostatic field must follow the direction of the electrostatic force. When we place a unit positive charge in an electric field, it either comes closer or goes far, depending upon the nature of the charge, by which the field is created. The direction of the electric field is given by the direction of motion of the unit positive charge.

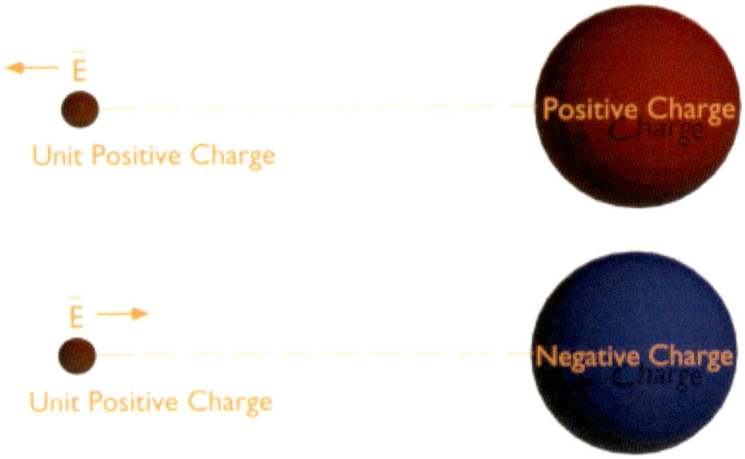

Representing Electric Fields (ESBPM)

We can represent the strength and direction of an electric field at a point using electric field lines.. In the following we will study what the electric fields look like around isolated charges.

Positive Charge Acting on a Test Charge

The magnitude of the force that a test charge experiences due to another charge is governed by Coulomb's law. In the diagram below, at each point around the positive charge, +Q+Q, we calculate the force a positive test charge, +q+q, would experience, and represent this force (a vector) with an arrow. The force vectors for some points around +Q+Q are shown in the diagram along with the positive test charge +q+q (in red) located at one of the points.

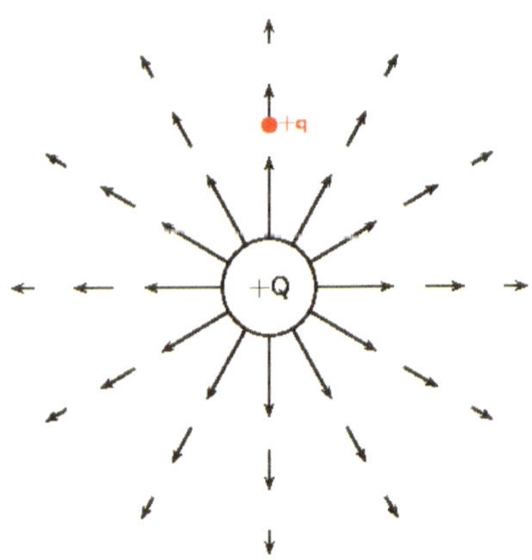

At every point around the charge \(+Q\), the positive test charge, \(+q\), will experience a force pushing it away. This is because both charges are positive and so they repel each other. We cannot draw an arrow at every point but we include enough arrows to illustrate what the field would look like. The arrows represent the force the test charge would experience at each point. Coulomb's law is an inverse-square law which means that the force gets weaker the greater the distance between the two charges. This is why the arrows get shorter further away from \(+Q\).

Negative Charge Acting on a Test Charge

For a negative charge, \(-Q\), and a positive test charge, \(+q\), the force vectors would look like:

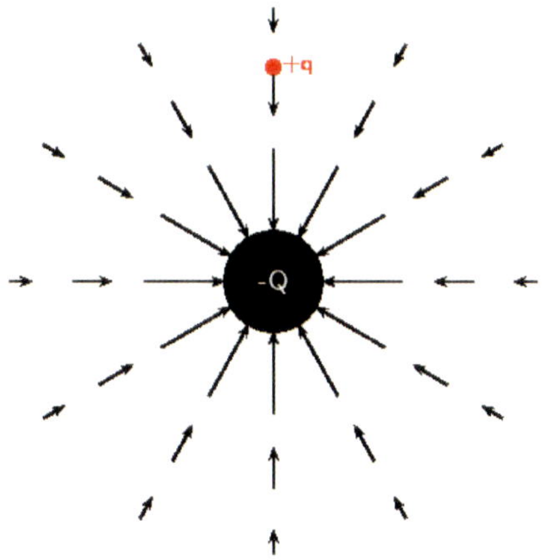

Notice that it is almost identical to the positive charge case. The arrows are the same lengths as in the previous diagram because the absolute magnitude of the charge is the same and so is the magnitude of the test charge. Thus the magnitude of the force is the same at the same points in space. However, the arrows point in the opposite direction because the charges now have opposite signs and attract each other.

Types of Electric Field

The electric fields are of four types:

- Uniform electric field
- Non-uniform electric field
- Static electric field
- Time varying electric field

1. Uniform Electric Field

When the electric field is constant at every point, then the field is called the uniform electric field. The constant field is obtained by placing the two conductor parallel to each other, and the potential difference between them remains same at every point.

Uniform Electric Field

2. Non-Uniform Electric Field

The field which is irregular at every point is called the non-uniform electric field. The non-uniform field has a different magnitude and directions.

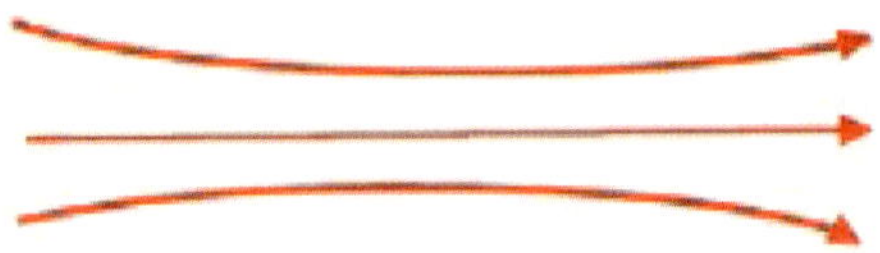

3. Static electric field

The strength of the electric field that does not vary with respect to time is called static electric field. This static electric field is produced by static electric charges.

4. Time varying electric field

The strength of the electric field that varies with respect to time is called time varying electric field.

Properties of Electric Field Lines

1. Electric field lines of force have a tendency to get separated from each other in the direction perpendicular to their lengths. They repel if they are of like charges.
2. Electric lines of force start from the positively charged surface of a body and end negatively charged surface of a body.
3. These lines of force are like elastic string, they comes to contract in length i.e. attract each other with respect to the opposite charges.
4. Closeness of lines of forces symbolizes more strength of electric field and vice versa.
5. Parallel lines indicate uniform field.
6. Two lines of forces never intersect each other.
7. Lines of force never pass through a conductor, i.e. field inside a conductor is always zero.
8. The tangential direction at any point on the lines of forces indicates the direction of the force acting on the positive charge at that point.

Electric Potential

Electric potential is the amount of electric potential energy that a unit charge has within theelectric field.

Electric potential energy is the energy that is needed to move a charge against an electric field. You need more energy to move a charge further in the electric field, but also more energy to move it through a stronger electric field.

Imagine that you have a huge negatively charged plate, with a little positively charged particle stuck to it through the electric force. There's an electric field around the plate that's pulling all positively charged objects toward it (while pushing other negatively charged objects away).

You take the positive particle, and start to pull it off the plate, against the pull of the electric field. It's hard work, because the electric force is pulling them together. If you let the positive particle go, it would snap back to the negative plate, pulled by the electric force. The energy that you used to move the particle away from the plate is stored in the particle as electrical potential energy. It is the potential that the particle has to move when it's let go.

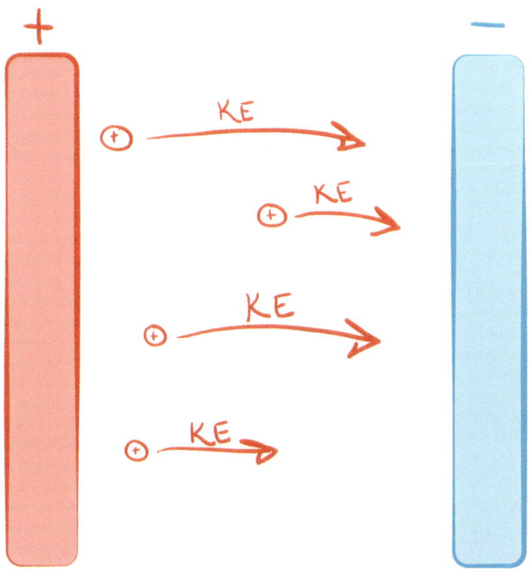

If you pulled the positive particle further away from the plate, you would have to use more energy, so the charge would have more electrical potential energy stored in it. If we doubled the charge on the plate, again, you would need more energy to move the positive particle. If we doubled the charge on the positive particle, you would need more energy to move it. You get the idea.

Imagine that instead of a negatively charged plate, our plate is positively charged. Our positive particle would be pushed away from the plate since they are both positively charged. This time, we have to put in energy to try to move the particle closer to the plate, instead of to pull it away. The closer we try to move it to the plate, the more energy we have to put in, so the more electrical potential energy the particle would have.

It may also be defined as, the amount of work done in moving a unit positive charge from a common reference point to a given point against the electric field of big positive charge. Any point outside the electric field of big positive charge can be considered as reference point. Electric potential is also called as the electric field potential or the electrostatic potential.

Electric potential can be mathematically written as,

$$\text{Electric potential} = \frac{W}{Q} = \frac{\text{Work}}{\text{Charge}}$$

Let us consider a big positive charge A and unit charge B as shown in the below figure. If a unit positive charge B is placed at a reference point outside the electric field of charge A, the force on charge B due to charge A is zero. Hence charge B has zero potential at this point. In order to move the charge B towards charge A work must be done against the electric field of charge A.

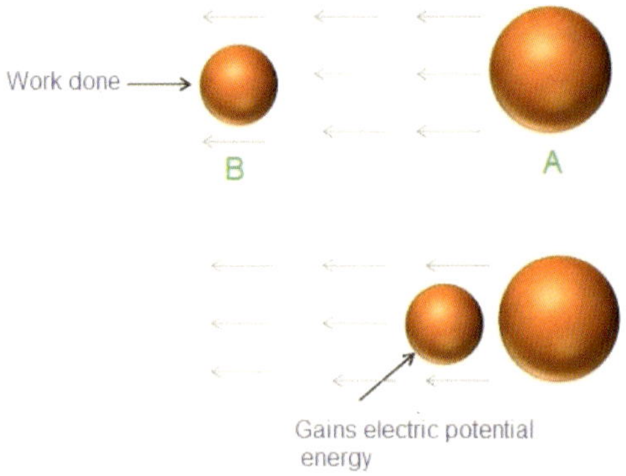

If the work is done on charge B against the electric field of charge A, the unit charge starts moving towards charge A. This work done on charge B is stored in the form of potential energy. The closer the unit charge to charge A, the higher will be the potential energy. This potential energy of unit positive charge within the electric field of charge A is called electric potential.

The SI unit of electric potential is volt. If one joule of work is done on unit positive charge (1C) to move from a reference point towards a big positive charge against the electric field, the electric potential of unit charge at that point is 1V.

Electric Potential Difference or Voltage

The difference in electric potential energy of a unit charge between two points within theelectric field is called electric potential difference.

It is also defined as; the amount of work done to move a unit positive charge from a point of lower potential to a point of higher potential against the static electric field. The point at which, unit charge has less potential energy is called lower potential. In the similar way, the point at which, unit charge has more potential energy is called higher potential. Electric potential difference is also called as voltage or electric pressure.

Consider two different points A and B in an electric field of positive charge Q as shown in figure (1). Here, point A is much closer to the charge Q than point B. Let a unit charge is placed at point B. This unit charge has electric potential energy of V_B.

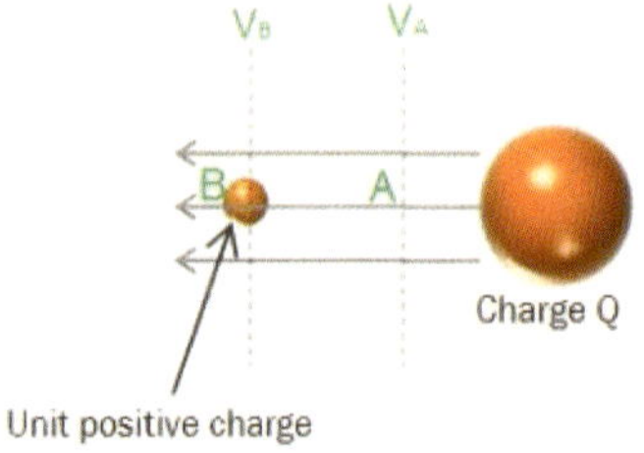

(1)

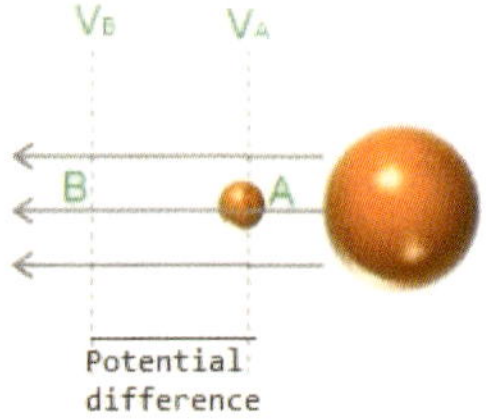

(2)

In order to move the unit positive charge from point B to point A work must be done on unit positive charge against the electric field. If the work is done on unit charge, the energy gets transferred to unit charge and it will start moving from point B to point A. Hence, at point Aunit charge gains electric potential energy as shown in figure. The amount of electric potential of unit charge at point A is V_A.

The amount of electric potential energy gained by unit charge from point B to Point A is called as potential difference. Therefore, the potential difference between point B and point A is $V_A - V_B$.

Electric potential difference can be mathematically written as ,

$$Electric\ potential\ difference\ (V) = V_A - V_B = \frac{work}{charge}$$

If one joule of work is done to move one coulomb of charge from one point (lower potential) to other (higher potential), the potential difference between these two points is 1V. In the similar way, If 5 joules of work is done to move one coulomb of charge from point of lower potential to point of higher potential, the potential difference between these two points is 5V. However, one joule of work is released as heat if one coulomb of charge moves from a point of higher potential to a point of lower potential.

The electric potential difference between two points is measured in volts. One volt is equal to one joule per coulomb. A voltmeter can be used to measure the potential difference between two points. Generally, ground or zero potential is taken as one common reference point.

Electric Potential in Circuits

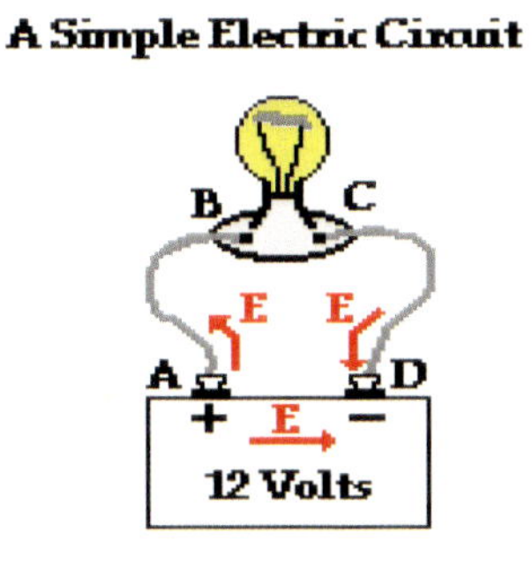

As we begin to discuss electric circuits, we will notice that a battery powered electric circuit has locations of high and low potential. Charge moving through the wires of the circuit will encounter changes in electric potential as it traverses the circuit. Within the electrochemical cells of the battery, there is an electric field established between the two terminals, directed from the positive terminal towards the negative terminal. As such, the movement of a positive test charge through the cells from the negative terminal to the positive terminal would require work, thus increasing the potential energy of every Coulomb of charge that moves along this path. This corresponds to a movement of positive charge against the electric field. It is for this reason that the positive terminal is described as the high potential terminal. Similar reasoning would lead one to conclude that the movement of positive charge through the wires from the positive terminal to the negative terminal would occur naturally. Such a movement of a positive test charge would be in the direction of the electric field and would not require work. The charge would lose potential energy as moves through the *external circuit* from the positive terminal to the negative terminal. The negative terminal is described as the low potential terminal. This assignment of high and low potential to the terminals of an electrochemical cell presumes the traditional convention that electric fields are based on the direction of movement of positive test charges.

In a certain sense, an electric circuit is nothing more than an energy conversion system. In the electrochemical cells of a battery-powered electric circuit, the chemical energy is used to do work on a positive test charge to move it from the low potential terminal to the high potential terminal. Chemical energy is transformed into electric potential energy within the *internal circuit* (i.e., the battery). Once at the high potential terminal, a positive test charge will then move through the external circuit and do work upon the light bulb or the motor or the heater coils, transforming its electric potential energy into useful forms for which the circuit was designed. The positive test charge returns to the negative terminal at a low energy and low potential, ready to repeat the cycle (or should we say *circuit*) all over again.

Electric Flux

Electric flux, property of an electric field that may be thought of as the number of electric lines of force (or electric field lines) that intersect a given area. Electric field lines are considered to originate on positive electric charges and to terminate on negative charges. Field lines directed into a closed surface are considered negative; those directed out of a closed surface are positive. If there is no net charge within a closed surface, every field line directed into the surface continues through the interior and is directed outward elsewhere on the surface. The negative flux just equals in magnitude the positive flux, so that the net, or total, electric flux is zero. If a net charge is contained inside a closed surface, the total flux through the surface is proportional to the enclosed charge, positive if it is positive, negative if it is negative.

The mathematical relation between electric flux and enclosed charge is known as Gauss's law for the electric field, one of the fundamental laws of electromagnetism. In the metre-kilogram-second system and the International System of Units (SI) the net flux of an electric field through any closed surface is equal to the enclosed charge, in units of coulombs, divided by a constant, called the permittivity of free space; in the centimetre gram-second system the net flux of an electric field through any closed surface is equal to the constant 4π times the enclosed charge, in electrostatic units (esu).

Formulation

An electric "charge," such as a single electron in space, has an electric field surrounding it. In pictorial form, this electric field is shown as a dot, the charge, radiating "lines of flux." These are called Gauss lines. Electric Flux Density is the amount of electric flux, the number of "lines," passing through a given area. Units are Gauss/square meter. Electric flux is proportional to the number of electric field lines going through a normally perpendicular surface. If the electric field is uniform, the electric flux passing through a surface of vector area S is

$$\Phi_E = \mathbf{E} \cdot \mathbf{S} = ES\cos\theta,$$

where E is the electric field (having units of V/m), E is its magnitude, S is the area of the surface, and θ is the angle between the electric field lines and the normal (perpendicular) to S.

For a non-uniform electric field, the electric flux $d\Phi_E$ through a small surface area dS is given by,

$$d\Phi_E = \mathbf{E} \cdot d\mathbf{S}$$

(the electric field, E, multiplied by the component of area perpendicular to the field). The electric flux over a surface S is therefore given by the surface integral:

$$\Phi_E = \iint_S \mathbf{E} \cdot d\mathbf{S}$$

where E is the electric field and $d\mathrm{S}$ is a differential area on the closed surface S with an outward facing surface normal defining its direction.

For a closed Gaussian surface, electric flux is given by:

$$\Phi_E = \oiint_S \mathbf{E} \cdot d\mathbf{S}$$

where,

E is the electric field,

S is any closed surface,

Q is the total electric charge inside the surface S,

ε_0 is the electric constant (a universal constant, also called the "permittivity of free space") ($\varepsilon_0 \approx 8.854\,187\,817... \times 10^{-12}$ farads per meter ($\mathrm{F \cdot m^{-1}}$)).

This relation is known as Gauss' law for electric field in its integral form and it is one of the four Maxwell's equations.

While the electric flux is not affected by charges that are not within the closed surface, the net electric field, E, in the Gauss' Law equation, can be affected by charges that lie outside the closed surface. While Gauss' Law holds for all situations, it is only useful for "by hand" calculations when high degrees of symmetry exist in the electric field. Examples include spherical and cylindrical symmetry.

Electrical flux has SI units of volt meters ($V\,m$), or, equivalently, newton meters squared per coulomb ($N\,m^2\,C^{-1}$). Thus, the SI base units of electric flux are $kg \cdot m^3 \cdot s^{-3} \cdot A^{-1}$.

Its dimensional formula is $[L^3MT^{-3}I^{-1}]$.

Electric Flux Density

The number of electric field lines or electric lines of force flowing perpendicularly through a surface area is called electric flux density.

Electric flux density can be mathematically defined as electric flux per surface area.

$$Electric\ flux\ density = \frac{Electric\ flux\,(\psi)}{Surface\ area\,(S)}$$

Electric flux is measured in Coulombs (C) and surface area is measured in square meters (m^2). Hence, the SI unit of electric flux density is coulomb per square meters (C/m^2). When we say that electric flux density in an electric field is $6C/m^2$, it means 6C of electric flux passes perpendicularly through a surface area of $1m^2$.

If the number of electric lines of force passing through a given surface area is high then the electric flux density is also high. Similarly, if the number of electric lines of force passing through a given surface area is low then the electric flux density is also low.

When the surface is placed perpendicular to the electric field, maximum number of electric field lines or electric lines of force will pass through the surface.

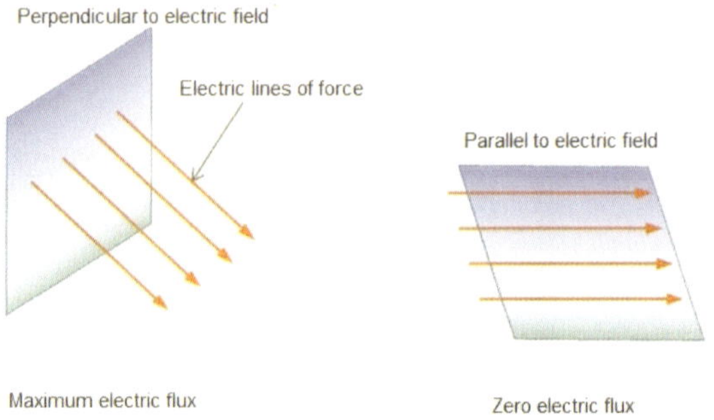

When the surface is placed parallel to the electric field, no electric lines of force will pass through the surface.

The concept of electric flux density can be easily understood by comparing it with the butterfly net. The amount of air moving through the net at any given time is the flux density. If the wind speed is high, then the flux through the net is also high. If this net size is made large, the air (flux) through the net gets increase even though the wind speed is same.

If the net is placed in a direction facing the wind, then maximum amount of air will pass through the net. On the other hand, if the net is placed parallel to the wind, then no wind will flows through the net.

The electric flux is considered as positive if the electric lines of force go away from a surface. On the other hand, it is considered as negative if the electric lines of force enter in to a surface.

Applications of Electric Flux

Electric Flux forms the basic foundation of electrostatics and has many applications like:

- Electric Flux helps in calculation of electric field
- Electric Flux simplifies evaluation of electric field in complex figures
- Gauss Theorem, one of the most useful theorems of electrostatics, is based mainly on electrostatics.

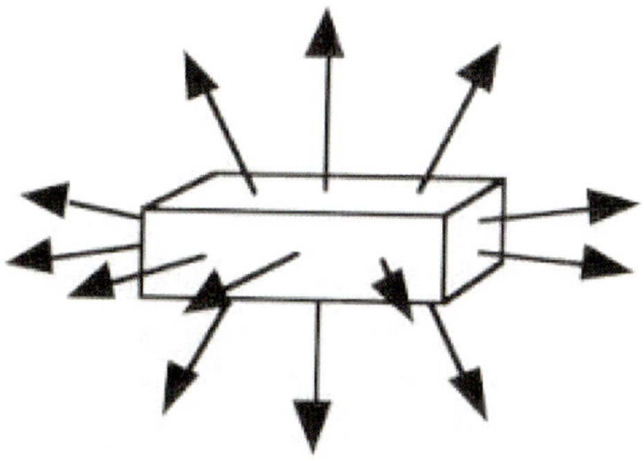

Direction of normal vector

Gauss's Law

Gauss's law, also known as Gauss's flux theorem, is a law relating the distribution of electric charge to the resulting electric field.

The law was formulated by Carl Friedrich Gauss in 1835, but was not published until 1867. It is one of the four Maxwell's equations which form the basis of classical electrodynamics, the other three being Gauss's law for magnetism, Faraday's law of induction, and Ampère's law with Maxwell's correction.

Carl Friedrich Gauss: Carl Friedrich Gauss, painted by Christian Albrecht Jensen

Gauss's law can be used to derive Coulomb's law, and vice versa. Note that since Coulomb's law only applies to stationary charges, there is no reason to expect Gauss's law to hold for moving charges based on this derivation alone. In fact, Gauss's law does hold for moving charges, and in this respect Gauss's law is more general than Coulomb's law.

In words, Gauss's law states that: The net outward normal electric flux through any closed surface is proportional to the total electric charge enclosed within that closed surface.

The law can be expressed mathematically using vector calculus in integral form and differential form, both are equivalent since they are related by the divergence theorem, also called Gauss's theorem. Each of these forms in turn can also be expressed two ways: In terms of a relation between the electric field E and the total electric charge, or in terms of the electric displacement field D and the free electric charge.

Gauss's law has a close mathematical similarity with a number of laws in other areas of physics, such as Gauss's law for magnetism and Gauss's law for gravity. In fact, any "inverse-square law" can be formulated in a way similar to Gauss's law: For example, Gauss's law itself is essentially equivalent to the inverse-square Coulomb's law, and Gauss's law for gravity is essentially equivalent to the inverse-square Newton's law of gravity.

The total charge enclosed in a closed surface is proportional to the total flux enclosed by the surface. If ϕ is total flux and ϵo is electric constant, the total electric charge Q enclosed by the surface is,

$$Q = \phi\ \epsilon o$$

The Gauss law formula is expressed by,

$$\phi = Q / \epsilon o$$

Where,

Q = total charge within the given surface,

ϵo = the electric constant.

Example 1

Find the total flux enclosed by the surface containing three charges q1, q2, and q3 having charge 6 C, 5 C and 4 C.

Solution:

Total charge,

$$Q = q1 + q2 + q3$$

$$= 6\text{ C} + 5\text{ C} + 4\text{ C}$$

$$= 15\text{ C}$$

The total flux is given by,

$$\phi = Q\ /\epsilon o$$

$\phi = 15\ C\ /\ 8.854 \times 10-12$

$\phi =\ 1.694\ Nm2\ /\ C-1$

Therefore, $\phi\ =\ \mathbf{1.694\ Nm2\ /\ C-1}$.

Example 2

Calculate the electric flux in a cylinder of length 5 cm, radius 3 cm having electric field intensity 2 N/C.

Solution:

Given:

Length l is 5 cm,

Radius r is 3 cm,

Electric field intensity is given by,

E = 2 N/C

Electric flux is given by,

$\phi =\ E\ (2\pi\pi\ r\ l)$

$\phi =\ 2\ N\ /\ C\ (2\pi \times 0.03\ m \times 0.05\ m)$

$\phi =\ 0.0188\ Nm2\ /\ C-1$.

Interpretation of Gauss' Law

What does this matter? Gauss' Law states that electric charge acts as sources or sinks for Electric Fields.

If you use the water analogy again, positive charge gives rise to flow out of a volume - this means positive electric charge is like a source (a faucet - pumping water into a region). Conversely, negative charge gives rise to flow into a volume - this means negative charge acts like a sink (fields flow into a region and terminate on the charge).

This gives us a lot of intuition about the way fields can physically act in any scenario. For instance, here are possible and impossible situations for the Electric Field, as decided by the universe in the Law of Gauss it setup:

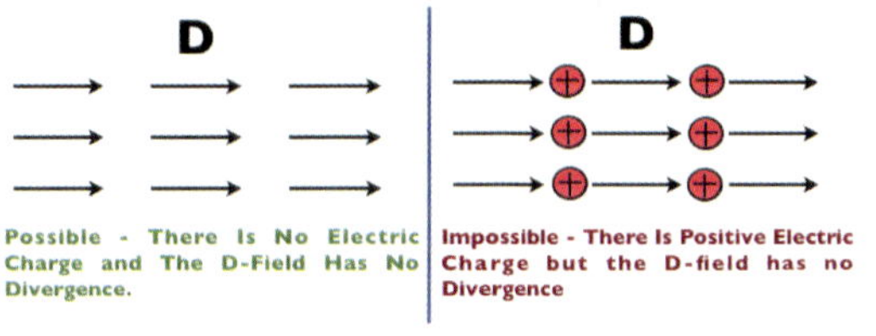

Example #1 of Gauss' Law: The D Field Must Have the Correct Divergence

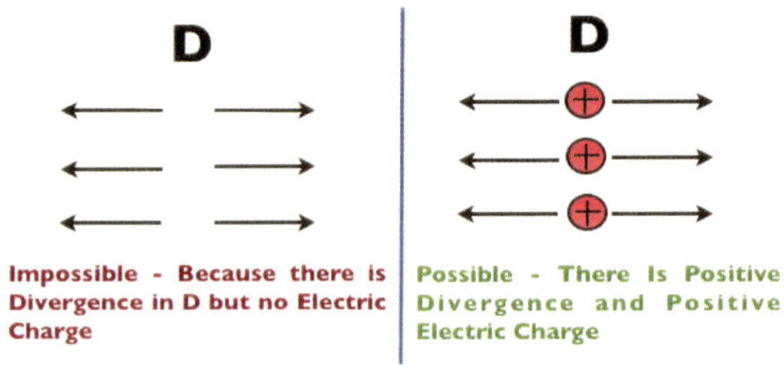

Example #2 of Gauss' Law: The Charges Dictate the Divergence of D

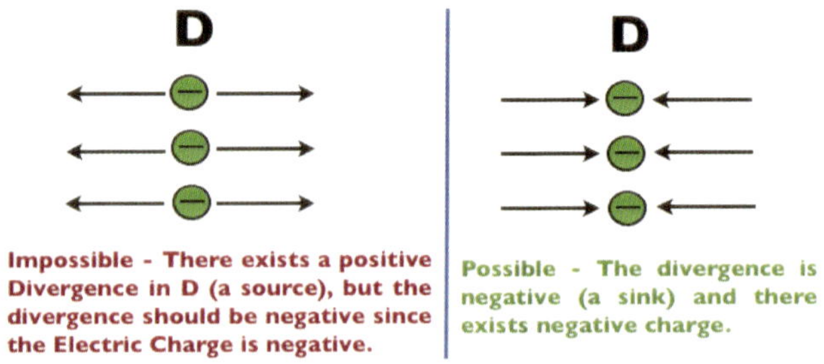

Example #3 of Gauss' Law: Negative Charge Indicates the Divergence of D should be negative

If you observe the way the D field must behave around charge, you may notice that Gauss' Law then is equivalent to the Force Equation for charges, which gives rise to the E field equation for point charges:

$$F=\frac{q_1 q_2}{4\pi\varepsilon_0 R^2}$$

$$|E|=\frac{q}{4\pi\varepsilon_0 R^2}$$

$$|D|=\frac{q}{4\pi R^2}$$

Above equation shows that charges exert a force on them, which means there exists E-fields that are away from positive charge and towards negative charge. This means opposite charges attract and negative charges repel. And since D and E are related by permittivity, we see that Gauss' Law is a more formal statement of the force equation for electric charges.

In summary, Gauss' Law means the following is true:

- D and E field lines diverge away from positive charges
- D and E field lines diverge towards negative charges
- D and E field lines start and stop on Electric Charges
- Opposite charges attract and negative charges repel
- The divergence of the D field over any region (volume) of space is exactly equal to the net amount of charge in that region.

Equation Involving the E Field

Integral Equation

Gauss's law in integral form is given below:

$$\int_V \nabla \cdot e\, dv = \oint_S e \cdot \hat{n}\, da = \frac{Q}{\varepsilon_0},$$

where:

- e is the electric field
- Q is the enclosed electric charge
- ε_0 is the electric permittivity of free space
- $\hat{n}$ is the outward pointing unit-normal

Flux is a measure of the strength of a field passing through a surface. Electric flux is defined in general as,

$$\Phi = \int_S e \cdot \hat{n}\; da.$$

We can think of electric field as flux density. Gauss's law tells us that the net electric flux through any closed surface is zero unless the volume bounded by that surface contains a net charge.

Differential Form

When considering a spatially extended charged body, we can think of its charge as being continously distributed throughout the body with density ρρ. The total charge is then given by the integral of the charge density over the volume of the body.

$$Q = \int_V \rho\, dv.$$

Using this definition and applying the divergence theorem to the left hand side of fisrt Gauss's law , we can rewrite the law as:

$$\int_V \nabla \cdot e\; dv = \int_V \frac{\rho}{\varepsilon_0}\, dv.$$

Since this equation must hold for any volume V , we can equate the integrands, giving the differential form of Gauss's law:

$$\nabla \cdot e = \frac{\rho}{\varepsilon_0}.$$

It can be shown that Gauss' law for electric fields is equivalent to Coulomb's law.

Gauss's Law in Matter

Gauss's law for electric fields is most easily understood by neglecting electric displacement (d). In matter, the dielectric permittivity may not be equal to the permittivity of free-space (i.e. $\varepsilon \neq \varepsilon_0$). In matter, the density of electric charges can be separated into a "free" charge density (ρf) and a "bounded" charge density (ρb), such that:

$$\rho = \rho_f + \rho_b$$

The free-charge density refers to charges which flow freely under the application of an electric field; i.e. they produce a current which is divergence-free. The bounded-charge density refers to electrical charges attributed to electrical polarization (p). By combining Eqs. fifth and sixth with our definition for electrical polarization, we find that:

$$\nabla \cdot d - \nabla \cdot p = \rho_f + \rho_b$$

By using the constitutive relationship d=εe and separating the previous equation into bounded and free contributions, we find that:

$$-\nabla \cdot p = \rho b$$

and

$$\nabla \cdot d = \rho_f$$

The above equation is the differential form of Gauss's equation in matter. Meanwhile, the integral form of Gauss's equations in matter is given by:

$$\int_V \nabla \cdot d \, dV = \oint_S d \cdot \hat{n} \, da = Q_f$$

where Q_f is the total enclosed free charge.

Equation Involving the D Field

Free, Bound, and Total Charge

The electric charge that arises in the simplest textbook situations would be classified as "free charge"—for example, the charge which is transferred in static electricity, or the charge on a capacitor plate. In contrast, "bound charge" arises only in the context of dielectric (polarizable) materials. (All materials are polarizable to some extent.) When such materials are placed in an external electric field, the electrons remain bound to their respective atoms, but shift a microscopic distance in response to the field, so that they're more on one side of the atom than the other. All these microscopic displacements add up to give a macroscopic net charge distribution, and this constitutes the "bound charge".

Although microscopically all charge is fundamentally the same, there are often practical reasons for wanting to treat bound charge differently from free charge. The result is that the more fundamental Gauss's law, in terms of E (above), is sometimes put into the equivalent form below, which is in terms of D and the free charge only.

Integral Form

This formulation of Gauss's law states the total charge form:

$$\Phi_D = Q_{\text{free}}$$

where Φ_D is the D-field flux through a surface S which encloses a volume V, and Q_{free} is the free charge contained in V. The flux Φ_D is defined analogously to the flux Φ_E of the electric field E thrugh S:

$$\Phi_D = \oiint_S \mathbf{D} \cdot \mathrm{d}\mathbf{A}$$

Differential Form

The differential form of Gauss's law, involving free charge only, states:

$$\nabla \cdot \mathbf{D} = \rho_{\text{free}}$$

where $\nabla \cdot \mathbf{D}$ is the divergence of the electric displacement field, and ρ_{free} is the free electric charge density.

Equivalence of Total and Free Charge Statements

Proof that the formulations of Gauss's law in terms of free charge are equivalent to the formulations involving total charge.

In this proof, we will show that the equation,

$$\nabla \cdot \mathbf{E} = \frac{\rho}{\varepsilon_0}$$

is equivalent to the equation

$$\nabla \cdot \mathbf{D} = \rho_{\text{free}}$$

Note that we are only dealing with the differential forms, not the integral forms, but that is sufficient since the differential and integral forms are equivalent in each case, by the divergence theorem.

We introduce the polarization density P, which has the following relation to E and D:

$$\mathbf{D} = \varepsilon_0 \mathbf{E} + \mathbf{P}$$

and the following relation to the bound charge:

$$\rho_{\text{bound}} = -\nabla \cdot \mathbf{P}$$

Now, consider the three equations:

$$\rho_{\text{bound}} = \nabla \cdot (-\mathbf{P})$$
$$\rho_{\text{free}} = \nabla \cdot \mathbf{D}$$
$$\rho = \nabla \cdot (\varepsilon_0 \mathbf{E})$$

The key insight is that the sum of the first two equations is the third equation. This completes the proof: The first equation is true by definition, and therefore the second equation is true if and only if the third equation is true. So the second and third equations are equivalent, which is what we wanted to prove.

Equation for linear materials

In homogeneous, isotropic, nondispersive, linear materials, there is a simple relationship between E and D:

$$\mathbf{D} = \varepsilon \mathbf{E}$$

where ε is the permittivity of the material. For the case of vacuum (aka free space), $\varepsilon = \varepsilon_0$. Under these circumstances, Gauss's law modifies to

$$\Phi_E = \frac{Q_{\text{free}}}{\varepsilon}$$

for the integral form, and

$$\nabla \cdot \mathbf{E} = \frac{\rho_{\text{free}}}{\varepsilon}$$

for the differential form.

Interpretations

In Terms of Fields of Force

Gauss's theorem can be interpreted in terms of the lines of force of the field as follows:

The flux through a closed surface is dependent upon both the magnitude and direction of the electric field lines penetrating the surface. In general a positive flux is defined by these lines leaving the surface and negative flux by lines entering this surface. This results in positive charges causing a positive flux and negative charges creating a negative flux. These electric field lines will extend to infinite decreasing in strength by a factor of one over the distance form the source of the charge squared. The larger the number of field lines emanating from a charge the larger the magnitude of the charge is, and the closer together the field lines are the greater the magnitude of the electric filed. This has the natural result of the electric field becoming weaker as one moves away from a charged particle, but the surface area also increases so that the net electric field exiting this particle will stay the same. In other words the closed integral of the electric field and the dot product of the derivative of the area will equal the net charge enclosed divided by permittivity of free space.

Relation to Coulomb's law

Deriving Gauss's Law from Coulomb's Law

Strictly speaking, Gauss's law cannot be derived from Coulomb's law alone, since Coulomb's law gives the electric field due to an individual point charge only. However, Gauss's law *can* be proven from Coulomb's law if it is assumed, in addition, that the electric field obeys the superposition principle. The superposition principle says that the resulting field is the vector sum of fields generated by each particle (or the integral, if the charges are distributed smoothly in space).

Outline of Proof

Coulomb's law states that the electric field due to a stationary point charge is:

$$\mathbf{E}(\mathbf{r}) = \frac{q}{4\pi\varepsilon_0}\frac{\mathbf{e}_r}{r^2}$$

where,

e_r is the radial unit vector,

r is the radius, |r|,

ε_0 is the electric constant,

q is the charge of the particle, which is assumed to be located at the origin.

Using the expression from Coulomb's law, we get the total field at r by using an integral to sum the field at r due to the infinitesimal charge at each other point s in space, to give,

$$\mathbf{E}(\mathbf{r}) = \frac{1}{4\pi\varepsilon_0}\int\frac{\rho(\mathbf{s})(\mathbf{r}-\mathbf{s})}{|\mathbf{r}-\mathbf{s}|^3}\mathrm{d}^3\mathbf{s}$$

where ρ is the charge density. If we take the divergence of both sides of this equation with respect to r, and use the known theorem,

$$\nabla\cdot\left(\frac{\mathbf{r}}{|\mathbf{r}|^3}\right) = 4\pi\delta(\mathbf{r})$$

where δ(r) is the Dirac delta function, the result is,

$$\nabla\cdot\mathbf{E}(\mathbf{r}) = \frac{1}{\varepsilon_0}\int\rho(\mathbf{s})\delta(\mathbf{r}-\mathbf{s})\mathrm{d}^3\mathbf{s}$$

Using the "sifting property" of the Dirac delta function, we arrive at,

$$\nabla\cdot\mathbf{E}(\mathbf{r}) = \frac{\rho(\mathbf{r})}{\varepsilon_0},$$

which is the differential form of Gauss' law, as desired.

Note that since Coulomb's law only applies to stationary charges, there is no reason to expect Gauss's law to hold for moving charges based on this derivation alone. In fact, Gauss's law does hold for moving charges, and in this respect Gauss's law is more general than Coulomb's law.

Deriving Coulomb's Law from Gauss's Law

Strictly speaking, Coulomb's law cannot be derived from Gauss's law alone, since Gauss's law does not give any information regarding the curl of E. However, Coulomb's law *can* be proven from Gauss's law if it is assumed, in addition, that the electric field from a point charge is spherically symmetric (this assumption, like Coulomb's law itself, is exactly true if the charge is stationary, and approximately true if the charge is in motion).

Outline of Proof

Taking S in the integral form of Gauss' law to be a spherical surface of radius r, centered at the point charge Q, we have

$$\mathbf{E}(\mathbf{r}) = \frac{Q}{4\pi\varepsilon_0}\frac{\hat{\mathbf{r}}}{r^2}$$

By the assumption of spherical symmetry, the integrand is a constant which can be taken out of the integral. The result is

$$4\pi r^2 \hat{\mathbf{r}} \cdot \mathbf{E}(\mathbf{r}) = \frac{Q}{\varepsilon_0}$$

where $\hat{\mathbf{r}}$ is a unit vector pointing radially away from the charge. Again by spherical symmetry, E points in the radial direction, and so we get

$$\mathbf{E}(\mathbf{r}) = \frac{Q}{4\pi\varepsilon_0}\frac{\hat{\mathbf{r}}}{r^2}$$

which is essentially equivalent to Coulomb's law. Thus the inverse-square law dependence of the electric field in Coulomb's law follows from Gauss' law.

Coulomb's Law

The electrostatic force was first studied in detail by Charles-Augustin de Coulomb around 1784. Through his observations he was able to show that the magnitude of the electrostatic force between two point-like charges is inversely proportional to the square of the distance between the charges. He also discovered that the magnitude of the force is proportional to the product of the charges. That is:

$$F \propto, \frac{Q_1 Q_2}{r^2},$$

where Q_1 and Q_2 are the magnitudes of the two charges respectively and r is the distance between them. The magnitude of the electrostatic force between two point-like charges is given by *Coulomb's law.*

Similarity of Coulomb's law to Newton's universal law of gravitation.

Notice how similar in form Coulomb's law is to Newton's universal law of gravitation between two point-like particles:

$$F_G = \frac{Gm_1m_2}{d^2},$$

where m_1 and m_2 are the masses of the two point-like particles, dd is the distance between them, and G is the gravitational constant. Both are inverse-square laws.

Both laws represent the force exerted by particles (point masses or point charges) on each other that interact by means of a field.

Worked Example 1: Coulomb's Law

Question

Two point-like charges carrying charges of $+3\times10^{-9}C$ C and $-5\times10^{-9}C$ are 2 m apart. Determine the magnitude of the force between them and state whether it is attractive or repulsive.

Solution

Step1: Determine what is required

We are required to determine the force between two point charges given the charges and the distance between them.

Step2 :Determine how to approach the problem

We can use Coulomb's law to calculate the magnitude of the force.

$$F = k\frac{Q_1Q_2}{r^2}$$

Step3 :Determine what is given

We are given:

- $Q_1 = +3\times10-9C$
- $Q_2 = -5\times10^{-9}\ C$
- $r = 2m$

We know that $k = 9{,}0\times10^9\ \text{N}\cdot\text{m}^2\cdot\text{C}^{-2}$.

We can draw a diagram of the situation.

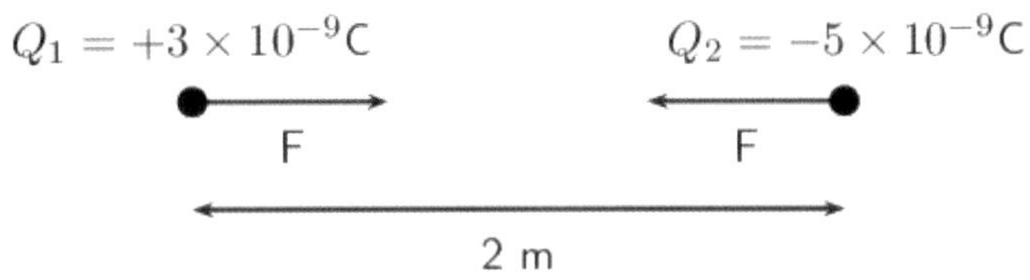

Step 4: Check units

All quantities are in SI units.

Step 5 :Determine the magnitude of the force

Using Coulomb's law we have

$$F = k\frac{Q_1 Q_2}{r^2}$$

$$= \frac{(9{,}0 \times 10^9)(3 \times 10^{-9})(5 \times 10^{-9})}{(2)^2}$$

$$= 3{,}37 \times 10^{-8}\ N$$

Thus the *magnitude* of the force is $= 3{,}37 \times 10^{-8}\ N$. However since the point charges have opposite signs, the force will be attractive.

Step 6 :Free body diagram

We can draw a free body diagram to show the forces. Each charge experiences a force with the same magnitude and the forces are attractive, so we have:

Next is another example that demonstrates the difference in magnitude between the gravitational force and the electrostatic force.

Worked Example 2: Coulomb's Law

Question

Determine the magnitudes of the electrostatic force and gravitational force between two electrons 10−1010−10 mm apart (i.e. the forces felt inside an atom) and state whether the forces are attractive or repulsive.

Solution

Step 1:Determine what is required

We are required to calculate the electrostatic and gravitational forces between two electrons, a given distance apart.

Step 2: Determine how to approach the problem

We can use:

$$F_e = k\frac{Q_1 Q_2}{r^2}$$

to calculate the electrostatic force and

$$F_g = \frac{Gm_1 m_2}{d^2}$$

to calculate the gravitational force.

Step 3 :Determine what is given

- $Q_1 = Q_2 = 1{,}6 \times 10^{-19}$ C (The charge on an electron)
- $m_1 = m_2 = 9{,}1 \times 10^{-31}$ kg (The mass of an electron)
- $r = d = 1 \times 10^{-10}\ m$

We know that:

- $k = 9{,}0 \times 10^9\ N \cdot m^2 \cdot C^{-2}$
- $G = 6{,}67 \times 10^{-11}\ N \cdot m^2 \cdot kg^{-2}$

All quantities are in SI units.

We can draw a diagram of the situation.

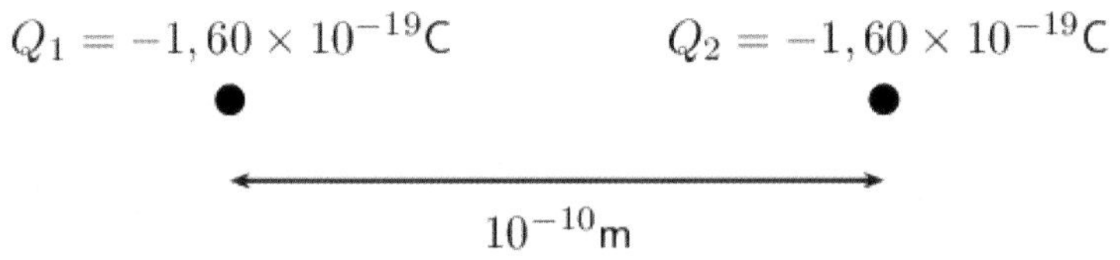

Step 4 :Calculate the electrostatic force

$$\begin{aligned} F_e &= k\frac{Q_1 Q_2}{r^2} \\ &= \frac{(9{,}0 \times 10^9)(1{,}60 \times 10^{-19})(1{,}60 \times 10^{-19})}{(10^{-10})^2} \\ &= 2{,}30 \times 10^{-8}\ N \end{aligned}$$

Hence the *magnitude* of the electrostatic force between the electrons is $2{,}30 \times 10^{-8}\ N$. Since electrons carry like charges, the force is repulsive.

Step 5 :Calculate the gravitational force

$$F_g = \frac{Gm_1m_2}{d^2}$$

$$= \frac{(6,67\times10^{-11})(9,11\times10^{-31})(9,11\times10^{-31})}{(10^{-10})^2}$$

$$= 5,54\times10^{-51}\ N$$

The magnitude of the gravitational force between the electrons is $5,54\times10^{-51}\ N$. Remember that the gravitational force is always an attractive force.

Notice that the gravitational force between the electrons is much smaller than the electrostatic force.

Worked Example 3: Coulomb's Law

Question

Three point charges are in a straight line. Their charges are $Q_1 = +2\times10^{-9}$ C, $Q_2 = +1\times10^{-9}\,C$ and $Q_3 = -3\times10^{-9}\,C$. The distance between Q_1 and Q_2 is 2×10^{-2} m and the distance between Q_2 and Q_3 is $4\times10^{-2}m$. What is the net electrostatic force on Q_2 due to the other two charges?

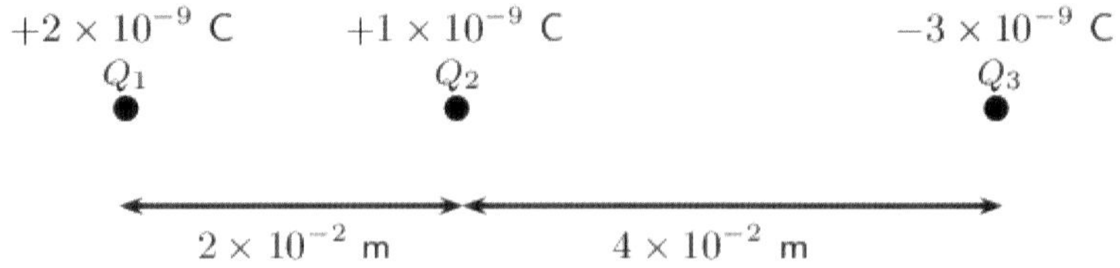

Solution

Step 1: Determine what is required

We need to calculate the net force on Q_2. This force is the sum of the two electrostatic forces - the forces between Q_1 on Q_2 and Q_3 on Q_2.

Step 2 :Determine how to approach the problem

- We need to calculate the two electrostatic forces on Q_2, using Coulomb's law.
- We then need to add up the two forces using our rules for adding vector quantities, because force is a vector quantity.

Step 3 :Determine what is given

We are given all the charges and all the distances.

Step 4 :Calculate the magnitude of the forces.

Force on Q_2 due to Q_1:

$$F_1 = k\frac{Q_1Q_2}{r^2}$$

$$=(9,0\times10^{9})\frac{(2\times10^{-9})(1\times10^{-9})}{(2\times10^{-2})^{2}}$$

$$=(9,0\times10^{9})\frac{(2\times10^{-9})(1\times10^{-9})}{(4\times10^{-4})}$$

$$=4,5\times10^{-5}\ N$$

Force on Q_2 due to Q_3:

$$F_3=k\frac{Q_2Q_3}{r^2}$$

$$=(9,0\times10^{9})\frac{(1\times10^{-9})(3\times10^{-9})}{(4\times10^{-2})^{2}}$$

$$=(9,0\times10^{9})\frac{(1\times10^{-9})(3\times10^{-9})}{(16\times10^{-4})}$$

$$=1,69\times10^{-5}\ N$$

Step 5 :Vector addition of forces

We know the force magnitudes but we need to use the charges to determine whether the forces are repulsive or attractive. It is helpful to draw the force diagram to help determine the final direction of the net force on Q_2. We choose the positive direction to be to the right (the positive xx-direction).

The force between Q_1 and Q_2 is repulsive (like charges). This means that it pushes Q_2 to the right, or in the positive direction.

The force between Q_2 and Q_3 is attractive (unlike charges) and pulls Q_2 to the right.

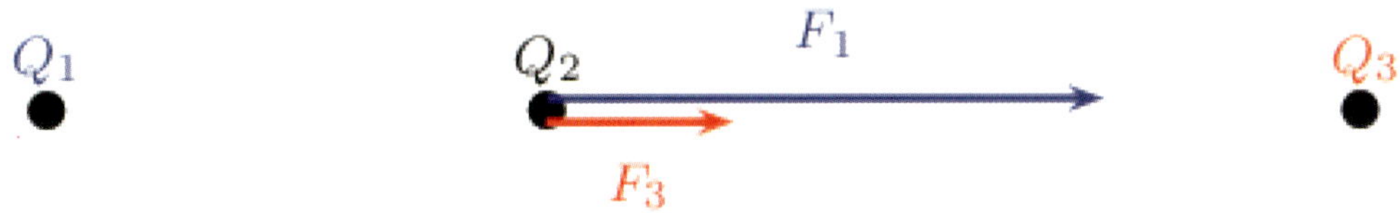

Therefore both forces are acting in the positive direction.

Therefore,

$$F_R=4,5\times10^{-5}\ N+1,69\times10^{-5}\ N$$
$$=6,19\times10^{-5}\ N$$

The resultant force acting on Q_2 is $6,19\times10^{-5}\ N$ to the right.

Worked Example 4: Coulomb's Law

Question

Three point charges form a right-angled triangle. Their charges are $Q_1 = 4\times10^{-9}\ \text{C} = 4\ \text{nC}$, $Q_2 = 6\times10^{-9}\,C = 6\,nC$ and $Q_3 = -3\times10^{-9}\,C = -3\,nC$. The distance between Q_1 and Q_2 is $5\times10^{-2}m$ and the distance between Q_1 and Q_3 is $3\times10^{-2}m$. What is the net electrostatic force on Q_1 due to the other two charges if they are arranged as shown?

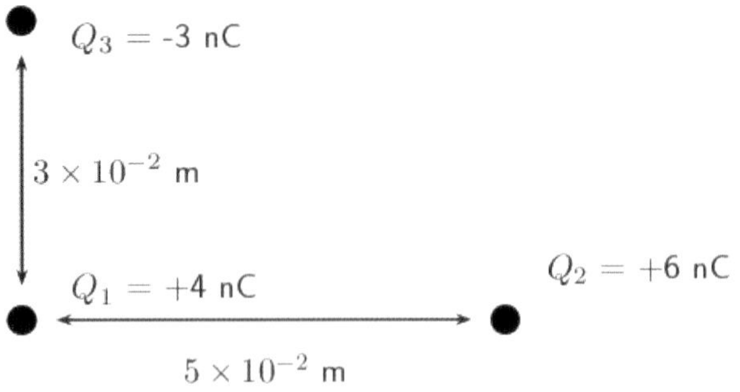

Solution

Step 1: Determine what is required

We need to calculate the net force on Q_1. This force is the sum of the two electrostatic forces - the forces of Q_2 on Q_1 and Q_3 on Q_1.

Step 2: Determine how to approach the problem

- We need to calculate, using Coulomb's law, the electrostatic force exerted on Q_1 by Q_2, and the electrostatic force exerted on Q_1 by Q_3.
- We then need to add up the two forces using our rules for adding vector quantities, because force is a vector quantity.

Step 3: Determine what is given

We are given all the charges and two of the distances.

Step 4: Calculate the magnitude of the forces.

The magnitude of the force exerted by Q_2 on Q_1, which we will call F_2, is:

$$
\begin{aligned}
F_2 &= k\frac{Q_1Q_2}{r^2} \\
&= (9{,}0\times10^9)\frac{(4\times10^{-9})(6\times10^{-9})}{(5\times10^{-2})^2} \\
&= (9{,}0\times10^9)\frac{(4\times10^{-9})(6\times10^{-9})}{(25\times10^{-4})} \\
&= 8{,}630\times10^{-5}\,N
\end{aligned}
$$

The magnitude of the force exerted by Q_3 on Q_1, which we will call F3F3, is:

$$F_3 = k\frac{Q_1Q_3}{r^2}$$
$$= (9{,}0\times10^9)\frac{(4\times10^{-9})(3\times10^{-9})}{(3\times10^{-2})^2}$$
$$= (9{,}0\times10^9)\frac{(4\times10^{-9})(3\times10^{-9})}{(9\times10^{-4})}$$
$$= 1{,}199\times10^{-4}\ N$$

Step 5: Vector addition of forces

This is a two-dimensional problem involving vectors. We have already solved many two-dimensional force problems and will use precisely the same procedure as before. Determine the vectors on the Cartesian plane, break them into components in the xx- and yy-directions, and then sum components in each direction to get the components of the resultant.

We choose the positive directions to be to the right (the positive xx-direction) and up (the positive y-direction). We know the magnitudes of the forces but we need to use the signs of the charges to determine whether the forces are repulsive or attractive. Then we can use a diagram to determine the directions.

The force between Q_1 and Q_2 is repulsive (like charges). This means that it pushes Q_1 to the left, or in the negative xx-direction.

The force between Q_1 and Q_3 is attractive (unlike charges) and pulls Q_1 in the positive y-direction.

We can redraw the diagram as a free-body diagram illustrating the forces to make sure we can visualise the situation:

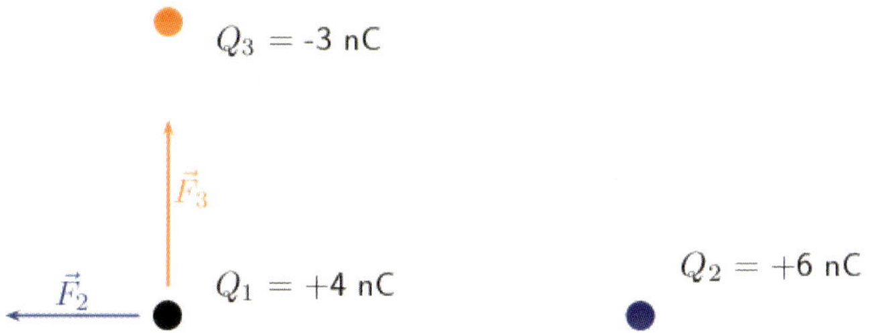

Step 6 : Resultant force

The magnitude of the resultant force acting on Q_1 can be calculated from the forces using Pythagoras' theorem because there are only two forces and they act in the x- and y-directions:

$$F_R^2 = F_2^2 22 + F_3^2 by\ Pythagoras'\ theorem$$
$$F_R = \sqrt{(8{,}630\times10^{-5})^2 + (1{,}199\times10^{-4})^2}$$
$$F_R = 1{,}48\times10^{-4}\ N$$

and the angle, θ_R made with the xx-axis can be found using trigonometry.

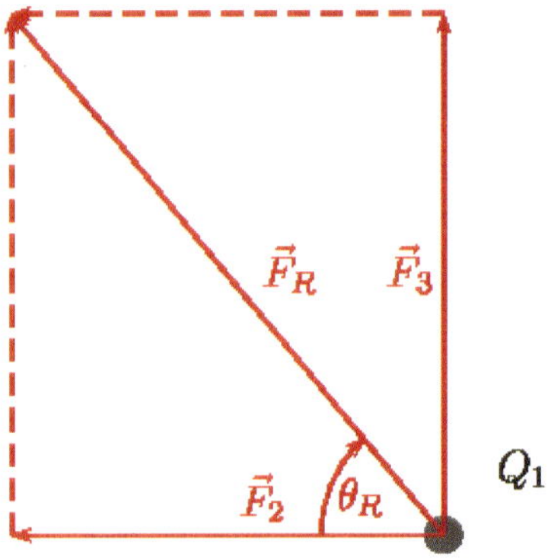

$$tan(\theta_R) = \frac{y - component}{x - component} =$$

$$tan(\theta_R) = \frac{1{,}199 \times 10^{-4}}{8{,}630 \times 10^{-5}}$$

$$\theta_R = tan^{-1}(\frac{1{,}199 \times 10^{-4}}{8{,}630 \times 10^{-5}}) =$$

$$\theta_R = 54{,}25° \, to \ 2 \ decimal \ places$$

The final resultant force acting on Q_1 is $1{,}48 \times 10^{-4} N$ acting at an angle of $54{,}25°$ to the negative x-axis or $25{,}75°$ to the positive xx-axis.

Statement of Coulomb's Law

First Law

Like charge particles repel each other and unlike charge particles attract each other.

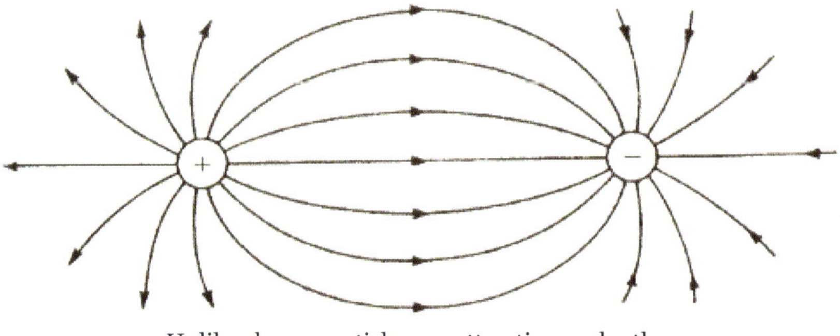

Unlike charge particles are attracting each other

Second Law

The force of attraction or repulsion between two electrically charged particles is directly proportional to the magnitude of their charges and inversely proportional to the square of the distance between them.

Formulas of Coulomb's Law

According to the Coulomb's second law,

$$Force\,(F) \propto \frac{Q_1 Q_2}{d^2}\;(OR)\;F = k\,\frac{Q_1 Q_2}{d^2}$$

Where,

1. 'F' is the repulsion or attraction force between two charged bodies.
2. 'Q_1' and 'Q_2' are the electrical charged of the bodies.
3. 'd' is distance between the two charged particles.
4. 'k' is a constant that depends on the medium in which charged bodies are presented. In S.I. system, as well as M.K.S.A. system $k = 1/4\pi\varepsilon$. Hence, the above equation becomes.

$$F = \frac{Q_1 Q_2}{4\pi \in d^2} = \frac{Q_1 Q_2}{4\pi \in_0 \in_r d^2}$$

The value of $\varepsilon_0 = 8.854 \times 10^{-12}\ C^2/Nm^2$.

$$Now,\ F = \frac{Q_1 Q_2}{4\pi * 8.854 * 10^{-12} * \in_r d^2} = 8.9878 * 10^9 \left[\frac{Q_1 Q_2}{\in_r d^2}\right]$$

Hence, Coulomb's law can be written for medium as,

$$F_{medium} = 9 * 10^9 \left[\frac{Q_1 Q_2}{\in_r d^2}\right]$$

Then, in air or vacuum $\in_r = 1$. Hence, Coulomb's law can be written for air medium as,

$$F_{air} = 9 * 10^9 \left[\frac{Q_1 Q_2}{d^2}\right]$$

The value of $\in_r$ would change depends on the medium. The expression for relative permittivity $\in_r$ is as follows;

$$\in_r = \frac{Force\ developed\ on\ air}{Force\ developed\ on\ medium} = \frac{F_{air}}{F_{medium}}$$

Principle of Coulomb's Law

Suppose if we have two charged bodies one is positively charged and one is negatively charged, then they will attract each other if they are kept at a certain distance from each other. Now if we increase the charge of one body keeping other unchanged, the attraction force is obviously increased. Similarly if we increase the charge of second body keeping first one unchanged, the attraction force between them is again increased. Hence, force between the charge bodies is proportional to the charge of either bodies or both.

$$\therefore F \propto Q_1 \,(OR)\, F \propto Q_2 \,(OR)\, F \propto Q_1 Q_2$$

Now, by keeping their charge fixed at Q_1 and Q_2 if you bring them nearer to each other the force between them increases and if you take them away from each other the force acting between them decreases. If the distance between the two charge bodies is d, it can be proved that the force acting on them is inversely proportional to d^2.

$$\therefore F \propto \frac{1}{d^2}$$

This development of force is not same for all mediums. As we discussed in the above formulas, εr would change for various medium. So, depends on the medium, creation of force can be varied.

Limitation of Coulomb's Law

1. Coulomb's law is valid, if the average number of solvent molecules between the two interesting charge particles should be large.
2. Coulomb's law is valid, if the point charges are at rest.
3. It is difficult to apply the Coulomb's law when the charges are in arbitrary shape. Hence, we cannot determine the value of distance 'd' between the charges when they are in arbitrary shape.

References

- Electrostatics, entry: newworldencyclopedia.org, Retrieved 28 March 2018
- Charge-electric-charge: whatis.techtarget.com, Retrieved 18 May 2018
- Properties-of-electric-charge, electromagnetics, electrostatics: physics-and-radio-electronics.com, Retrieved 28 June 2018
- Electric-flux-and-gausss-law, boundless-physics: lumenlearning.com, Retrieved 18 June 2018
- Gauss-law-formula: byjus.com, Retrieved 08 July 2018
- Coulombs-law: electrical4u.com, Retrieved 19 July 2018

5

Magnetostatics

The study of magnetic fields produced by steady currents falls under the domain of magnetostatics. In order to completely understand magnetostatics, it is necessary to understand the basics of magnetic field, magnetic moment, magnetization, magnetic flux, Biot-Savart law, Gauss's law for magnetism and Ampère's circuital law, which have been carefully analyzed in this chapter.

Magnetostatics is a branch of electromagnetic studies involving magnetic fields produced by steady non-time varying currents.

The center of Magnetostatics is calculating the Magnetic Field produced by steadily moving accumulations of charge. Thereafter, one can analyse the magnetostatic force on charged particles. This is basically the application of the law of Biot and Savart. However, just like life, things are not quite that simple. Just like the study of Coulomb's Law, the mathematics is quite cumbersome due to the vector nature of the Magnetic Field and its non zero curl. Therefore, the study of Magnetostatics is *really* about defining a vector magnetic potential 'A' and using this potential in order to evaluate the field and the force it produces. Much time and effort is devoted to the magnetic potential and applying a power series expansion to it.

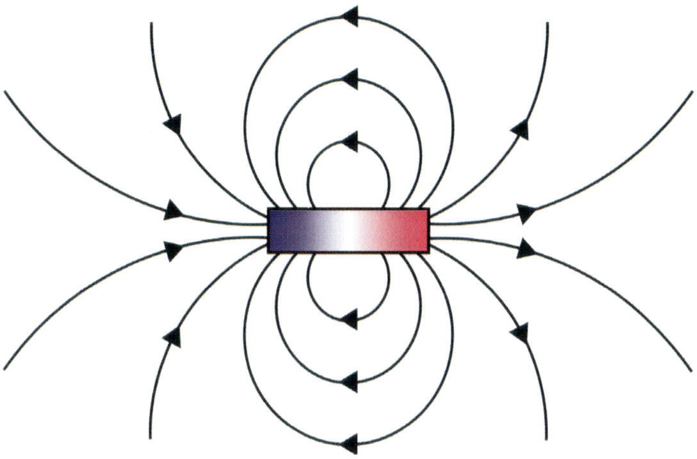

To realize steady currents, a large number of charges must be involved so that collection of charges can be regarded as a continuous fluid. In non-time varying cases $(\partial/\partial t = 0)$, the charge conservation principle requires that,

$$\frac{\partial \rho}{\partial t} + \nabla . \mathrm{J} = \nabla . \mathrm{J} = 0, \text{ in steady state.}$$

This implies that a steady current is transverse. A static magnetic Öeld can also be produced by a nonuniform magnetization (magnetic dipole density in the case of a collection of magnetic dipoles) M (A/m) which produces a magnetization current,

$$J_M = \nabla \times M, \left(A\,m^{-2}\right).$$

The magnetization current is transverse (divergence-free), since $\nabla . J_M = \nabla .(\nabla \times M) \equiv 0$.

The magnetic field B exerts a force perpendicular to the velocity of charged particles. The force to act on a charge q is,

$$F = q\ (E + v \times B).$$

The magnetic field does not do any work on charged particles since the rate of work $v.\ F$ is independent of B,

$$v\,.\,F = qv\ .\,E.$$

(Note the $v\,.(v \times B\,) = 0$.) In contrast to the electric field which is a true (or polar) vector, the magnetic field is a pseudo (or axial) vector, or more precisely, constitutes a pseudotensor,

$$B^{ij} = \begin{bmatrix} 0 & -B_z & B_y \\ B_z & 0 & -B_z \\ -B_y & B_x & 0 \end{bmatrix}.$$

The magnetic field due to a prescribed current distribution can be calculated using the BiotSavartís law. This law is generic because all other laws in magnetostatics, such as the Ampereís law and vanishing divergence of the magnetic field $\nabla \cdot B = 0$, 0; follow from it. Calculation of the magnetic field is facilitated by introducing a magnetic vector potent A, A; which is related to the magnetic field through $B = \nabla \times A$. A: This expression is consistent with $\nabla \cdot\ B = 0$, 0; since $\nabla \cdot\ (\nabla \times A) \equiv 0$ identically. It also follows that $\nabla \cdot A = 0$ 0 in static cases.

One important objective of magnetostatics is to derive formulae for the self-inductance and mutual inductance for given current configurations. Use of the vector potential greatly facilitates calculation of the magnetic flux.

$$\Psi = \int B \cdot dS = \int \nabla \times A \cdot dS = \oint A \cdot dI$$

and inductance,

$$L = \frac{\Psi}{I}.$$

Applications

Magnetostatics as a Special Case of Maxwell's Equations

Starting from Maxwell's equations and assuming that charges are either fixed or move as a steady current , the equations separate into two equations for the electric field and two for the magnetic field. The fields are independent of time and each other. The magnetostatic equations, in both differential and integral forms, are shown in the table.

Name	Form	
	Partial differential	Integral
Gauss's law for magnetism	$\nabla \cdot \mathbf{B} = 0$	$\oint_S \mathbf{B} \cdot \mathrm{d}\mathbf{S} = 0$
Ampère's law	$\nabla \times \mathbf{H} = \mathbf{J}$	$\oint_C \mathbf{H} \cdot \mathrm{d}\mathbf{l} = I_{\mathrm{enc}}$

Where ∇ with the dot denotes divergence, and B is the magnetic flux density, the first integral is over a surface S with oriented surface element $d\mathbf{S}$. Where ∇ with the cross denotes curl, J is the current density and H is the magnetic field intensity, the second integral is a line integral around a closed loop C with line element $\mathbf{l}$. The current going through the loop is I_{enc}.

The quality of this approximation may be guessed by comparing the above equations with the full version of Maxwell's equations and considering the importance of the terms that have been removed. Of particular significance is the comparison of the $\mathbf{J}$ term against the $\partial\mathbf{D}/\partial t$ term. If the $\mathbf{J}$ term is substantially larger, then the smaller term may be ignored without significant loss of accuracy.

Re-introducing Faraday's Law

A common technique is to solve a series of magnetostatic problems at incremental time steps and then use these solutions to approximate the term $\partial\mathbf{B}/\partial t$. Plugging this result into Faraday's Law finds a value for $\mathbf{E}$ (which had previously been ignored). This method is not a true solution of Maxwell's equations but can provide a good approximation for slowly changing fields.

Solving for the Magnetic Field

Current Sources

If all currents in a system are known (i.e., if a complete description of the current density $\mathbf{J}(\mathbf{r})$ is available) then the magnetic field can be determined, at a position r, from the currents by the Biot–Savart equation:[:174]

$$\mathbf{B}(\mathbf{r}) = \frac{\mu_0}{4\pi}\int \frac{\mathbf{J}(\mathbf{r}')\times(\mathbf{r}-\mathbf{r}')}{|\mathbf{r}-\mathbf{r}'|^3}\mathrm{d}^3\mathbf{r}'.$$

This technique works well for problems where the medium is a vacuum or air or some similar material with a relative permeability. This includes air-core inductors and air-core transformers. One advantage of this technique is that, if a coil has a complex geometry, it can be divided into sections and the integral evaluated for each section. Since this equation is primarily used to solve linear problems, the contributions can be added. For a very difficult geometry, numerical integration may be used.

For problems where the dominant magnetic material is a highly permeable magnetic core with relatively small air gaps, a magnetic circuit approach is useful. When the air gaps are large in comparison to the magnetic circuit length, fringing becomes significant and usually requires a finite element calculation. The finite element calculation uses a modified form of the magnetostatic equations above in order to calculate magnetic potential. The value of B can be found from the magnetic potential.

The magnetic field can be derived from the vector potential. Since the divergence of the magnetic flux density is always zero,

$$\mathbf{B} = \nabla \times \mathbf{A},$$

and the relation of the vector potential to current is:

$$\mathbf{A}(\mathbf{r}) = \frac{\mu_0}{4\pi} \int \frac{\mathbf{J}(\mathbf{r}')}{|\mathbf{r} - \mathbf{r}'|} \mathrm{d}^3\mathbf{r}'.$$

Magnetization

Strongly magnetic materials (i.e., ferromagnetic, ferrimagnetic or paramagnetic) have a magnetization that is primarily due to electron spin. In such materials the magnetization must be explicitly included using the relation ,

$$\mathbf{B} = \mu_0(\mathbf{M} + \mathbf{H}).$$

Except in metals, electric currents can be ignored. Then Ampère's law is simply ,

$$\nabla \times \mathbf{H} = 0.$$

This has the general solution ,

$$\mathbf{H} = -\nabla \Phi_M,$$

where Φ_M is a scalar potential.[:192] Substituting this in Gauss's law gives ,

$$\nabla^2 \Phi_M = \nabla \cdot \mathbf{M}.$$

Thus, the divergence of the magnetization, $\nabla \cdot \mathbf{M}$,has a role analogous to the electric charge in electrostatics and is often referred to as an effective charge density ρ_M.

The vector potential method can also be employed with an effective current density ,

$$\mathbf{J}_\mathbf{M} = \nabla \times \mathbf{M}.$$

Magnetic Field

A magnetic field is a picture that we use as a tool to describe how the magnetic force is distributed in the space around and within something magnetic.

Most of us have some familiarity with everyday magnetic objects and recognize that there can be forces between them. We understand that magnets have two poles and that depending on the orientation of two magnets there can be attraction (opposite poles) or repulsion (similar poles). We recognize that there is some region extending around a magnet where this happens. The magnetic field describes this region.

There are two different ways that a magnetic field is typically illustrated:

1. The magnetic field is described mathematically as a *vector field*. This vector field can be plotted directly as a set of many vectors drawn on a grid. Each vector points in the direction that a compass would point and has length dependent on the strength of the magnetic force.

Arranging many small compasses in a grid pattern and placing the grid in a magnetic field illustrates this technique. The only difference here is that a compass doesn't indicate the strength of a field.

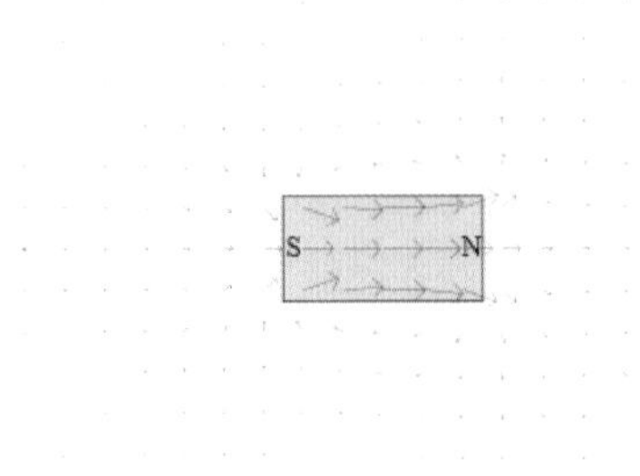

Vector field plot for a bar magnet.

2. An alternative way to represent the information contained within a vector field is with the use of *field lines*. Here we dispense with the grid pattern and connect the vectors with smooth lines. We can draw as many lines as we want.

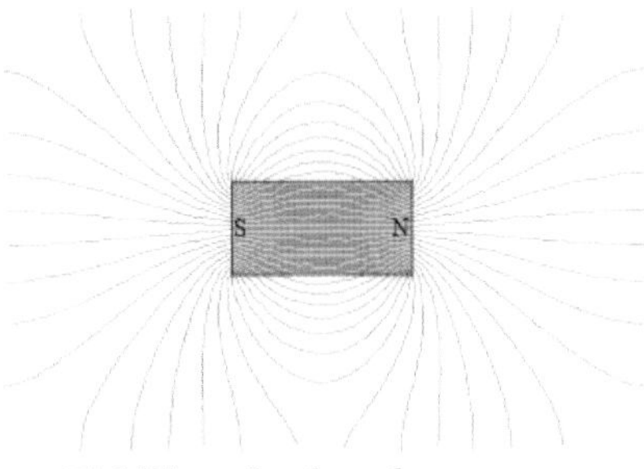

Field line plot for a bar magnet

The field-line description has some useful properties:

- Magnetic field lines never cross.
- Magnetic field lines naturally bunch together in regions where the magnetic field is the strongest. This means that the density of field lines indicates the strength of the field.
- Magnetic field lines don't start or stop anywhere, they always make closed loops and will continue inside a magnetic material (though sometimes they are not drawn this way).
- We require a way to indicate the direction of the field. This is usually done by drawing arrowheads along the lines. Sometimes arrowheads are not drawn and the direction must be indicated in some other way. For historical reasons the convention is to label one region 'north' and another 'south' and draw field lines only from these 'poles'. The field is assumed to follow the lines from north to south. 'N' and 'S' labels are usually placed on the ends of a magnetic field source, although strictly this is arbitrary and there is nothing special about these locations.

- Field lines can be visualized quite easily in the real world. This is commonly done with iron filings dropped on a surface near something magnetic. Each filing behaves like a tiny magnet with a north and south pole. The filings naturally separate from each other because similar poles repel each other. The result is a pattern that resembles field lines. While the general pattern will always be the same, the exact position and density of lines of filings depends on how the filings happened to fall, their size and magnetic properties.

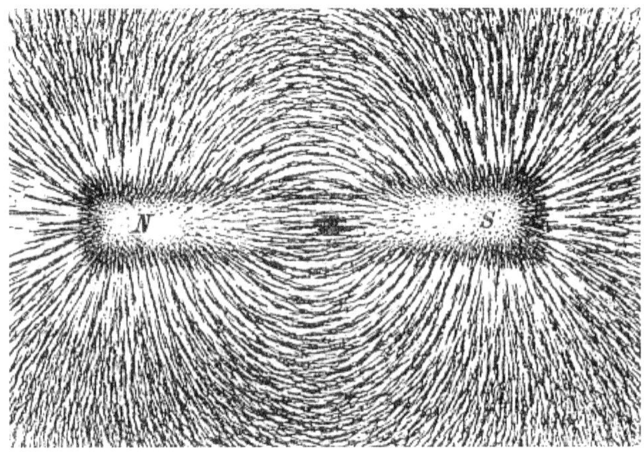

Magnetic field lines around a bar magnet visualized using iron filings.

Measuring Magnetic Fields

Because a magnetic field is a vector quantity, there are two aspects we need to measure to describe it; the strength and direction.

The direction is easy to measure. We can use a magnetic compass which lines up with the field. Magnetic compasses have been used for navigation (using the earth's magnetic field) since the 11th century.

Interestingly, measuring the strength is considerably more difficult. Practical *magnetometers* only came available in the 19th century. Most of these magnetometers work by exploiting the force an electron feels as it moves through a magnetic field.

Very accurate measurement of small magnetic fields has only been practical since the discovery in 1988 of *giant magnetoresistance* in specially layered materials. This discovery in fundamental physics was quickly applied to the magnetic hard-disk technology used for storing data in computers. This lead to a thousand-fold increase in data storage capacity in just a few years immediately following the implementation of the technology (0.1 to 100 Gbit/inch^2 between 1991 and 2003). In 2007 Albert Fert and Peter Grünberg were awarded the Nobel Prize in Physics for this discovery.

In the SI system, the magnetic field is measured in tesla (symbol T, amed after Nikola Tesla). The Tesla is defined in terms of how much force is applied to a moving charge due to the field. A small refrigerator magnet produces a field of around 0.001 T and the earth's field is about $5 \cdot 10^{-5\,T}$ An alternative measurement is also often used, the Gauss (symbol G) There is a simple conversion factor, 1 T=$10^{4\,G}$. Gauss is often used because 1 Tesla is a very large field.

In equations the magnitude of the magnetic field is given the symbol B. You may also see a quantity called the magnetic field strength which is given the symbol H. Both B and H have the same units, but H takes into account the effect of magnetic fields being concentrated by magnetic materials. For simple problems taking place in air you won't need to worry about this distinction.

Origin of the Magnetic Field

Magnetic fields occur whenever charge is in motion. As more charge is put in more motion, the strength of a magnetic field increases.

Magnetism and magnetic fields are one aspect of the electromagnetic force, one of the four fundamental forces of nature.

There are two basic ways which we can arrange for charge to be in motion and generate a useful magnetic field:

1. We make a current flow through a wire, for example by connecting it to a battery. As we increase the current (amount of charge in motion) the field increases proportionally. As we move further away from the wire, the field we see drops off proportionally with the distance. This is described by Ampere's law. Simplified to tell us the magnetic field at a distance r from a long straight wire carrying current I the equation is,

$$B = \frac{\mu_0 I}{2\pi r}$$

Here μ_0 is a special constant known as the *permeability* of free space. $\mu_0 = 4\pi . 10^{-7}\ \mathrm{T.m/A}$.

Some materials have the ability to concentrate magnetic fields, this is described by those materials having higher *permeability*.

Since the magnetic field is a vector, we also need to know the direction. For conventional current flowing through a straight wire this can be found by the *right-hand-grip-rule*. To use this rule imagine gripping your right hand around the wire with your thumb pointing in the direction of the current. The fingers show the direction of the magnetic field which wraps around the wire.

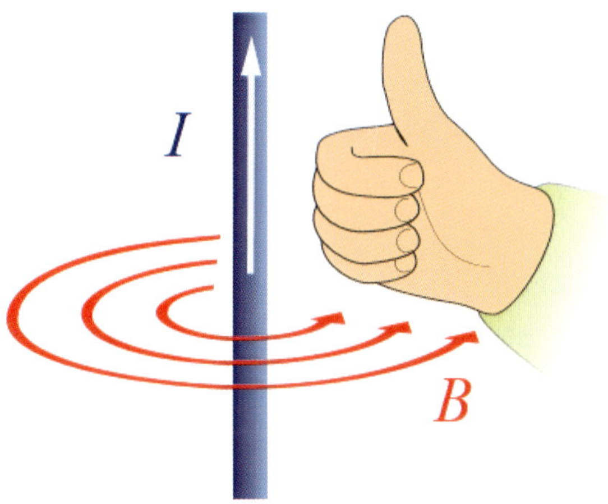

Right-hand-grip rule used to find the direction of the magnetic field (B) based on the direction of a current (I).

2. We can exploit the fact that electrons (which are charged) appear to have some motion around the nuclei of atoms. This is how permanent magnets work. As we know from experience, only some 'special' materials can be made into magnets and some magnets are much stronger than others. So some specific conditions must be required:

   - Although atoms often have many electrons, they mostly 'pair up' in such a way that the overall magnetic field of a pair cancels out. Two electrons paired in this way are said to have *opposite spin.* So if we want something to be magnetic we need atoms that have one or more unpaired electrons with the same spin. Iron for example is a 'special' material that has four such electrons and therefore is good for making magnets out of.
   - Even a tiny piece of material contains billions of atoms. If they are all randomly orientated the overall field will cancel out, regardless of how many unpaired electrons the material has. The material has to be stable enough at room temperature to allow an overall preferred orientation to be established. If established permanently then we have a permanent magnet, also known as a *ferromagnet.*
   - Some materials can only become sufficiently well ordered to be magnetic when in the presence of an external magnetic field. The external field serves to line all the electron spins up, but this alignment disappears once the external field is removed. These kinds of materials are known as *paramagnetic.*

     The metal of a refrigerator door is an example of a paramagnet. The refrigerator door itself is not magnetic, but behaves like a magnet when a refrigerator magnet is placed on it. Both then attract each other strongly enough to easily keep in place a shopping list, sandwiched between the two.

Magnetic Field Basics

B and H

In common usage, physicists refer to the magnetic field in "empty" space by using the symbol B and use a second term, H (magnetic field strength), to characterize the magnetic field that passes through a magnetically responsive material, such as iron, which contributes an internal magnetic field. Physicists also refer to H as *auxiliary magnetic field* or *magnetizing field.* Among electrical engineers H is known as the *magnetic field intensity* or *magnetic field strength,* and B is known as *magnetic flux density,magnetic induction,* or simply *magnetic field,* as used by physicists.

Units

The magnetic field B has the SI units of teslas (T), equivalent to webers per square meter (Wb/m^2) or volt seconds per square meter (V s/m^2). (The weber is the SI unit of magnetic flux defined as the amount of flux required to induce in a surrounding conducting loop an electromotive force of 1 volt if the flux is reduced to zero over one second.) In cgs units, B has units of gauss (G), where 1T = 10,000G. The vector field H is measured in Amperes/meter (A/m) in SI or oersted (Oe) in cgs units.

Vector Fields

A magnetic field belongs mathematically to the class of vector fields, which are defined at each point in their region of occurrence by both a magnitude and a direction. The magnetic field B, whose magnitude is measured in Teslas and the magnetic field strength H, whose magnitude is measured in Amperes/meter, both therefore are more properly represented by a vector (looking like an arrow) and the full field is represented by many arrows whose directions collectively show the curve of the magnetic field extending between the two magnetic poles.

Permanent Magnets and Magnetic Poles

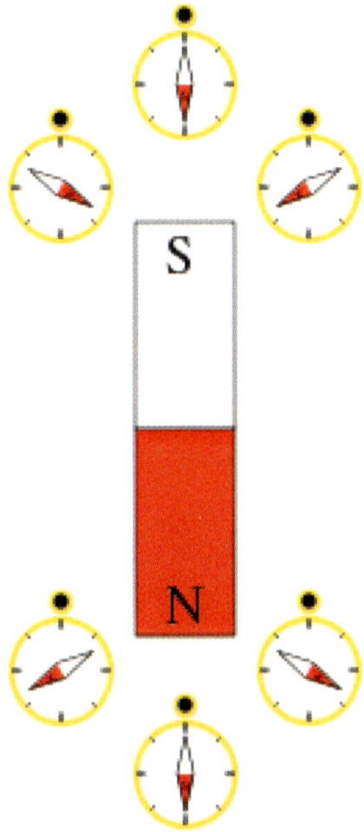

The direction of the magnetic field near the poles of a magnet is revealed by placing compasses nearby. As seen here, the magnetic field points towards a magnet's south pole and away from its north pole.

Permanent magnets are objects that produce their own persistent magnetic fields. All permanent magnets have both a north and a south pole. (Magnetic poles always come in north-south pairs.) Like poles repel and opposite poles attract. The magnetism in a permanent magnet arises from properties of the atoms (in particular the electrons) that compose it. Each atom acts like a little individual magnet. If these magnets line up, they combine to create a macroscopic magnetic effect.

If allowed to twist freely, a magnet will turn to point in the direction of the magnetic field at its location. A compass is a small magnet that uses this effect to point in the direction of the local magnetic field. By definition, the direction of the magnetic field at a point is the direction that the north pole of a magnet would want to point.

If a compass is placed near the north pole of a magnet then it will point away from that pole—like poles repel. In other words, the magnetic field points away from a magnet near its north pole. The opposite occurs if we place the compass near a magnet's south pole; the magnetic field points towards the magnet near its south pole. Not all magnetic fields are describable in terms of poles, though. A straight current-carrying wire, for instance, produces a magnetic field that points neither towards nor away from the wire, but encircles it instead.

Visualizing the Magnetic Field

The strength and direction of the magnetic field due to an object varies from position to position. Mapping out this magnetic field is simple in principle. First, measure the strength and direction of the magnetic field at a large number of points. Then mark each location with an arrow (called a vector) pointing in the direction of the magnetic field with a length proportional to the strength of the magnetic field. This is a valid and useful way of marking out and visualizing the magnetic field of an object. It has the unfortunate consequence, though, of cluttering up a graph even when using a small number of points. An alternative method of visualizing the magnetic field is to use "magnetic field lines."

Magnetic B Field Lines

Various physical phenomena have the effect of displaying magnetic field lines. For example, iron filings placed in a magnetic field will line up in such a way as to visually show the orientation of the magnetic field. Another place where magnetic fields are visually displayed is in the polar auroras, in which visible streaks of light line up with the local direction of Earth's magnetic field (due to plasma particle dipole interactions). In these phenomena, lines or curves appear that follow along the direction of the local magnetic field.

These field lines provide us with a way to depict or draw the magnetic field (or any other vector field). Technically, field lines are a set of lines through space whose direction at any point is the direction of the local magnetic field, and whose density is proportional to the magnitude of the local magnetic field. Note that when a magnetic field is depicted with field lines, it is *not* meant to imply that the field is only nonzero along the drawn-in field lines. (The use of iron filings to display a field presents something of an exception to this picture: The magnetic field is in fact much larger along the "lines" of iron, due to the large permeability of iron relative to air.) Rather, the field is typically smooth and continuous everywhere, and can be estimated at *any* point (whether on a field line or not) by looking at the direction and density of the field lines nearby. The choice of which field lines to draw in such a depiction is arbitrary, apart from the requirement that they be spaced out so that their density approximates the magnitude of the local field. The level of detail at which the magnetic field is depicted can be increased by increasing the number of lines.

Field lines are a useful way to represent any vector field and can often be used to reveal sophisticated properties of that field quite simply. One important property of the magnetic B field that can be verified with field lines is that it always makes complete loops. Magnetic field lines neither start nor end (although they can extend to or from infinity). To date no exception to this rule has been found.

Even when it appears that a magnetic field has an end (such as when it leaves near a north pole or enters near a south pole of a magnet) in reality it does not. In the case of the permanent magnet the field lines complete the loop inside of the magnet traveling from the south to the north pole. (To see that this must be true imagine placing a compass inside of the magnet. The north pole of the compass will point toward the north pole of the magnet since magnets stacked on each other point in the same direction.)

Since magnetic field lines always come in loops, magnetic poles always come in N and S pairs. If a magnetic field line enters a magnet somewhere it has to leave the magnet somewhere else; it is not allowed to have an end point. For this reason as well, cutting a magnet in half will result in two separate magnets each with both a north and a south pole.

Field lines are also a good tool for visualizing magnetic forces. When dealing with magnetic fields in ferromagnetic substances like iron, and in plasmas, the magnetic forces can be understood by imagining that the field lines exert a tension, (like a rubber band) along their length, and a pressure perpendicular to their length on neighboring field lines. The "unlike" poles of magnets attract because they are linked by many field lines, while "like" poles repel because the field lines between them don't meet, but run parallel, pushing on each other.

Earth's Magnetic Field

Because of Earth's magnetic field, a compass placed anywhere on Earth will turn so that the "north pole" of the magnet inside the compass points roughly north, toward Earth's north magnetic pole in northern Canada. This is the traditional definition of the "north pole" of a magnet, although other equivalent definitions are also possible. One confusion that arises from this definition is that if Earth itself is considered as a magnet, the *south* pole of that magnet would be the one nearer the north magnetic pole, and vice-versa. (Opposite poles attract and the north pole of the compass magnet is attracted to the north magnetic pole.) The north magnetic pole is so named not because of the polarity of the field there but because of its geographical location.

The figure to the right is a sketch of Earth's magnetic field represented by field lines. The magnetic field at any given point does not point straight toward (or away) from the poles and has a significant up/down component for most locations. (In addition, there is an East/West component as Earth's magnetic poles do not coincide exactly with Earth's geological pole.) The magnetic field is as if there were a magnet deep in Earth's interior.

Earth's magnetic field is probably due to a dynamo that produces electric currents in the outer liquid part of its core. Earth's magnetic field is not constant: Its strength and the location of its poles vary. The poles even periodically reverse direction, in a process called geomagnetic reversal.

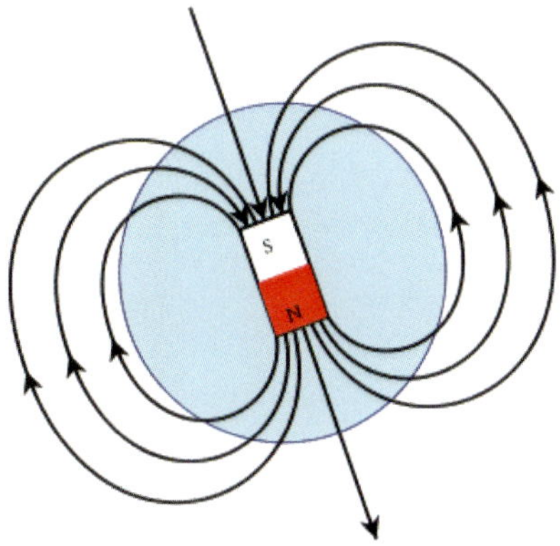

A sketch of Earth's magnetic field representing the source of Earth's magnetic field as a magnet.

The north pole of earth is near the top of the diagram, the south pole near the bottom. Notice that the south pole of that magnet is deep in Earth's interior below Earth's North Magnetic Pole. Earth's magnetic field is produced in the outer liquid part of its core due to a dynamo that produce electrical currents there.

Effects of the Magnetic Field, B

A magnetic field has many effects on materials and on individual particles. All of these effects can be expressed due to its affects on elementary charges and magnetic dipoles. There are four elementary ways that a magnetic field can affect a charge or a magnetic dipole.

1. Sideways force on a moving charge or current
2. Torque on a magnetic dipole
3. Force on a magnetic dipole due to a non-uniform B
4. Force on a charge due to a changing B

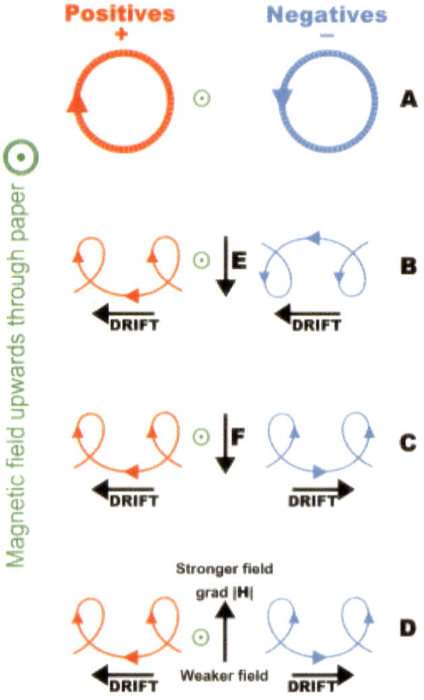

Charged particle drifts in a homogenous magnetic field. (A) No disturbing force (B) With an electric field, E (C) With an independent force, F (for example, gravity) (D) In an inhomogeneous magnetic field, grad H

Force Due to a Magnetic Field on a Moving Charge

Force on a Charged Particle

A charged particle moving in a magnetic B field will feel a *sideways* force that is proportional to the strength of the magnetic field, the component of the velocity that is perpendicular to the magnetic field and the charge of the particle. This force is known as the Lorentz Force. The force is always perpendicular to both the velocity of the particle and the magnetic field that created it. Neither a stationary particle nor one moving in the direction of the magnetic field lines will experience a force. For that reason, charged particles move in a circle (or more generally, helix) around magnetic field lines; this is called cyclotron motion. Because the magnetic field is always perpendicular to the motion, the magnetic fields can do no work on a charged particle; a magnetic field alone cannot speed up or slow down a charged particle. It can and does, however, change the particle's direction, even to the extent that a force applied in one direction can cause the particle to drift in a perpendicular direction.

Force on Current-carrying Wire

The force on a current carrying wire is similar to that of a moving charge as expected since a charge carrying wire is a collection of moving charges. A current carrying wire will feel a sideways force in the presence of a magnetic field. The Lorentz force on a macroscopic current is often referred to as the *Laplace force*.

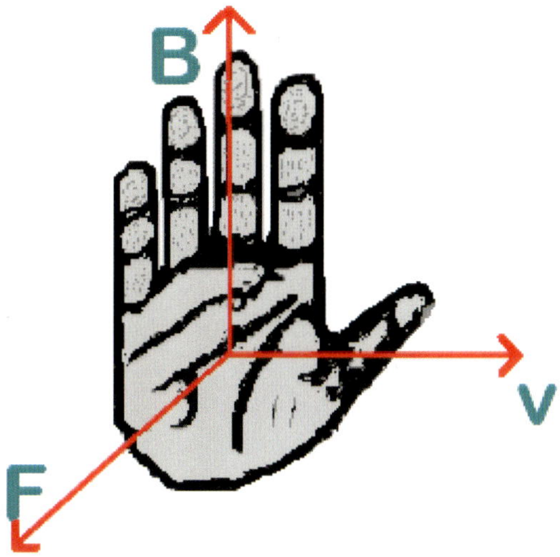

The right-hand rule: For a conventional current or moving positive charge in the direction of the thumb of the right hand and the magnetic field along the direction of the fingers (pointing away from palm) the force on the current will be in a direction out of the palm. The direction of the force is reversed for a negative charge.

Direction of Force

The direction of force on a positive charge or a current is determined by the right-hand rule. See the figure on the right. Using the right hand and pointing the thumb in the direction of the moving positive charge or positive current and the fingers in the direction of the magnetic field the resulting force on the charge will point outwards from the palm. The force on a negative charged particle is in the opposite direction. If both the speed and the charge are reversed then the direction of the force remains the same. For that reason a magnetic field measurement (by itself) cannot distinguish whether there is a positive charge moving to the right or a negative charge moving to the left. (Both of these will produce the same current.) On the other hand, a magnetic field combined with an electric field *can* distinguish between these.

An alternative, similar trick to the right hand rule is Fleming's left hand rule.

Torque on a Magnetic Dipole

A magnet placed in a magnetic field will feel a torque that will try to align the magnet with the magnetic field. The torque on a magnet due to an external magnetic field is easy to observe by placing two magnets near each other while allowing one to rotate. This magnetic torque is the basis for how compasses work. It is used to define the direction of the magnetic field.

The magnetic torque also provides the driving torque for simple electric motors. A magnet (called a rotor) placed on a rotating shaft will feel a strong torque if like poles are placed near its own poles. If the magnet that caused the rotation—called the stator—is constantly being flipped such that it always has like poles close to the rotor then the rotor will generate a torque that is transferred to the shaft. The polarity of the rotor can easily be flipped if it is an electromagnet by flipping the direction of the current through its coils.

Force on a Magnetic Dipole Due to a Non-uniform B

The most commonly experienced effect of the magnetic field is the force between two magnets: Like poles repel and opposites attract. One can, in fact, express this force in terms of the pole locations and strengths (or more generally, pole distributions) in the two magnets attracting and repulsing each other. This model is called the "Gilbert model" and produces both the correct force between two magnets, and the correct B field outside of the magnets, but the wrong magnetic field *inside* the magnets. (Although the Gilbert model is useful in certain contexts as a mathematical model, the idea of "poles" does not accurately reflect what physically happens inside a magnet; see ferromagnetism.)

A more physically accurate picture would be based on the fundamental fact that a magnetic dipole experiences a force, when placed in a *non-uniform* external magnetic field. (In a uniform field, it will experience a torque but no force.) The south pole of one magnet is attracted to the north pole of another magnet because of the specific way in which each of the microscopic dipoles in either magnet responds to the non-uniform field of the other magnet.

The force on a magnetic dipole does not depend directly on the strength or direction of the magnetic B field, but only on how these vary with location. A magnet will move to maximize the magnetic field in the direction of its magnetic moment.

Care should be taken to distinguish the magnetic force on a magnetic dipole from the magnetic force on a moving charge. The magnetic force on a charge only occurs when the charge is moving and is in a sideways direction. It is felt for both uniform and non-uniform magnetic fields. The magnetic force on a dipole, on the other hand, is present only in non-uniform (in space) fields and is in the direction that increases the component of the magnetic field in the direction parallel to the dipole's magnetic moment. Neither does the force on a magnetic dipole depend on its speed (except at velocities approaching the speed of light).

Electric Force Due to a Changing B

If the magnetic field in an area is varying with time it generates an electric field that forms closed loops around that area. A conducting wire that forms a closed loop around the area will have an induced voltage generated by this changing magnetic field. This effect is represented mathematically as Faraday's Law and forms the basis of many generators. Care must be taken to understand that the changing magnetic field is a source for an *extended* electric field. The changing magnetic field does not only create an electric field at that location; rather it generates an electric field that forms closed loops around the location where the magnetic field is changing.

Mathematically, Faraday's law is most often represented in terms of the change of magnetic flux with time. The magnetic flux is the property of a closed loop (say of a coil of wire) and is the product of the area times the magnetic field that is normal to that area. Engineers and physicists often use magnetic flux as a convenient physical property of a loop(s). They then express the magnetic B field as the magnetic flux per unit area. It is for this reason that the B field is often referred to as the "magnetic flux density." This approach has the benefit of making certain calculations easier such as in magnetic circuits. It is typically not used outside of electrical circuits, though, because the magnetic B field truly is the more "fundamental" quantity in that it directly connects all of electrodynamics in the simplest manner.

Sources of Magnetic Fields

Magnetic fields can be created in a number of different ways. All of these ways are based on three elementary ways to create a magnetic B field.

1. Electrical currents (moving charges)
2. Magnetic dipoles
3. Changing electric field

These sources are thought to affect the virtual particles that compose the field.

Electrical Currents (Moving Charges)

All moving charges produce a magnetic field. The magnetic field of a moving charge is very complicated but is well known. It forms closed loops around a line that is pointing in the direction the charge is moving. The magnetic field of a current on the other hand is much easier to calculate.

Magnetic Field of a Steady Current

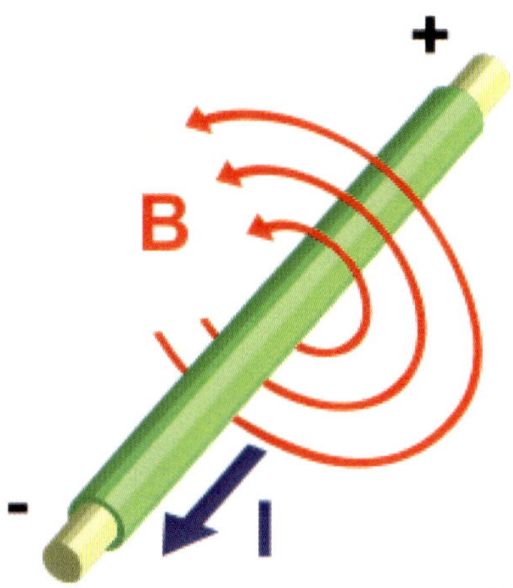

Current *(I)* through a wire produces a magnetic field (B) around the wire.
The field is oriented according to the right hand grip rule.

The magnetic field generated by a *steady current* (a continual flow of charges, for example through a wire, which is constant in time and in which charge is neither building up nor depleting at any point), is described by the Biot-Savart law. This is a consequence of Ampere's law, one of the four Maxwell's equations that describe electricity and magnetism. The magnetic field lines generated by a current carrying wire form concentric circles around the wire. The direction of the magnetic field of the loops is determined by the right hand grip rule. The strength of the magnetic field decreases with distance from the wire.

A current carrying wire can be bent in a loop such that the field is concentrated (and in the same direction) inside of the loop. The field will be weaker outside of the loop. Stacking many such loops to form a solenoid (or long coil) can greatly increase the magnetic field in the center and decrease the magnetic field outside of the solenoid. Such devices are called electromagnets and are extremely important in generating strong and well controlled magnetic fields. An infinitely long solenoid will have a uniform magnetic field inside of the loops and no magnetic field outside. A finite length electromagnet will produce essentially the same magnetic field as a uniform permanent magnet of the same shape and size. An electromagnet has the advantage, though, that you can easily vary the strength (even creating a field in the opposite direction) simply by controlling the input current. One important use is to continually switch the polarity of a stationary electromagnet to force a rotating permanent magnet to continually rotate using the fact that opposite poles attract and like poles repel. This can be used to create an important type of electrical motor.

Magnetic Dipoles

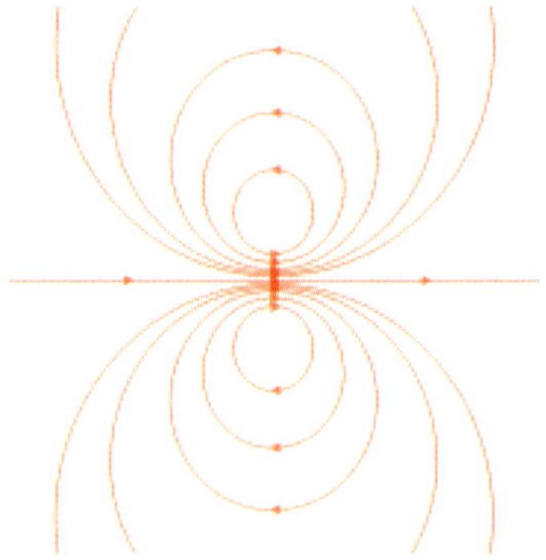

Magnetic field lines around a "magnetostatic dipole" the magnetic dipole itself is in the center and is seen from the side.

The magnetic field due to a permanent magnet is well known. But, what causes the magnetic field of a permanent magnet? The answer again is that the magnetic field is essentially created due to currents. But this time it is due to thc cumulativc cffcct of many small "currents" of electrons "orbiting" the nuclei of the magnetic material. Alternatively it is due to the structure of the electron itself which, in some sense, can be thought of as forming a tiny loop of current. (The true nature of the electron's magnetic field is relativistic in nature, but this model often works.) Both of these tiny loops are modeled in terms of what is called the magnetic dipole. The dipole moment of that dipole can be defined as the current times the area of the loop, then an equation for the magnetic field due to that magnetic dipole can be derived. (See the above image for what that magnetic field looks like.) Magnetic field of a larger magnet can be calculated by adding up the magnetic fields of many magnetic dipoles.

Changing Electric Field

The final known source of magnetic fields is a changing electric field. Just as a changing magnetic field generates an electric field so does a changing electric field generate a magnetic field.

(These two effects bootstrap together to form electromagnetic waves, such as light.) Similar to the way magnetic field lines form close loops around a current a time varying electric field generates a magnetic field that forms closed loops around the region where the electric field is changing. The strength of this magnetic field is proportional to the time rate of the change of the electric field (which is called the displacement current). The fact that a changing electric field creates a magnetic field is known as Maxwell's correction to Ampere's Law.

Magnetic Monopole (Hypothetical)

The magnetic monopole is a hypothetical particle (it may or may not exist). A magnetic monopole would have, as its name suggests, only one pole. In other words, it would possess "magnetic charge" analogous to electric charge. Positive magnetic charge would correspond to an isolated north pole, and negative magnetic charge would correspond to an isolated south pole.

Modern interest in this concept stems from particle theories, notably Grand Unified Theories and superstring theories, that predict either the existence or the possibility of magnetic monopoles. These theories and others have inspired extensive efforts to search for monopoles. Despite these efforts, no magnetic monopole has been observed to date. (Two experiments produced candidate events that were initially interpreted as monopoles, but these are now regarded to be inconclusive.)

Definition and Mathematical Properties of B

There are several different but physically equivalent ways to define the magnetic B field. In principle any of the above effects due to the magnetic field or any of the sources of the magnetic field can be used to define its magnitude and the direction. Its direction at a given point can be thought of as being the direction that a *hypothetical* freely rotating small test dipole would rotate to point if it *were* placed at that point. Its magnitude is defined (in SI units) in terms of the voltage induced per unit area on a current carrying loop in a uniform magnetic field normal to the loop when the magnetic field is reduced to zero in a unit amount of time. The SI unit of magnetic field is the Tesla.

The magnetic field vector is a pseudovector (also called an axial vector). (This is a technical statement about how the magnetic field behaves when you reflect the world in a mirror.) This fact is apparent from many of the definitions and properties of the field; for example, the magnitude of the field is proportional to the torque on a dipole, and torque is a well-known pseudovector.

Maxwell's Equations

As discussed above, the magnetic field is a vector field. (The magnetic field at each point in space and time is represented by its own vector.) As a vector field, the magnetic field has two important mathematical properties. These properties, along with the corresponding properties of the electric field, make up Maxwell's Equations.

The first is that the magnetic B field never starts nor ends at a point. Whatever magnetic field lines enter a region has to eventually leave that region. This is mathematically equivalent to saying that the divergence of the magnetic B is zero. (Such vector fields are called solenoidal vector fields.)

This property is called Gauss' law for magnetism and is one of Maxwell's Equations. It is also equivalent to the statement that there are no magnetic monopoles.

The second mathematical property of the magnetic field is that it always loops around the source that creates it. This source could be a current, a magnet, or a changing electric field, but it is always within the loops of magnetic field they create. Mathematically, this fact is described by the Ampère-Maxwell equation.

Measuring the Magnetic B Field

There are many ways of measuring the magnetic B field, many of which use the effects described above. Devices used to measure the local magnetic field are called magnetometers. Important magnetometers include using a rotating coil, Hall effect magnetometers, NMR magnetometer, SQUID magnetometer, and a fluxgate magnetometer. The magnetic fields of distant astronomical objects can be determined by noting their effects on local charged particles. For instance, electrons spiraling around a field line will produce synchotron radiation which is detectable in radio waves.

Hall Effect

Because the Lorentz force is charge-sign-dependent, it results in charge separation when a conductor with current is placed in a transverse magnetic field, with a buildup of opposite charges on two opposite sides of conductor in the direction normal to the magnetic field, and the potential difference between these sides can be measured.

The Hall effect is often used to measure the magnitude of a magnetic field as well as to find the sign of the dominant charge carriers in semiconductors (negative electrons or positive holes).

SQUID Magnetometer

Superconductors are materials with both distinctive electric properties (perfect conductivity) and magnetic properties (such as the Meissner effect, in which many superconductors can perfectly expel magnetic fields). Due to these properties, it turns out that loops that incorporate superconducting material and their Josephson junctions can function as very sensitive magnetometers, called SQUIDs.

The H Field

The term "magnetic field" can also be used to describe the magnetic H field. The magnetic H field is similar to B in that it is a vector field, but its units are often different. In SI units, B and H are measured in teslas (T) and amperes per meter (A/m), respectively; or, in cgs units, in gauss (G) and oersteds (Oe), respectively. Outside of magnetizable materials, the two fields are identical (apart from possibly a constant conversion factor), but inside a magnetic material they can differ substantially. Quantitatively, the fields are related by the following equations:

$$\mathrm{B} = \mu_0 (\mathrm{H} + \mathrm{M}) \quad \text{(SI units)}$$

$$\mathrm{B} = \mathrm{H} + 4\pi \mathrm{M} \quad \text{(cgs units)},$$

where M is magnetization density of any magnetic material.

Physical Interpretation of the H Field

When magnetic materials are present, the total magnetic field B is caused by two different types of currents which need to be distinguished: free current and bound current. Free currents are the ordinary currents in wires and other conductors, that can be controlled and measured. Bound currents are the tiny circular currents inside atoms that are responsible for the magnetization M of magnetic materials. Although the actual source of the magnetic field in electron orbitals of atoms is complex, the magnetic properties of a material can be accounted for by assuming it is divided into tiny blocks, each of which has a current flowing around its outside surface, perpendicular to the magnetic field axis. As an example of bound current consider a uniform permanent bar magnet. A piece of iron is formed of many tiny regions called magnetic domains, each of which is a magnetic dipole, essentially a tiny loop of current. In a bar magnet, most of these dipoles have their poles lined up, creating a large magnetic field. If we add up the currents of all these tiny loops we will find that the currents cancel in the interior of the material, but add up along the sides of the bar. (This current loops around the sides and not at the poles.) No one charge makes the complete trip around the magnet (each charge is bound to its tiny loop) but the net effect is exactly equivalent to a real current that flows around the outside surface of the magnet, perpendicular to the magnetic field axis. (If the magnetization is not uniform then a bound current will flow through the bulk of the magnetic material as well.)

The magnetic H is useful because it treats these two types of currents differently. The free currents it treats in the normal fashion and therefore has the same form as the magnetic B field it would generate. The magnetic H fields treats the field inside of a magnetic material (due to that magnetic material) in a manner similar to the Gilbert model. (By subtracting the magnetization from the B field we are essentially converting the bound current sources to Gilbert-like magnetic charges at the poles.) Unlike the magnetic B, which always forms closed loops, the field due to the magnetic charges flow outward (or inward depending on the sign of the magnetic charge) in both directions from the poles. And while the magnetic field is exactly the same on the outside of the magnetic material for both models the magnetic fields inside are quite different.

Putting both sources together we see that the magnetic H field is the same as the magnetic B field to a multiplicative constant outside of magnetic materials, but is completely different from the magnetic B field inside a magnetic material. The advantage of this hybrid H field is that these sources are treated so differently that we can often pick out one source from the other. For example a line integral of the magnetic H field in a closed loop will yield the total free current in the loop (and not the bound current). This is unlike the magnetic B field where a similar integral will yield the sum of both the free and the bound current. If one wants to isolate the contribution due to the bound currents then a surface integral of H over any closed surface will pick out the "magnetic charges" at the poles.

Sources of the H Field

Unlike the magnetic B field that only has a current source such that the magnetic B field loops around currents, the magnetic H field has two types of sources. The first source of magnetic H field are the *free* currents for which H loop around similar to the way B field loops around the total current. The second source of the magnetic H field are "magnetic charges" near the poles of the magnetic material. More precisely, these "magnetic charges" are calculated as $-\nabla \cdot M$.

Uses of the H Field

Energy Stored in Magnetic Fields

In order to create a magnetic field we need to do work to establish a free current. If one were to ask how much energy does it take to create a specific magnetic field using a particular free current then one needs to distinguish between the free and the bound currents. It is the free current that one is "pushing" on. The bound currents are freeloaders. They create a magnetic field that the free current has to work against without doing any of the work. If one was to calculate the energy of creating a magnetic field B we need to have a way of separating out the free current. The magnetic B cannot be used to determine this free current since B does not distinguish between bound and free current.

The magnetic H field does treat the two sources differently. Therefore it is useful in calculating the energy needed to create a magnetic field with a free current in the presence of magnetic materials. In this case the energy density needed, assuming a linear relationship between H and B, has the form of:

$$u = \frac{\mathrm{H.B}}{2}.$$

If there are no magnetic materials around then we can replace H with $\frac{\mathrm{B}}{\mu_{o,}}$

$$u = \frac{\mathrm{B} \cdot \mathrm{B}}{2\mu_o}.$$

Magnetic Circuits

A second use for H is in magnetic circuits where inside a linear material $\mathrm{B} = \mu \mathrm{H}$. Here, μ is the permeability of the material. This is similar in form to Ohm's Law $\mathrm{J} = \sigma \mathrm{E}$, where J is the current density, σ is the conductance and E is the Electric field. Extending this analogy we derive the counterpoint to the macroscopic Ohm's law ($\frac{V}{R} = I$) as:

$$\Phi = \frac{F}{R_m},$$

where $\Phi = \int \mathrm{B} \cdot d\mathrm{A}$ is the magnetic flux in the circuit, $F = \int \mathrm{H} \cdot dl$ is the magnetomotive force applied to the circuit, and R_m is the reluctance of the circuit. Here the reluctance R_m is a quantity similar in nature to resistance for the flux.

Using this analogy it is straight-forward to calculate the magnetic flux of complicated magnetic field geometries, by using all the available techniques of circuit theory.

History of B and H

The difference between the B and the H vectors can be traced back to Maxwell's 1855 paper entitled, *On Faraday's Lines of Force*. It is later clarified in his concept of a sea of molecular vortices that appears in his 1861 paper, On Physical Lines of Force—1861. Within that context, H represented pure vorticity (spin), whereas B was a weighted vorticity that was weighted for the density of the vortex sea. Maxwell considered magnetic permeability μ to be a measure of the density of the vortex sea. Hence, the relationship,

(1) Magnetic induction current causes a magnetic current density

$$B = \mu H$$

was essentially a rotational analogy to the linear electric current relationship,

(2) Electric convection current

$$J = \rho v$$

where ρ is electric charge density. B was seen as a kind of magnetic current of vortices aligned in their axial planes, with H being the circumferential velocity of the vortices. With μ representing vortex density, we can now see how the product of μ with vorticity H leads to the term magnetic flux density which we denote as B.

The electric current equation can be viewed as a convective current of electric charge that involves linear motion. By analogy, the magnetic equation is an inductive current involving spin. There is no linear motion in the inductive current along the direction of the B vector. The magnetic inductive current represents lines of force. In particular, it represents lines of inverse square law force.

The extension of the above considerations confirms that where B is to H, and where J is to ρ, then it necessarily follows from Gauss's law and from the equation of continuity of charge that E is to D. ie. B parallels with E, whereas H parallels with D.

Rotating Magnetic Fields

The rotating magnetic field is a key principle in the operation of alternating-current motors. A permanent magnet in such a field will rotate so as to maintain its alignment with the external field. This effect was conceptualized by Nikola Tesla, and later utilized in his, and others', early AC (alternating-current) electric motors. A rotating magnetic field can be constructed using two orthogonal coils with 90 degrees phase difference in their AC currents. However, in practice such a system would be supplied through a three-wire arrangement with unequal currents. This inequality would cause serious problems in standardization of the conductor size and so, in order to overcome it, three-phase systems are used where the three currents are equal in magnitude and have 120 degrees phase difference. Three similar coils having mutual geometrical angles of 120 degrees will create the rotating magnetic field in this case. The ability of the three-phase system to create a rotating field, utilized in electric motors, is one of the main reasons why three-phase systems dominate the world's electrical power supply systems.

Because magnets degrade with time, synchronous motors and induction motors use short-circuited rotors (instead of a magnet) following the rotating magnetic field of a multicoiled stator. The short-circuited turns of the rotor develop eddy currents in the rotating field of the stator, and these currents in turn move the rotor by the Lorentz force.

Special Relativity and Electromagnetism

Magnetic fields played an important role in helping to develop the theory of special relativity.

Moving Magnet and Conductor Problem

Imagine a moving conducting loop that is passing by a stationary magnet. Such a conducting loop will have a current generated in it as it passes through the magnetic field. But why? It is answering this seemingly innocent question that led Albert Einstein to develop his theory of special relativity.

A stationary observer would see an unchanging magnetic field and a moving conducting loop. Since the loop is moving all of the charges that make up the loop are also moving. Each of these charges will have a sideways, Lorentz force, acting on it which generates the current. Meanwhile, an observer on the moving reference frame would see a *changing* magnetic field and *stationary* charges. (The loop is not moving in this observers reference frame. The magnet is.) This changing magnetic field generates an *electric* field.

The stationary observer claims there is *only* a magnetic field that creates a *magnetic force* on a moving charge. The moving observer claims that there is both a magnetic and an electric field but all of the force is due to the *electric* field. Which is true? Does the electric field exist or not? The answer, according to special relativity, is that both observers are right from their reference frame. A pure magnetic field in one reference can be a mixture of magnetic and electric field in another reference frame.

Electric and Magnetic Fields Different Aspects of the Same Phenomenon

According to special relativity, electric and magnetic forces are part of a single physical phenomenon, electromagnetism; an electric force perceived by one observer will be perceived by another observer in a different frame of reference as a mixture of electric and magnetic forces. A magnetic force can be considered as simply the relativistic part of an electric force when the latter is seen by a moving observer.

More specifically, rather than treating the electric and magnetic fields as separate fields, special relativity shows that they naturally mix together into a rank-2 tensor, called the electromagnetic tensor. This is analogous to the way that special relativity "mixes" space and time into spacetime, and mass, momentum and energy into four-momentum.

Magnetic Field Shape Descriptions

Schematic quadrupole magnet("*four-pole*") magnetic field. There are four steel pole tips, two opposing magnetic north poles and two opposing magnetic south poles.

- An azimuthal magnetic field is one that runs east-west.
- A meridional magnetic field is one that runs north-south. In the solar dynamo model of the Sun, differential rotation of the solar plasma causes the meridional magnetic field to stretch into an azimuthal magnetic field, a process called the *omega-effect*. The reverse process is called the alpha-effect.
- A dipole magnetic field is one seen around a bar magnet or around a charged elementary particle with nonzero spin.
- A quadrupole magnetic field is one seen, for example, between the poles of four bar magnets. The field strength grows linearly with the radial distance from its longitudinal axis.
- A solenoidal magnetic field is similar to a dipole magnetic field, except that a solid bar magnet is replaced by a hollow electromagnetic coil magnet.
- A toroidal magnetic field occurs in a doughnut-shaped coil, the electric current spiraling around the tube-like surface, and is found, for example, in a tokamak.
- A poloidal magnetic field is generated by a current flowing in a ring, and is found, for example, in a tokamak.
- A radial magnetic field is one in which the field lines are directed from the center outwards, similar to the spokes in a bicycle wheel. An example can be found in a loudspeaker transducers (driver).
- A helical magnetic field is corkscrew-shaped, and sometimes seen in space plasmas such as the Orion Molecular Cloud.

Magnetic Moment

The magnetic moment of a magnet is the quantity that finds the force that a magnet is capable of exerting on electric currents and the torque that the magnetic field ill exert on it. Magnetic moments are exhibited by numerous things including a loop of electric current, a bar magnet, an electron, a molecule or a planet.

Magnetic moment is also a vector which as it has magnitude as well as a direction. The direction of the magnetic moment points from the south to North Pole of a magnet. The magnetic field produced by a magnet is proportional to its magnetic moment as well. The term magnetic moment refers to the system's magnetic dipole moment. The dipole component of an object's magnetic field is symmetric about the direction of its magnetic dipole moment, and decreases as the inverse cube of the distance from the object.

Magnetic moment, which is also called a magnetic dipole moment, is a measure of the object's tendency to align with a magnetic field. Its positive direction depends on the way the object responds to the magnetic field. The objects have a tendency to place themselves in such a way that the magnetic moment vector becomes parallel to the magnetic field lines. Magnetic moment is basically generated in two ways including:

- The Motion of Electric Charge
- Spin Angular Momentum

For example, a loop of wire with a current running through it will have a magnetic moment proportional to the current and area of the loop pointing in the direction of your right thumb if your fingers are curling in the direction of the current.

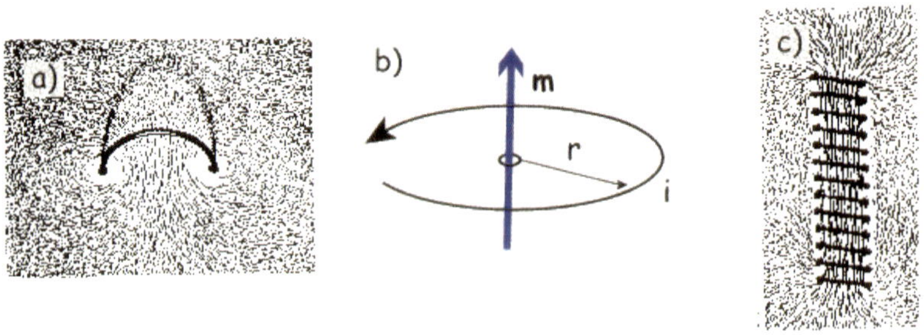

The above figure shows how the magnetic field is generated by the current flowing in the loop. The (b) part of the figure shows a current loop with current i and area producing a magnetic moment m. the last figure depicts the magnetic fields of loops arranged as a solenoid which shows the total contribution of the individual loops.

External Magnetic Field produced by a Magnetic Dipole Moment

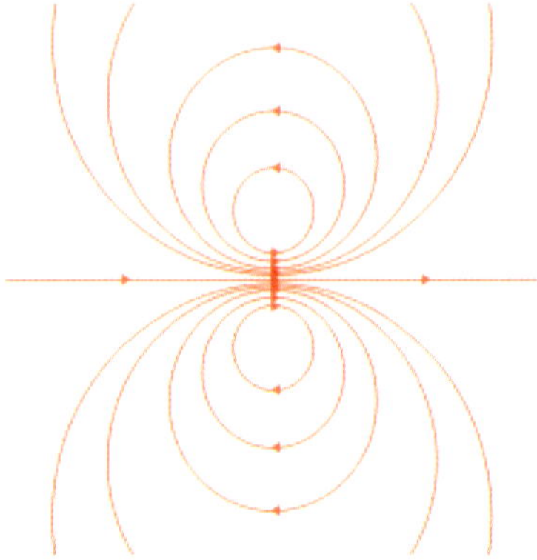

Any system that has a net magnetic dipole moment m produces a dipolar magnetic field. Although, the net magnetic field produced by the system may exhibit higher order multipole components, but they fall out very rapidly with distance and at last, only the dipolar component will dominate the magnetic field of the system at distances far away from it.

Internal Magnetic Field of a Dipole

Both the models for a dipole give the same result for the magnetic field far from the source. But the predictions inside the source region are different. The magnetic field between poles is in the opposite direction to the magnetic moment. it is apparent form the figure that the limits of these fields must also be different as the sourecs ultimately reduce to zero size.

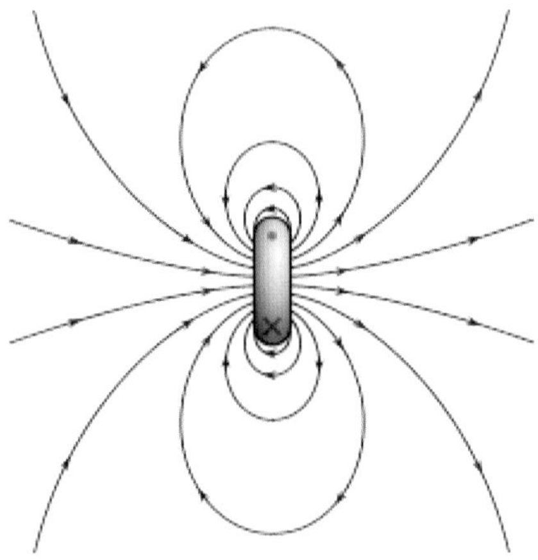

Magnetic dipole in a Uniform Field Magnetic Field

Consider a uniform magnetic field of strength B. Let a magnetic dipole be suspended in it such that its axis makes an angle θ with the field as shown in figure. If 'm' is the strength of each pole, the two poles N and S experience two equal and opposite forces 'mB'. These forces constitute a couple which tends to rotate the dipole. Suppose the couple exerts a torque of magnitude τ.

τ = (either force) × (perpendicular distance between the two forces)

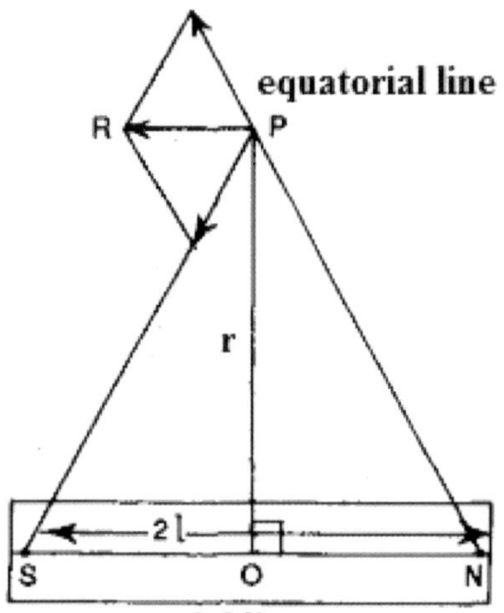

$$= mB \times AC = mB \times 2l \sin\theta$$

$$\text{Or } \tau = MB \sin\theta$$

Here, M = 2ml = magnetic moment of the dipole

If B = 1 and θ = 90°

$$\tau = M$$

Magnetic moment M of the magnetic dipole is defined as the moment of the couple acting on it when it is placed at right angles to a uniform field of unit strength.

In such case,

Force on each pole = m × 1

Perpendicular distance between two forces = NS = 2l

So, $\tau = m \times 2l$

Thus, magnetic moment of a magnetic dipole is also defined as the product of its pole strength and the magnetic length.

Magnetic moment of a magnet is a vector quantity and is directed along SN inside the magnet.

$$\vec{M} = M\hat{l}$$

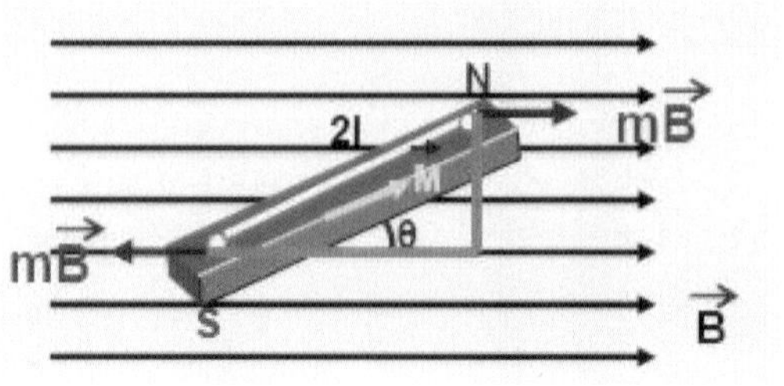

Here, $\hat{l}$ = unit vector along SN.

Since $\vec{M}$ and $\vec{B}$, both, are vectors the torque acting on a magnet suspended in the magnetic field can be expressed as the cross-product of $\vec{M}$ and $\vec{B}$ as follows.

$$\vec{\tau} = \vec{M} \times \vec{B}$$

The direction of $\vec{\tau}$ can be obtained by applying right hand thumb rule.

Units of M

Units of M can be obtained from the relation

$$\tau = MB \sin\theta$$

So, $M = \tau / B \sin\theta$

Therefore, M can be measured in terms of joule per tesla (JT^{-1}) or (NmT^{-1})

Since ,1 tesla = Wb/m^2

Thus, Unit of M = $J/(Wb/m^2) = Jm^2/Wb$

Since 1 J = 1 Nm

Therefore, Unit of M = $1\ Nm\ T^{-1} = 1\ Nm/(Wb/m^2) = 1\ Nm^3\ Wb^{-1}$

Since unit of pole strength = Am

Thus, Unit of M = (Am) m = Am^2

Work done in Rotating a Magnetic Dipole in a Magnetic Field

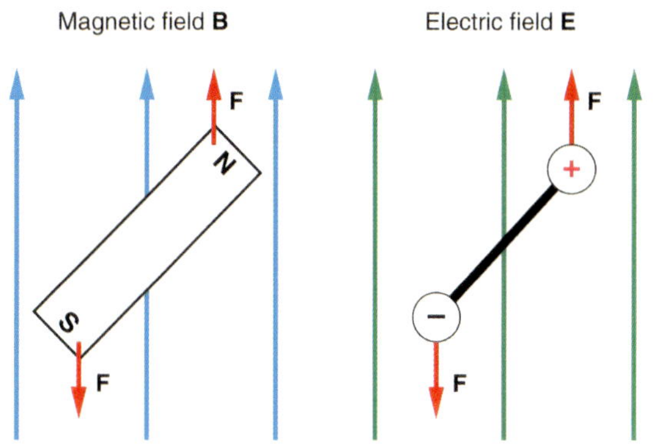

When a magnetic dipole of magnetic moment M is oriented in a magnetic field of strength, B, making an angle ‘θ’. With its lines of force, it experiences a torque τ given by

$$\vec{\tau} = \vec{M} \times \vec{B}$$

Or

$$|\vec{\tau}| = M\ B \sin\theta$$

Let $d\vec{\theta}$ be the small abgular displacement given to the dipole, work done dW is given by

$$dW = \vec{\tau}.d\vec{\theta} = \tau d\theta$$

Since the angle between $\vec{\tau}$ and $d\vec{\theta}$ is zero.

So, dW = MB sin θ dθ

Work done in displacing the dipole from an angle θ_1 to θ_2 is

$$W = \int_0^W dW = \int_{\theta_1}^{\theta_2} M\ B \sin\theta d\theta$$

$W = MB\,[(-\cos\theta_2) - (-\cos\theta_1)]$

Or, $W = MB\,(\cos\theta_1 - \cos\theta_2)$

Potential Energy of a Magnetic Dipole in a Magnetic Field

Potential energy of a magnetic dipole, in a magnetic field, is defined as the amount of work done in rotating the dipole from zero potential energy position to any desired position.

It is convenient to choose the zero potential energy position to be the one when the dipole is at right angles to the lines of force of magnetic field.

So, $\theta_1 = 90^\circ$ and $\theta_2 = \theta$

Making these substitution in equation (3), we get,

Potential energy = W = MB (cos 90° – cos θ)

Or, W = – MB cosθ

In vector form, $W = -\vec{M}.\vec{B}$

Special cases:

(a) When θ = 0°, cos θ = 1

So, W = – MB

(b) When θ = 90°, cos θ = 0

So, W = – MB(0) = 0

(c) When θ = 180°, cos θ = -1

So, W = – MB (-1) = MB

θ = 0° position corresponds to the position of dipole aligning itself such that its magnetic moment vector $\vec{M}$ is parallel to field. In this case, potential energy is – MB, a minimum value. We know that every body in state of equilibrium must possess minimum potential energy. That is why the magnetic dipole exists in equilibrium when $\vec{M}$ is parallel to $\vec{B}$.

Current Loop as a Magnetic Dipole

Ampere found that the distribution of magnetic lines of force around a finite current carrying solenoid is similar to that produced by a bar magnet. This is evident from the fact that a compass needle when moved around these two bodies show similar deflections. After noting the close resemblance between these two, Ampere demonstrated that a simple current loop behaves like a bar magnet and put forward that all the magnetic phenomena is due to circulating electric current. This is Ampere's hypothesis.

The magnetic induction at a point along the axis of a circular coil carrying current is,

$$B = \mu_0 nIa^2/2(a^2+x^2)^{3/2}$$

The direction of this magnetic field is along the axis and is given by right hand rule. For points which are far away from the centre of the coil, $x>>a$, a^2 is small and it is neglected. Hence for such points,

$$B = \mu_0 nIa^2/2x^3$$

If we consider a circular loop, n = 1, its area $A = \pi a^2$

$$\text{So, } B = \mu_0 IA/2\pi x^3$$

The magnetic induction at a point along the axial line of a short bar magnet is,

$$B = (\mu_0/4\pi)\,(2M/x^3)$$

$$\text{Or, } B = (\mu_0/2\pi)\,(M/x^3)$$

Comparing equations first and second, we find that,

$$M = IA$$

Hence a current loop is equivalent to a magnetic dipole of moment M = IA

The magnetic moment of a current loop is defined as the product of the current and the loop area. Its direction is perpendicular to the plane of the loop.

The Magnetic Dipole Moment of a Revolving Electron

According to Neil Bohr's atom model, the negatively charged electron is revolving around a positively charged nucleus in a circular orbit of radius r. The revolving electron in a closed path constitutes an electric current. The motion of the electron in anticlockwise direction produces conventional current in clockwise direction.

Since the electron is negatively charged, conventional current flows in a direction opposite to its motion. The magnetic moments associated with the two types of motion of electron in an orbit are shown in figure. An electron revolving in an orbit of radius 'r' is equivalent to a magnetic shell of magnetic moment 'M' given by,

$$M = i \times A$$

Here 'i' is the current to which the rotating electron is equivalent to and A is the area of the orbit. If 'τ' is the time period of rotation of electron.

$$i = \text{charge/period} = e/\tau = e/(2\pi/\omega)$$

or $i = e\omega/2\pi$

Here, 'ω' is the angular velocity of the electron.

$$\text{So, } M = (e\omega/2\pi) \times (\pi r^2) = e\omega r^2/2$$

According to Bohr's theory, an electron can revolve in an orbit in which its angular momentum is an integral multiple of $h/2\pi$ where 'h' is Planck's constant.

$$mr^2\omega = n\,(h/2\pi)$$

$$\text{Or, } r^2\omega = n\,(h/2\pi m)$$

Substituting for $r^2\omega$ in equation fourth, we get,

$$M = n\,(eh/4\pi m)$$

Since an atom may have a large number of electrons, the resultant magnetic moment of the atom is the vector sum of the magnetic moments of various electrons.

The term $eh/4\pi m$ is called Bohr's magneton. It is the smallest value of magnetic moment which an electron possess. Any atom can possess magnetic moment which is an integral multiple of Bohr's magneton. Thus, the magnetic moment is also quantised in the atomic and sub-atomic level.

In addition to the magnetic moment due to its orbital motion, the electron possesses magnetic moment due to its spin. Hence the resultant magnetic moment of an electron is the vector sum of its orbital magnetic moment and its spin magnetic moment.

Calculation of Magnetic Moment of a Bar Magnet

(a) If a magnet of length l and magnetic moment M is bent in the form of a semicircular are then its new magnetic moment will be $M' = 2M/\pi$

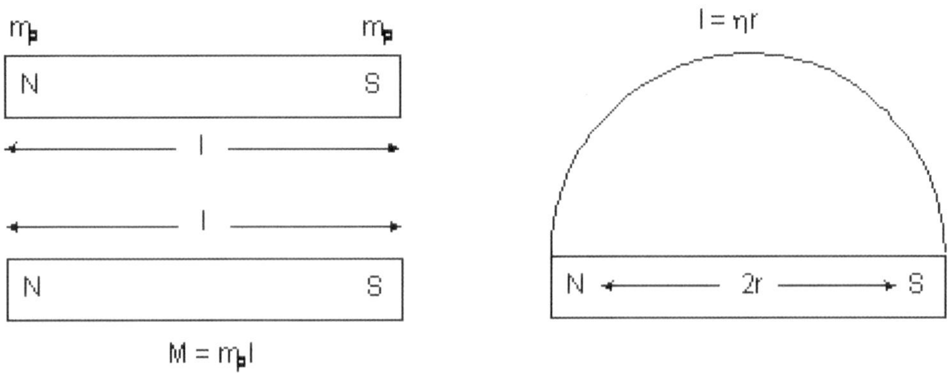

(b) The magnetic moment of an electron due to its orbital motion is $1\mu_B$ whereas that due to its spin motion it is $\mu_B/2$.

i.e. $M_{orbital} = 1\left(\frac{eh}{4\pi m_l}\right) = \left(\frac{\bar{e}h}{2m_l}\right)$

and $M_{spin} = s\left(\frac{eh}{4\pi m_s}\right) = s\left(\frac{\bar{e}h}{2m_s}\right) = s\mu_B = \frac{\mu_B}{2}$

Here μ_B = Bohr magneton

(i) The value of Bohr magneton $\mu_B = eh / 4\pi m$

(ii) $\mu B = 0.93 \times 10^{-23}$ Amp-m^2

(c) Other formula for magnetic moment includes:

(i) $M = ni\, r^2$

(ii) $M = eVr/2 = er^2\omega/2 = er^2 2\pi f/2 = er^2\pi/T$

(iii) e/mJ

(iv) $M = n\omega_B$

(d) How to calculate resultant magnetic moment:

(i) When two bar magnets are lying mutually perpendicular to each other, then

$M = \sqrt{M_1^2 + M_2^2} = \sqrt{2}m_p l$

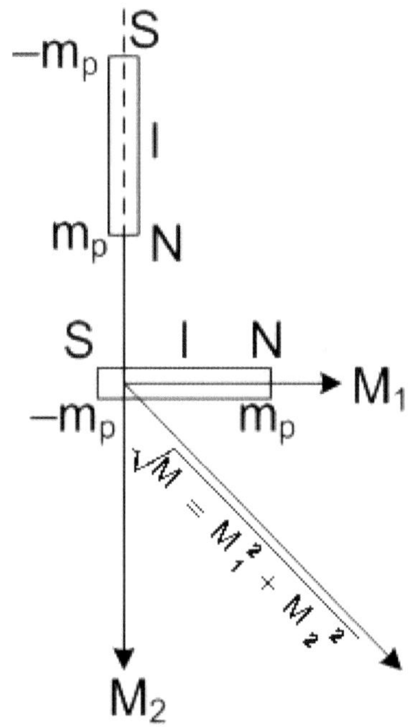

(ii) When two coils, each of radius r and carrying current i, are lying concentrically with their planes at right angles to each other, then

$$M=\sqrt{M_1^2+M_2^2} = \sqrt{2}i\pi r^2 \text{ if } M_1 = M_2$$

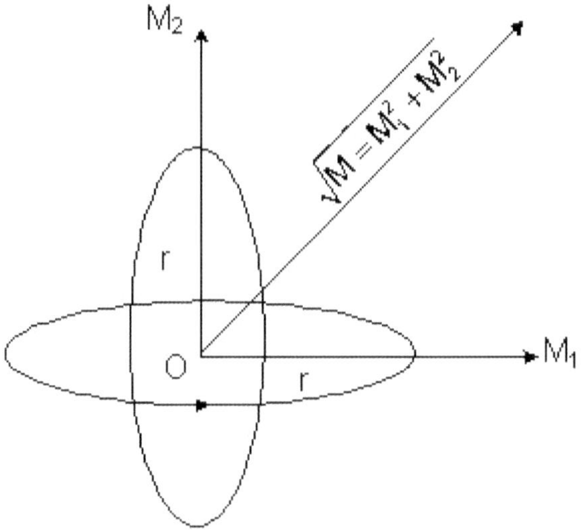

Illustration: *A square loop OABCO of side l carries a current i. Find the magnetic moment of the loop.*

Solution: Magnetic moment of the loop can be written as,

M = i (BC x CO), where the letters in bold denote the vectors

BC = -lk

$$\overline{CO} = -\ell\cos 60^\circ\,\hat{i} - \ell\sin 60^\circ\,\hat{j} = -\frac{\ell}{2}\hat{i} - \frac{\sqrt{3}}{2}\ell\hat{j}$$

$$\therefore \overline{M} = i\left[\left(-\ell\hat{k}\right)\times\left(-\frac{\ell}{2}\hat{i} - \frac{\sqrt{3}\ell}{2}\hat{j}\right)\right]$$

or $$\overline{M} = \frac{i\ell^2}{2}\left(\hat{j} - \sqrt{3}\,\hat{i}\right) \quad \text{Ans}$$

Illustration:Find the magnitude of magnetic moment of the current carrying loop ABCDEFA. Each side of the loop is 10 cm long and current in the loop is i = 2.0 A.

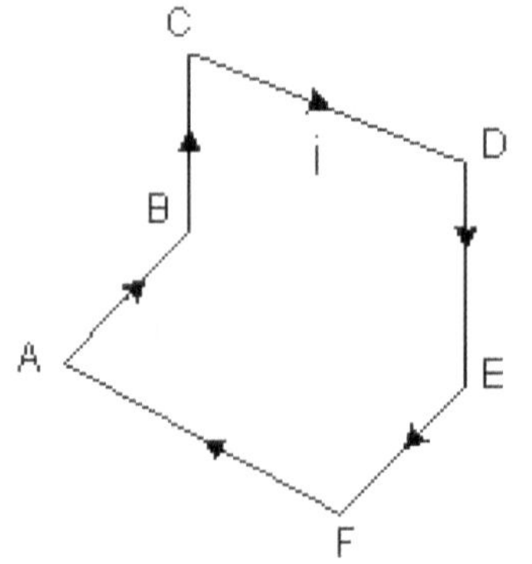

Solution: By assuming two equal and opposite currents in BE, two current carrying loop (ABEFA and BCDEB) are formed. Their magnetic moments are equal in magnitude but perpendicular to each other. Hence

$$M_{net} = \sqrt{M^2 + M^2} = \sqrt{2}\,M$$

Where $M = ia = (2.0)(0.1)(0.1) = 0.02\ \text{A-m}^2$

$$M_{net} = (\sqrt{2})(0.02)\text{A–m}^2 = 0.028\text{A-m}^2$$

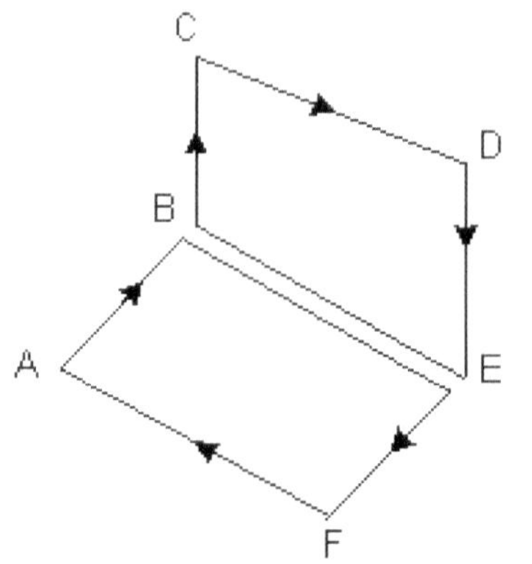

Illustration: Two identical bar magnets each of length L and pole–strength m are placed at right angles to each other with the north pole of one touching the south pole of the other. Evaluate the magnetic moment of the system.

Solution: As magnetic moment is a vector, so we know

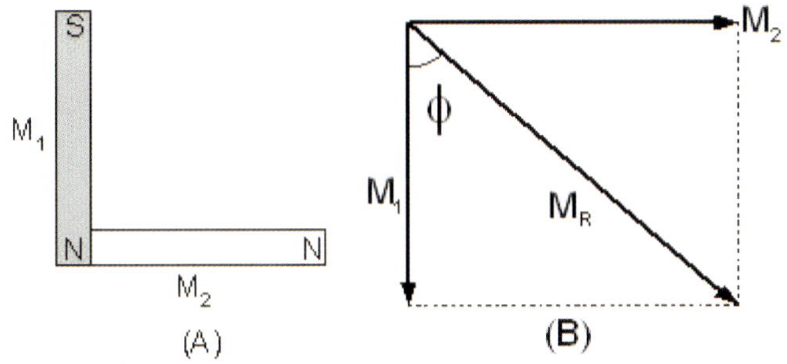

$$M_R = (M_1^2 + M_2^2 + 2M_1M_2 \cos\theta)^{1/2} \text{ with } \tan\Phi\ M_2 \sin\theta / M_1 + M_2 \cos\theta$$

and as here $M_1 = M_2 = mL$ and $\theta = 90°$, so

$$M_R = (M^2 + M^2 + 2MM \cos 90)^{1/2} = (\sqrt{2})\ mL$$

And, $\tan\Phi\ M\sin 90/M + M\cos 90 = 1$, i.e., $\Phi = \tan^{-1}(1) = 45°$

The JEE asks tricky questions on magnetic moment and resultant magnetic moment. It is crucial for the JEE aspirants to master various formulae like the resultant magnetic moment formula and the resultant of two magnetic dipole as questions on these topics are often asked.

Relation to Magnetization

The magnetic moment is a quantity that describes the magnetic strength of an entire object. Sometimes, though, it is useful or necessary to know how much of the net magnetic moment of the object is produced by a particular portion of that magnet. Therefore, it is useful to define the magnetization field **M** as:

$$\mathbf{M} = \frac{\mathbf{m}_{\Delta V}}{V_{\Delta V}},$$

where $\mathbf{m}_{\Delta V}$ and $V_{\Delta V}$ are the magnetic dipole moment and volume of a sufficiently small portion of the magnet ΔV. This equation is often represented using derivative notation such that ,

$$\mathbf{M} = \frac{d\mathbf{m}}{dV},$$

where dm is the elementary magnetic moment and dV is the volume element. The net magnetic moment of the magnet **m** therefore is

$$\mathbf{m} = \iiint \mathbf{M} dV,$$

where the triple integral denotes integration over the volume of the magnet. For uniform magnetization (where both the magnitude and the direction of **M** is the same for the entire magnet (such as a straight bar magnet) the last equation simplifies to:

$$\mathbf{m} = \mathbf{M}V,$$

where V is the volume of the bar magnet.

The magnetization is often not listed as a material parameter for commercially available ferromagnetic materials, though. Instead the parameter that is listed is residual flux density (or remanence), denoted $\mathbf{B}_r$. The formula needed in this case to calculate $\mathbf{m}$ in (units of A•m²) is:

$$\mathbf{m} = \frac{1}{\mu_o}\mathbf{B}_r V,$$

where:

- $\mathbf{B}_r$ is the residual flux density, expressed in teslas.
- V is the volume (m³) of the magnet.
- $\mu_o = 4\pi \cdot 10^{-7}$ H/m is the permeability of vacuum.

Models

The preferred classical explanation of a magnetic moment has changed over time. Before the 1930s, textbooks explained the moment using hypothetical magnetic point charges. Since then, most have defined it in terms of Ampèrian currents. In magnetic materials, the cause of the magnetic moment are the spin and orbital angular momentum states of the electrons, and varies depending on whether atoms in one region are aligned with atoms in another.

Magnetic Pole Model

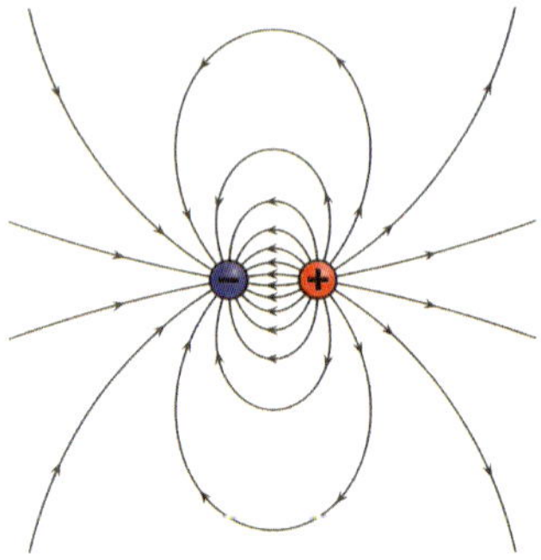

An electrostatic analog for a magnetic moment: two opposing charges separated by a finite distance.

The sources of magnetic moments in materials can be represented by poles in analogy to electrostatics. This is sometimes known as the Gilbert model. In this model, a small magnet is modeled by a pair of magnetic poles of equal magnitude but opposite polarity. Each pole is the source of magnetic force which weakens with distance. Since magnetic poles always come in pairs, their forces partially cancel each other because while one pole pulls, the other repels. This cancellation is greatest when the poles are close to each other i.e. when the bar magnet is short. The magnetic force produced by a bar magnet, at a given point in space, therefore depends on two factors: the strength p of its poles (*magnetic pole strength*), and the vector l separating them. The magnetic dipole moment m is related to the fictitious poles as

$$\mathbf{m} = pl.$$

It points in the direction from South to North pole. The analogy with electric dipoles should not be taken too far because magnetic dipoles are associated with angular momentum (see Relation to angular momentum). Nevertheless, magnetic poles are very useful for magnetostatic calculations, particularly in applications to ferromagnets. Practitioners using the magnetic pole approach generally represent the magnetic field by the irrotational field H, in analogy to the electric field E.

Amperian Loop Model

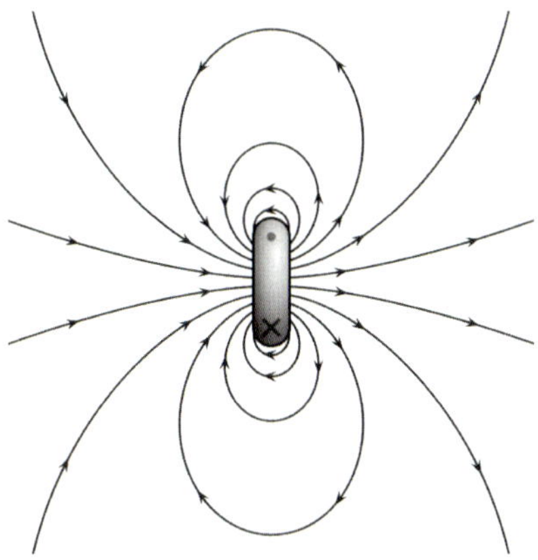

The Amperian loop model: A current loop (ring) that goes into the page at the x and comes out at the dot produces a B-field (lines). The north pole is to the right and the south to the left.

After Hans Christian Ørsted discovered that electric currents produce a magnetic field and André-Marie Ampère discovered that electric currents attract and repel each other similar to magnets, it was natural to hypothesize that all magnetic fields are due to electric current loops. In this model developed by Ampère, the elementary magnetic dipole that makes up all magnets is a sufficiently small amperian loop of current I. The dipole moment of this loop is

$$\mathbf{m} = I\mathbf{S},$$

where S is the area of the loop. The direction of the magnetic moment is in a direction normal to the area enclosed by the current consistent with the direction of the current using the right hand rule.

Localized Current Distributions

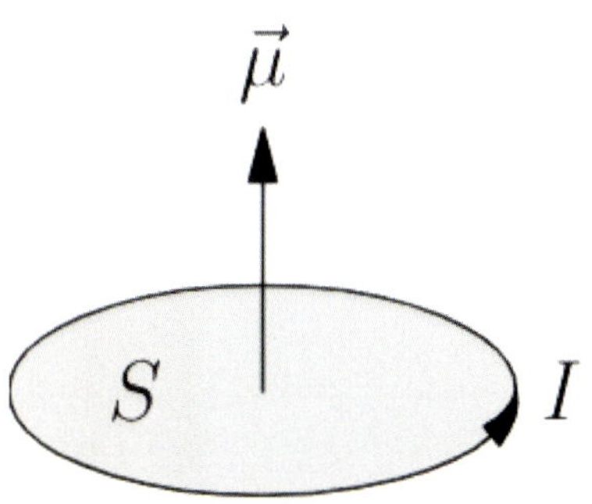

Moment m of a planar current having magnitude I and enclosing an area S

The magnetic dipole moment can be calculated for a localized (does not extend to infinity) current distribution assuming that we know all of the currents involved. Conventionally, the derivation starts from a multipole expansion of the vector potential. This leads to the definition of the magnetic dipole moment as:

$$\mathbf{m} = \tfrac{1}{2}\iiint_V \mathbf{r} \times \mathbf{j}\, dV,$$

where × is the vector cross product, r is the position vector, and j is the electric current density and the integral is a volume integral. When the current density in the integral is replaced by a loop of current I in a plane enclosing an area S then the volume integral becomes a line integral and the resulting dipole moment becomes

$$\mathbf{m} = I\mathbf{S},$$

which is how the magnetic dipole moment for an Amperian loop is derived.

Practitioners using the current loop model generally represent the magnetic field by the solenoidal field B, analogous to the electrostatic field D.

Magnetic moment of a solenoid

Image of a solenoid

A generalization of the above current loop is a coil, or solenoid. Its moment is the vector sum of the moments of individual turns. If the solenoid has N identical turns (single-layer winding) and vector area S,

$$\mathbf{m} = NI\mathbf{S}.$$

Quantum mechanical model

When calculating the magnetic moments of materials or molecules on the microscopic level it is often convenient to use a third model for the magnetic moment that exploits the linear relationship between the angular momentum and the magnetic moment of a particle. While this relation is straight forward to develop for macroscopic currents using the amperian loop model , neither the magnetic pole model nor the amperian loop model truly represents what is occurring at the atomic and molecular levels. At that level quantum mechanics must be used. Fortunately, the linear relationship between the magnetic dipole moment of a particle and its angular momentum still holds; although it is different for each particle. Further, care must be used to distinguish between the intrinsic angular momentum (or spin) of the particle and the particle's orbital angular momentum.

Effects of an External Magnetic Field

Torque on a Moment

The torque $\hat{\mathbf{o}}$ on an object having a magnetic dipole moment $\mathbf{m}$ in a uniform magnetic field $\mathbf{B}$ is:

$$\hat{\mathbf{o}} = \mathbf{m} \times \mathbf{B}.$$

This is valid for the moment due to any localized current distribution provided that the magnetic field is uniform. For non-uniform B the equation is also valid for the torque about the center of the magnetic dipole provided that the magnetic dipole is small enough.

An electron, nucleus, or atom placed in a uniform magnetic field will precess with a frequency known as the Larmor frequency.

Force on a Moment

A magnetic moment in an externally produced magnetic field has a potential energy U:

$$U = -\mathbf{m} \cdot \mathbf{B}$$

In a case when the external magnetic field is non-uniform, there will be a force, proportional to the magnetic field gradient, acting on the magnetic moment itself. There are two expressions for the force acting on a magnetic dipole, depending on whether the model used for the dipole is a current loop or two monopoles (analogous to the electric dipole). The force obtained in the case of a current loop model is

$$\mathbf{F}_{\text{loop}} = \nabla(\mathbf{m} \cdot \mathbf{B}).$$

In the case of a pair of monopoles being used (i.e. electric dipole model), the force is

$$\mathbf{F}_{\text{dipole}} = (\mathbf{m} \cdot \nabla)\mathbf{B}.$$

And one can be put in terms of the other via the relation

$$\mathbf{F}_{\text{loop}} = \mathbf{F}_{\text{dipole}} + \mathbf{m} \times (\nabla \times \mathbf{B}).$$

In all these expressions m is the dipole and B is the magnetic field at its position. Note that if there are no currents or time-varying electrical fields $\nabla \times B = 0$ and the two expressions agree.

Magnetism

In addition, an applied magnetic field can change the magnetic moment of the object itself; for example by magnetizing it. This phenomenon is known as magnetism. An applied magnetic field can flip the magnetic dipoles that make up the material causing both paramagnetism and ferromagnetism. Too, the magnetic field can affect the currents that create the magnetic fields (such as the atomic orbits) which causes diamagnetism.

Effects on Environment

Magnetic Field of a Magnetic Moment

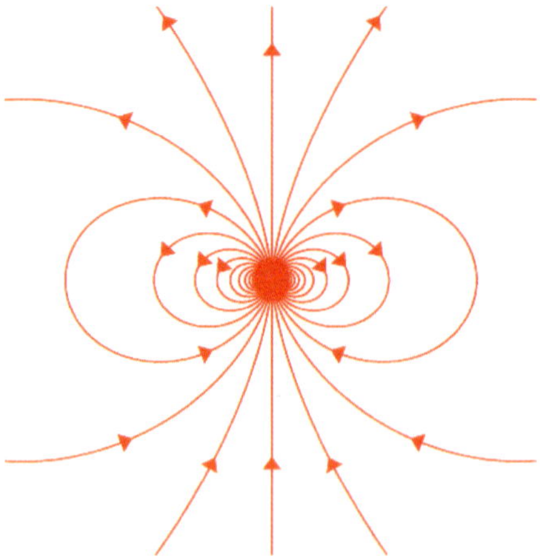

Magnetic field lines around a "magnetostatic dipole". The magnetic dipole itself is located in the center of the figure, seen from the side, and pointing upward.

Any system possessing a net magnetic dipole moment m will produce a dipolar magnetic field in the space surrounding the system. While the net magnetic field produced by the system can also have higher-order multipole components, those will drop off with distance more rapidly, so that only the dipole component will dominate the magnetic field of the system at distances far away from it.

The magnetic field of a magnetic dipole depends on the strength and direction of a magnet's magnetic moment $\mathbf{m}$ but drops off as the cube of the distance such that:

$$\mathbf{H}(\mathbf{r}) = \frac{1}{4\pi}\left(\frac{3\mathbf{r}(\mathbf{m}\cdot\mathbf{r})}{|\mathbf{r}|^5} - \frac{\mathbf{m}}{|\mathbf{r}|^3}\right),$$

where $\mathbf{H}$ is the magnetic field produced by the magnet and $\mathbf{r}$ is a vector from the center of the magnetic dipole to the location where the magnetic field is measured. The inverse cube nature of this equation is more readily seen by expressing the location vector $\mathbf{r}$ as the product of its magnitude times the unit vector in its direction ($\mathbf{r} = |\mathbf{r}|\mathbf{n}$) so that:

$$\mathbf{H}(\mathbf{r}) = \frac{1}{4\pi}\frac{3\mathbf{n}(\mathbf{n}\cdot\mathbf{m}) - \mathbf{m}}{|\mathbf{r}|^3}.$$

The equivalent equations for the magnetic $\mathbf{B}$-field are the same except for a multiplicative factor of $\mu_0 = 4\pi\times10^{-7}$ H/m, where μ_0 is known as the vacuum permeability. For example:

$$\mathbf{B}(\mathbf{r}) = \frac{\mu_0}{4\pi}\frac{3\mathbf{n}(\mathbf{n}\cdot\mathbf{m}) - \mathbf{m}}{|\mathbf{r}|^3}.$$

Forces between Two Magnetic Dipoles

The force exerted by a dipole loop with moment m_1 on another with moment m_2 is

$$\mathbf{F} = \nabla(\mathbf{m}_2 \cdot \mathbf{B}_1),$$

where B_1 is the magnetic field due to moment m_1. The result of calculating the gradient is

$$\mathbf{F}(\mathbf{r},\mathbf{m}_1,\mathbf{m}_2) = \frac{3\mu_0}{4\pi|\mathbf{r}|^4}\left(\mathbf{m}_2(\mathbf{m}_1\cdot\hat{\mathbf{r}}) + \mathbf{m}_1(\mathbf{m}_2\cdot\hat{\mathbf{r}}) + \hat{\mathbf{r}}(\mathbf{m}_1\cdot\mathbf{m}_2) - 5\hat{\mathbf{r}}(\mathbf{m}_1\cdot\hat{\mathbf{r}})(\mathbf{m}_2\cdot\hat{\mathbf{r}})\right),$$

where $\hat{r}$ is the unit vector pointing from magnet 1 to magnet 2 and r is the distance. An equivalent expression is

$$\mathbf{F} = \frac{3\mu_0}{4\pi|\mathbf{r}|^4}\left((\hat{\mathbf{r}}\times\mathbf{m}_1)\times\mathbf{m}_2 + (\hat{\mathbf{r}}\times\mathbf{m}_2)\times\mathbf{m}_1 - 2\hat{\mathbf{r}}(\mathbf{m}_1\cdot\mathbf{m}_2) + 5\hat{\mathbf{r}}((\hat{\mathbf{r}}\times\mathbf{m}_1)\cdot(\hat{\mathbf{r}}\times\mathbf{m}_2))\right).$$

The force acting on m_1 is in the opposite direction.

Torque of One Magnetic Dipole on Another

The torque of magnet 1 on magnet 2 is

$$\tau = m_2 \times B_1.$$

Relation to Angular Momentum

The magnetic moment has a close connection with angular momentum called the *gyromagnetic effect*. This effect is expressed on a macroscopic scale in the Einstein-de Haas effect, or "rotation by magnetization," and its inverse, the Barnett effect, or "magnetization by rotation." Further, a torque applied to a relatively isolated magnetic dipole such as an atomic nucleus can cause it to precess (rotate about the axis of the applied field). This phenomenon is used in nuclear magnetic resonance.

Viewing a magnetic dipole as current loop brings out the close connection between magnetic moment and angular momentum. Since the particles creating the current (by rotating around the loop) have charge and mass, both the magnetic moment and the angular momentum increase with the rate of rotation. The ratio of the two is called the gyromagnetic ratio or γ so that:

$$\mathbf{m} = \gamma\mathbf{L},$$

where $\mathbf{L}$ is the angular momentum of the particle or particles that are creating the magnetic moment.

In the amperian loop model, which applies for macroscopic currents, the gyromagnetic ration is one half of the charge-to-mass ratio. This can be shown as follows. The angular momentum of a moving charged particle is defined as:

$$\mathbf{L} = \mathbf{r}\times\mathbf{p} = \mu\mathbf{r}\times\mathbf{v},$$

where μ is the mass of the particle and v is the particles velocity. The angular momentum of the very large number of charged particles that make up a current therefore is:

$$\mathbf{L} = \iiint_V \mathbf{r} \times (\rho \mathbf{v}) \mathrm{d}V,$$

where ρ is the mass density of the moving particles. By convention the direction of the cross product is given by the right hand grip rule.

This is similar to the magnetic moment created by the very large number of charged particles that make up that current:

$$\mathbf{m} = \tfrac{1}{2} \iiint_V \mathbf{r} \times (\rho_Q \mathbf{v}) \mathrm{d}V,$$

where $\mathbf{j} = \rho_Q \mathbf{v}$ and ρ_Q is the charge density of the moving charged particles.

Comparing the two equations results in:

$$\mathbf{m} = \frac{e}{2\mu} \mathbf{L},$$

where e is the charge of the particle and μ is the mass of the particle.

Even though atomic particles cannot be accurately described as orbiting (and spinning) charge distributions of uniform charge-to-mass ratio, this general trend can be observed in the atomic world so that:

$$\mathbf{m} = g \frac{e}{2\mu} \mathbf{L},$$

where the g-factor depends on the particle and configuration. For example the g-factor for the magnetic moment due to an electron orbiting a nucleus is one while the g-factor for the magnetic moment of electron due to its intrinsic angular momentum (spin) is a little larger than 2. The g-factor of atoms and molecules must account for the orbital and intrinsic moments of its electrons and possibly the intrinsic moment of its nuclei as well.

In the atomic world the angular momentum (spin) of a particle is an integer (or half-integer in the case of spin) multiple of the reduced Planck constant $\hbar$. This is the basis for defining the magnetic moment units of Bohr magneton (assuming charge-to-mass ratio of the electron) and nuclear magneton (assuming charge-to-mass ratio of the proton).

Atoms, Molecules, and Elementary Particles

Fundamentally, contributions to any system's magnetic moment may come from sources of two kinds: motion of electric charges, such as electric currents; and the intrinsic magnetism of elementary particles, such as the electron.

Contributions due to the sources of the first kind can be calculated from knowing the distribution of all the electric currents (or, alternatively, of all the electric charges and their velocities) inside the system, by using the formulas below. On the other hand, the magnitude of each elementary particle's intrinsic magnetic moment is a fixed number, often measured experimentally to a great precision. For example, any electron's magnetic moment is measured to be -9.284764×10^{-24} J/T. The direction of the magnetic moment of any elementary particle is entirely determined by the direction of its spin, with the negative value indicating that any electron's magnetic moment is antiparallel to its spin.

The net magnetic moment of any system is a vector sum of contributions from one or both types of sources. For example, the magnetic moment of an atom of hydrogen-1 (the lightest hydrogen isotope, consisting of a proton and an electron) is a vector sum of the following contributions:

1. the intrinsic moment of the electron,
2. the orbital motion of the electron around the proton,
3. the intrinsic moment of the proton.

Similarly, the magnetic moment of a bar magnet is the sum of the contributing magnetic moments, which include the intrinsic and orbital magnetic moments of the unpaired electrons of the magnet's material and the nuclear magnetic moments.

Magnetic Moment of an Atom

For an atom, individual electron spins are added to get a total spin, and individual orbital angular momenta are added to get a total orbital angular momentum. These two then are added using angular momentum coupling to get a total angular momentum. For an atom with no nuclear magnetic moment, the magnitude of the atomic dipole moment is then,

$$m_{\text{Atom}} = g_J \mu_B \sqrt{j(j+1)}$$

where j is the total angular momentum quantum number, g_J is the Landé g-factor, and μ_B is the Bohr magneton. The component of this magnetic moment along the direction of the magnetic field is then,

$$m_{\text{Atom}}(z) = -m g_J \mu_B$$

where m is called the magnetic quantum number or the *equatorial* quantum number, which can take on any of $2j + 1$ values:

$$-j, -(j-1) \cdots 0 \cdots +(j-1), +j\,.$$

The negative sign occurs because electrons have negative charge.

Due to the angular momentum, the dynamics of a magnetic dipole in a magnetic field differs from that of an electric dipole in an electric field. The field does exert a torque on the magnetic dipole tending to align it with the field. However, torque is proportional to rate of change of angular momentum, so precession occurs: the direction of spin changes. This behavior is described by the Landau–Lifshitz–Gilbert equation:

$$\frac{1}{\gamma}\frac{d\mathbf{m}}{dt} = \mathbf{m}\times\mathbf{H}_{\text{eff}} - \frac{\lambda}{\gamma m}\mathbf{m}\times\frac{d\mathbf{m}}{dt}$$

where γ is the gyromagnetic ratio, m is the magnetic moment, λ is the damping coefficient and H_{eff} is the effective magnetic field (the external field plus any self-induced field). The first term describes precession of the moment about the effective field, while the second is a damping term related to dissipation of energy caused by interaction with the surroundings.

Magnetic Moment of an Electron

Electrons and many elementary particles also have intrinsic magnetic moments, an explanation of which requires a quantum mechanical treatment and relates to the intrinsic angular momentum of the particles as discussed in the article Electron magnetic moment. It is these intrinsic magnetic moments that give rise to the macroscopic effects of magnetism, and other phenomena, such as electron paramagnetic resonance.

The magnetic moment of the electron is

$$\mathbf{m}_S = -\frac{g_S \mu_B \mathbf{S}}{\hbar},$$

where μ_B is the Bohr magneton, S is electron spin, and the g-factor g_S is 2 according to Dirac's theory, but due to quantum electrodynamic effects it is slightly larger in reality: 2.00231930436. The deviation from 2 is known as the anomalous magnetic dipole moment.

Again it is important to notice that m is a negative constant multiplied by the spin, so the magnetic moment of the electron is antiparallel to the spin. This can be understood with the following classical picture: if we imagine that the spin angular momentum is created by the electron mass spinning around some axis, the electric current that this rotation creates circulates in the opposite direction, because of the negative charge of the electron; such current loops produce a magnetic moment which is antiparallel to the spin. Hence, for a positron (the anti-particle of the electron) the magnetic moment is parallel to its spin.

Magnetic Moment of a Nucleus

The nuclear system is a complex physical system consisting of nucleons, i.e., protons and neutrons. The quantum mechanical properties of the nucleons include the spin among others. Since the electromagnetic moments of the nucleus depend on the spin of the individual nucleons, one can look at these properties with measurements of nuclear moments, and more specifically the nuclear magnetic dipole moment.

Most common nuclei exist in their ground state, although nuclei of some isotopes have long-lived excited states. Each energy state of a nucleus of a given isotope is characterized by a well-defined magnetic dipole moment, the magnitude of which is a fixed number, often measured experimentally to a great precision. This number is very sensitive to the individual contributions from nucleons, and a measurement or prediction of its value can reveal important information about the content of the nuclear wave function. There are several theoretical models that predict the value of the magnetic dipole moment and a number of experimental techniques aiming to carry out measurements in nuclei along the nuclear chart.

Magnetic Moment of a Molecule

Any molecule has a well-defined magnitude of magnetic moment, which may depend on the molecule's energy state. Typically, the overall magnetic moment of a molecule is a combination of the following contributions, in the order of their typical strength:

- magnetic moments due to its unpaired electron spins (paramagnetic contribution), if any
- orbital motion of its electrons, which in the ground state is often proportional to the external magnetic field (diamagnetic contribution)
- the combined magnetic moment of its nuclear spins, which depends on the nuclear spin configuration.

Examples of Molecular Magnetism

- The dioxygen molecule, O_2, exhibits strong paramagnetism, due to unpaired spins of its outermost two electrons.
- The carbon dioxide molecule, CO_2, mostly exhibits diamagnetism, a much weaker magnetic moment of the electron orbitals that is proportional to the external magnetic field. The nuclear magnetism of a magnetic isotope such as ^{13}C or ^{17}O will contribute to the molecule's magnetic moment.
- The dihydrogen molecule, H_2, in a weak (or zero) magnetic field exhibits nuclear magnetism, and can be in a para- or an ortho- nuclear spin configuration.
- Many transition metal complexes are magnetic. The spin-only formula is a good first approximation for high-spin complexes of first-row transition metals.

Number of unpaired electrons	**Spin-only moment (μ_B)**
1	1.73
2	2.83
3	3.87
4	4.90
5	5.92

Elementary Particles

In atomic and nuclear physics, the Greek symbol μ represents the magnitude of the magnetic moment, often measured in Bohr magnetons or nuclear magnetons, associated with the intrinsic spin of the particle and/or with the orbital motion of the particle in a system. Values of the intrinsic magnetic moments of some particles are given in the table below:

Intrinsic magnetic moments and spins of some elementary particles		
Particle	**Magnetic dipole moment (10^{-27} J·T^{-1})**	**Spin quantum number (dimensionless)**
electron (e^-)	−9284.764	1/2
proton (H^+)	−0 014.106067	1/2
neutron (n)	0 00−9.66236	1/2
muon (μ^-)	0 0−44.904478	1/2
deuteron ($^2H^+$)	−0 004.3307346	1
triton ($^3H^+$)	−0 015.046094	1/2
helion ($^3He^{2+}$)	0 0−10.746174	1/2
alpha particle ($^4He^{2+}$)	−0 000	0

Magnetization

Magnetization is the density of magnetic dipole moments that are induced in a magnetic material when it is placed near a magnet. The magnetic effects of a material can also be induced by passing an electrical current through the material; the magnetic effect is caused by the motion of electrons in atoms, or the spin of the electrons or the nuclei.

Magnetization is also known as magnet polarization.

All matter is built up out of atoms, and each atom consists of electrons in motion. The currents associated with this motion are termed *atomic currents*. Each atomic current is a tiny closed circuit of atomic dimensions, and may therefore be appropriately described as a magnetic dipole. If the atomic currents of a given atom all flow in the same plane then the atomic dipole moment is directed normal to the plane (in the sense given by the right-hand rule), and its magnitude is the product of the total circulating current and the area of the current loop. More generally, if $\mathrm{j}(\mathrm{r})$ is the atomic current density at the point r then the magnetic moment of the atom is,

$$\mathrm{m} = \frac{1}{2}\int \mathrm{r} \times \mathrm{j} d^3 \mathrm{r},$$

where the integral is over the volume of the atom. If there are N such atoms or molecules per unit volume then the *magnetization* M (*i.e.*, the magnetic dipole moment per unit volume) is given by $\mathrm{M} = N\mathrm{m}$. More generally,

$$\mathrm{M}(\mathrm{r}) = \sum_i N_i \langle \mathrm{m}_i \rangle,$$

where $\langle \mathrm{m}_i \rangle$ is the average magnetic dipole moment of the ith type of molecule in the vicinity of point r, and N_i is the average number of such molecules per unit volume at r.

Consider a general medium which is made up of molecules which are polarizable and possess a net magnetic moment. It is easily demonstrated that any circulation in the magnetization field $\mathrm{M}(\mathrm{r})$ gives rise to an effective current density j_m in the medium. In fact,

$j_m = \nabla \times M$. This current density is called the magnetization current density, and is usually distinguished from the true current density, j_t, which represents the convection of free charges in the medium. In fact, there is a third type of current called a polarization current, which is due to the apparent convection of bound charges. It is easily demonstrated that the polarization current density, j_p, is given by,

$$j_p = \frac{\partial P}{\partial t}.$$

Thus, the total current density, j, in the medium takes the form

$$j = j_t + \nabla \times M + \frac{\partial P}{\partial t}.$$

It must be emphasized that all terms on the right-hand side of the above equation represent real physical currents, although only the first term is due to the motion of real charges (over more than atomic dimensions).

The differential form of Ampère's law is

$$\nabla \times B = \mu_0 j + \mu_0 \epsilon_0 \frac{\partial E}{\partial t},$$

which can also be written

$$\nabla \times B = \mu_0 j_t + \mu_0 \nabla \times M + \mu_0 \frac{\partial D}{\partial t},$$

where use has been made of the definition $D = \epsilon_0 E + P$. The above expression can be rearranged to give

$$\nabla \times H = j_t + \frac{\partial D}{\partial t},$$

where,

$$H = \frac{B}{\mu_0} - M$$

is termed the *magnetic intensity*, and has the same dimensions as M (*i.e.*, magnetic dipole moment per unit volume). In a steady-state situation, Stokes' theorem tell us that

$$\oint_C H \cdot dl = \int_S j_t \cdot dS.$$

In other words, the line integral of H around some closed curve is equal to the flux of true current through any surface attached to that curve. Unlike the magnetic field B (which specifies the force $q v \times B$ acting on a charge q moving with velocity v), or the magnetization M (the magnetic dipole moment per unit volume), the magnetic intensity H has no clear physical meaning. The only reason for introducing it is that it enables us to calculate fields in the presence of magnetic materials without first having to know the distribution of magnetization currents. However, this is only possible if we possess a constitutive relation connecting B and H.

Magnetic Flux

The number of magnetic lines of forces set up in a magnetic circuit is called Magnetic Flux. It is analogous to electric current I in an electric circuit. Its SI unit is Weber (Wb) and its CGS unit is Maxwell. It is denoted by φ_m. The magnetic flux measures through flux meter. The fluxmeter has to measure coil which measures the variation of voltage to measure the flux.

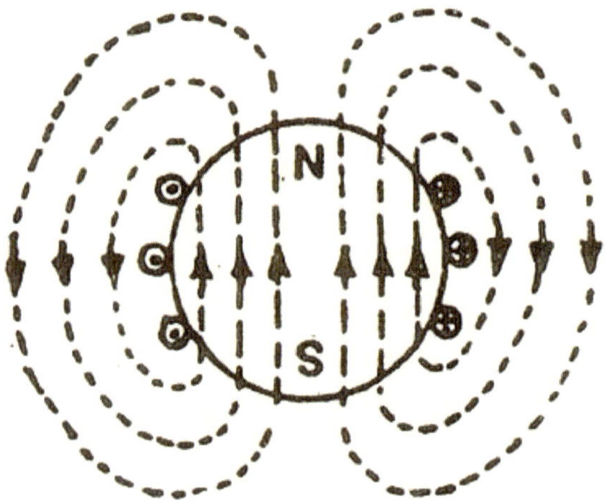

Net numbers of lines passing through the surface are called magnetic lines of forces.

$$\varphi_B = B.S$$

If the magnetic field is constant than the magnetic flux passing through a surface (S) is

$$\varphi_B = B.S\,COS\theta$$

where

B – magnitude of magnetic field

S – area of surface

θ – angle between the magnetic field lines and perpendicular distance normal to the surface area

Magnetic flux for closed surface

$$\varphi_B = \iint s B.ds = 0$$

Magnetic flux for open surface,

$$E = \int (\mathrm{E} + \mathrm{v} \times \mathrm{B})\, \mathrm{dt}$$

$$E = d\varphi_{\mathrm{B}} \over \mathrm{dt}$$

Where

E – electromotive force

v – velocity of the boundary

E – electric field

B – magnetic field

$\varnothing_{\mathrm{B}}$ – magnetic flux through the open surface

dl – infinitesimal vector element

The magnetic flux through a closed surface is always zero, but in the open surface, it is not zero.

Properties of Magnetic Flux

1. Magnetic flux of a filed is considered as the total number of magnetic lines of force in the field. These are also called magnetic flux lines.
2. Each magnetic flux line is closed loop.
3. Each magnetic flux line starts from north pole of a magnet and comes to the south pole through the field and continues from south pole to north pole in the body of the magnet.
4. No two flux lines cross each other.
5. Two similar lines of force travel side by side but repeal each other.
6. The lines of force are stretched like elastic cord.

Magnetic Flux Density

The number of magnetic lines of force passing through a unit area surface perpendicular to the magnetic field is called magnetic flux density. If total φ Weber flux perpendicularly through a surface of area A m^2, Magnetic flux density of the field would be,

$$B = \frac{\phi}{A} \quad Wb/m^2$$

We generally represent magnetic flux density by capital letter B.

Measure Magnetic Flux

The SI unit of magnetic flux is the Weber (named after German physicist and co-inventor of the telegraph Wilhelm Weber) and the unit has the symbol Wb.

Because the magnetic flux is just a way of expressing the magnetic field in a given area, it can be measured with a *magnetometer* in the same way as the magnetic field. For example, suppose a small magnetometer probe is moved around (without rotating) inside a 0.5 m² area near a large sheet of magnetic material and indicates a constant reading of 5 mT. The magnetic flux through the area is then $(5{\cdot}10^{-3}\ \mathrm{T}){\cdot}(0.5\ \mathrm{m}^2)=0.0025$ Wb. In the event that the magnetic field reading changes with position, it would be necessary to find the average reading.

A related term that you may come across is the *magnetic flux density*. This is measured in Wb/ m2. Because we are dividing flux by area we could also directly state the units of flux density in Tesla. In fact, the term magnetic flux density is often used synonymously with the magnitude of the magnetic field.

Exercise 2:

Figure below shows a map of a non-uniform magnetic field measured near a sheet of magnetic material. If the green line represents a loop of wire, what is the magnetic flux through the loop?

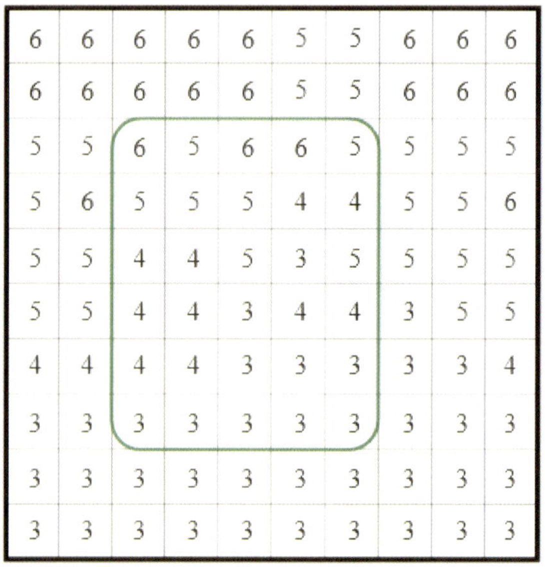

All numbers represent mT out of the page.
Grid lines drawn with 1 cm spacing.

A map of magnetic field measurements around a loop of wire (green).

To find the magnetic flux we need to know the area of the loop and the magnitude of the magnetic field intersecting it. In this case the field is not uniform so we need to take an average over the area enclosed by the coil.

$$B_{avg} = \frac{123\ \mathrm{mT}}{30}$$
$$= 4.1\ \mathrm{mT}$$
$$\Phi = AB_{avg}$$
$$= (0.05\,\mathrm{m}.\,0.06\,\mathrm{m}).(4.1\ \mathrm{mT})$$
$$= 0.0123\,\mathrm{m\,Wb}$$

Why is this useful?

There are a couple of reasons why the description of magnetic flux can be more useful than that of a magnetic field directly.

1. When a coil of wire is moved through a magnetic field a voltage is generated which depends on the magnetic flux through the area of the coil. This is described by *Faraday's law* and is explored in our topic on Faraday's law. Electric motors and generators apply Faraday's law to coils which rotate in a magnetic field as depicted in figure below. In this example the flux changes as the coil rotates. The description of magnetic flux allows engineers to easily calculate the voltage generated by an electric generator even when the magnetic field is complicated.

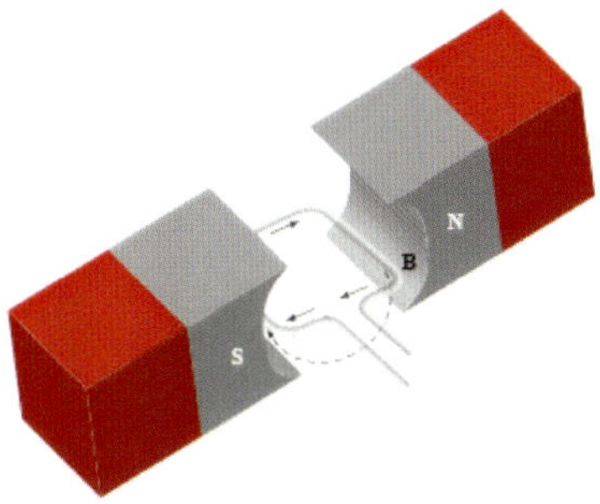

Simplified diagram of a rotating coil in an electric generator (public domain).

2. Although we have thus-far only concerned ourselves with magnetic flux measured for a simple flat test-area, we can make our test-area a surface of any shape we like. In-fact, we can use a *closed surface* such as a sphere which encloses a region of interest. Closed surfaces are particularly interesting to physicists because of *Gauss's law for magnetism*. Because magnets always have two poles there is no possibility (as far as we know) that there is a magnetic monopole inside a closed surface. This means that the net magnetic flux through such a closed surface is always zero and therefore all the magnetic field lines going into the closed surface are exactly balanced by field lines coming out. This fact is useful for simplifying magnetic field problems.

Magnetic Flux Around a Current-carrying Wire

Exercise

Figure below shows a square loop of wire placed near a current carrying wire. Using the dimensions shown in the figure, find the magnetic flux through a coil. If you don't know how to calculate the magnetic field around a wire, review our topic on the magnetic field. Hint: it may be useful to plot the magnetic field vs vertical distance from the wire.

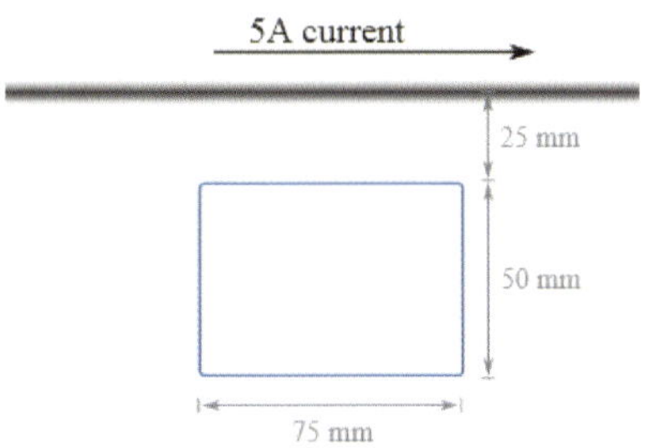

Magnetic flux through a loop near a straight current carrying wire.

Recall from on the magnetic field, the equation for magnetic field at a distance r from a long straight wire carrying current I is

$$B = \frac{\mu_0 I}{2\pi r}.$$

Here μ_0 is the permeability of free space. $\mu_0 = 4\pi.10^{-7}$ T. m/A.

If we plot the magnetic field as a function of vertical distance we can use the area under the curve representing the area intersected by the coil to find the flux. There are many ways to evaluate this; the most accurate is to use calculus. However, a simple graphical evaluation is sufficient for many purposes. Figure shows a plot with the gray area representing the area intersected by the coil.

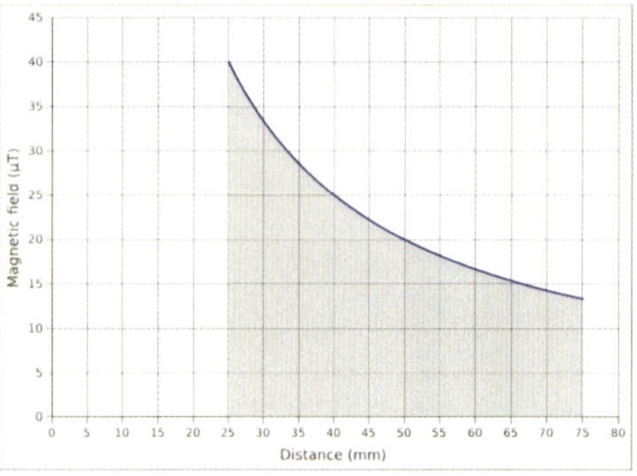

Plot of magnetic field variation with distance in exercise 1.

Counting the gray squares in figure above, we have approximately 44.5 squares. So the total area is,

$$44.5.(5.10^{-3}\text{ m}).(5.10^{-6}\text{ T}) = 1.1125.10^{-6}\text{ Tm}$$

The area directly gives us magnetic field times distance but we need magnetic field times area. The field is not varying along the horizontal direction so we can simply multiply by the width of the coil.

$$\Phi = (1.1125.10^{-6}\text{ Tm}).(75.10^{-3}\text{ m})$$
$$\simeq 8.10^{-8}\text{ Wb}$$

Magnetic Flux Through a Closed Surface

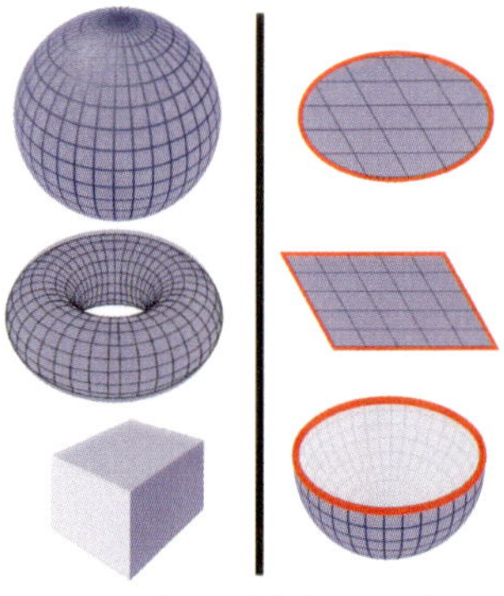

Some examples of closed surfaces (left) and open surfaces (right). Left: Surface of a sphere, surface of a torus, surface of a cube. Right: Disk surface, square surface, surface of a hemisphere. (The surface is blue, the boundary is red.)

Gauss's law for magnetism, which is one of the four Maxwell's equations, states that the total magnetic flux through a closed surface is equal to zero. (A "closed surface" is a surface that completely encloses a volume(s) with no holes.) This law is a consequence of the empirical observation that magnetic monopoles have never been found.

In other words, Gauss's law for magnetism is the statement:

$$\Phi = \oiint B \,.\, dS = 0$$

for any closed surface S.

Magnetic Flux Through an Open Surface

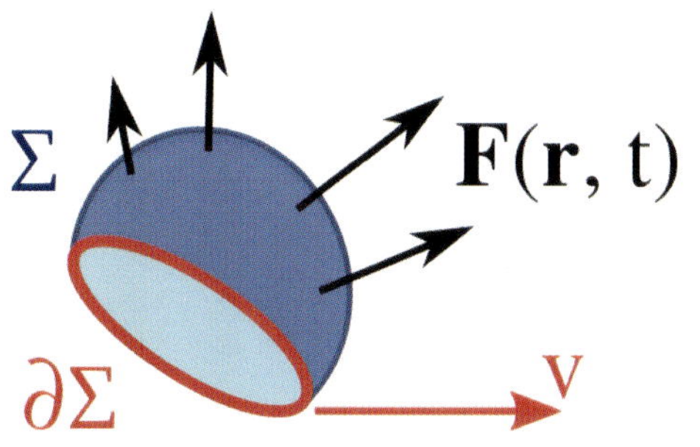

For an open surface Σ, the electromotive force along the surface boundary, ∂Σ, is a combination of the boundary's motion, with velocity v, through a magnetic field B (illustrated by the generic F field in the diagram) and the induced electric field caused by the changing magnetic field.

While the magnetic flux through a closed surface is always zero, the magnetic flux through an open surface need not be zero and is an important quantity in electromagnetism.

Changing Magnetic Flux

For example, a change in the magnetic flux passing through a loop of conductive wire will cause an electromotive force, and therefore an electric current, in the loop. The relationship is given by Faraday's law:

$$\mathcal{E} = \oint_{\partial\Sigma} (\mathbf{E} + \mathbf{v} \times \mathbf{B}) \cdot d\ell = -\frac{d\Phi_B}{dt},$$

where

$\mathcal{E}$ is the electromotive force (EMF),

Φ_B is the magnetic flux through the open surface Σ,

∂Σ is the boundary of the open surface Σ; note that the surface, in general, may be in motion and deforming, and so is generally a function of time. The electromotive force is induced along this boundary.

$d\ell$ is an infinitesimal vector element of the contour ∂Σ,

v is the velocity of the boundary ∂Σ,

E is the electric field,

B is the magnetic field.

The two equations for the EMF are, firstly, the work per unit charge done against the Lorentz force in moving a test charge around the (possibly moving) surface boundary $\partial\Sigma$ and, secondly, as the change of magnetic flux through the open surface Σ. This equation is the principle behind an electrical generator.

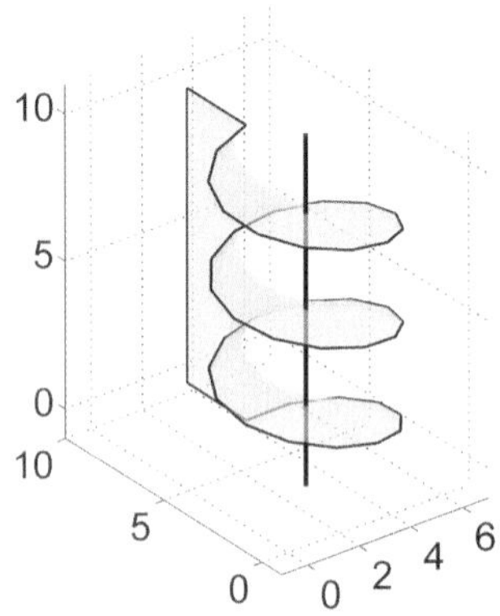

Area defined by an electric coil with three turns.

Comparison with Electric Flux

By way of contrast, Gauss's law for electric fields, another of Maxwell's equations, is

$$\Phi_E = \oiint_S \mathbf{E} \cdot d\mathbf{S} = \frac{Q}{\epsilon_0}$$

where

E is the electric field,

S is any closed surface,

Q is the total electric charge inside the surface S,

ε_0 is the electric constant (a universal constant, also called the "permittivity of free space").

Note that the flux of E through a closed surface is *not* always zero; this indicates the presence of "electric monopoles", that is, free positive or negative charges.

Biot–Savart Law

Biot-Savart law, in physics, a fundamental quantitative relationship between an electric current and the magnetic field it produces, based on the experiments in 1820 of the French scientists Jean-Baptiste Biot and Félix Savart.

An electric current flowing in a conductor, or a moving electric charge, produces a magnetic field, or a region in the space around the conductor in which magnetic forces may be detected. The value of the magnetic field at a point in the surrounding space may be considered the sum of all the contributions from each small element, or segment, of a current-carrying conductor. The Biot-Savart law states how the value of the magnetic field at a specific point in space from one short segment of current-carrying conductor depends on each factor that influences the field. In the first place, the value of the magnetic field at a point is directly proportional to both the value of the current in the conductor and the length of the current-carrying segment under consideration. The value of the field depends also on the orientation of the particular point with respect to the segment of current. If the line from the point to the short segment of current makes an angle of 90° with the current segment or lies straight out from it, the field is greatest. As this angle gets smaller, the field of the current segment diminishes, becoming zero when the point lies on a line of which the current element itself is a segment. In addition, the magnetic field at a point depends upon how far the point is from the current element. At twice the distance, the magnetic field is four times smaller, or the value of the magnetic field is inversely proportional to the square of the distance from the current element that produces it.

The Biot-Savart law is applied in a specific case by adding up the contributions to the magnetic field at a given point from the whole series of short current segments that constitute a specific conductor of whatever shape. For instance, with a very long straight wire carrying current, the value of the magnetic field at a point nearby is just directly proportional to the value of the current and inversely proportional to the perpendicular distance from the wire to the given point. *Compare* Ampère's law.

After, through observations and calculations they had derived a mathematical expression, which shows, the magnetic flux density of which dB, is directly proportional to the length of the element dl, the current I, the sine of the angle and θ between direction of the current and the vector joining a given point of the field and the current element and is inversely proportional to the square of the distance of the given point from the current element, r. This is Biot Savart law statement.

$$Hence, dB \propto \frac{Idl \sin\theta}{r^2} \quad or\ dB = k\frac{Idl \sin\theta}{r^2}$$

Where, k is a constant, depends upon the magnetic properties of the medium and system of the units employed. In SI system of unit,

$$k = \frac{\mu_0 \mu_r}{4\pi}$$

Therefore, final Biot Savart law derivation is,

$$dB = \frac{\mu_0 \mu_r}{4\pi} \times \frac{I\,dl \sin\theta}{r^2}$$

Let us consider a long wire carrying a current I and also consider a point p in the space. The wire is presented in the picture below, by red color. Let us also consider an infinitely small length of the wire dl at a distance r from the point P as shown. Here, r is a distance vector which makes an angle θ with the direction of current in the infinitesimal portion of the wire.

If you try to visualize the condition, you can easily understand the magnetic field density at the point P due to that infinitesimal length dl of the wire is directly proportional to current carried by this portion of the wire. As the current through that infinitesimal length of wire is same as the current carried by the whole wire itself, we can write,

$$dB \propto I$$

It is also very natural to think that the magnetic field density at that point P due to that infinitesimal length dl of wire is inversely proportional to the square of the straight distance from point P to center of dl. Mathematically we can write this as,

$$dB \propto \frac{1}{r^2}$$

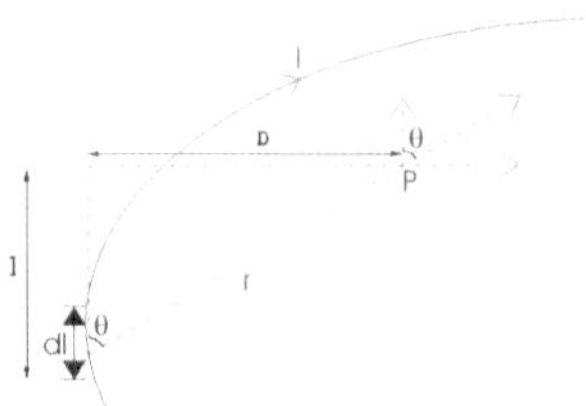

Lastly, magnetic field density at that point P due to that infinitesimal portion of wire is also directly proportional to the actual length of the infinitesimal length dl of wire. As θ be the angle between distance vector r and direction of current through this infinitesimal portion of the wire, the component of dl directly facing perpendicular to the point P is dlsinθ,

$$\textit{Hence, } dB \propto dl \sin\theta$$

Now, combining these three statements, we can write,

$$dB \propto \frac{I \cdot dl \cdot \sin\theta}{r^2}$$

This is the basic form of Biot Savart's Law Now, putting the value of constant k in the above expression, we get

$$dB = k\frac{I \cdot dl \cdot \sin\theta}{r^2}$$

$$\Rightarrow dB = \frac{\mu_0 \mu_r}{4\pi} \times \frac{I\, dl \sin\theta}{r^2}$$

Here, μ_0 used in the expression of constant k is absolute permeability of air or vacuum and it's value is $4\pi 10^{-7}$ W_b/ A-m in SI system of units. μ_r of the expression of constant k is relative permeability of the medium. Now, flux density(B) at the point P due to total length of the current carrying conductor or wire can be represented as,

$$B = \int dB \Rightarrow dB = \int \frac{\mu_0 \mu_r}{4\pi} \times \frac{I\, dl \sin\theta}{r^2} = \frac{I \mu_0 \mu_r}{4\pi} \int \frac{\sin\theta}{r^2} dl$$

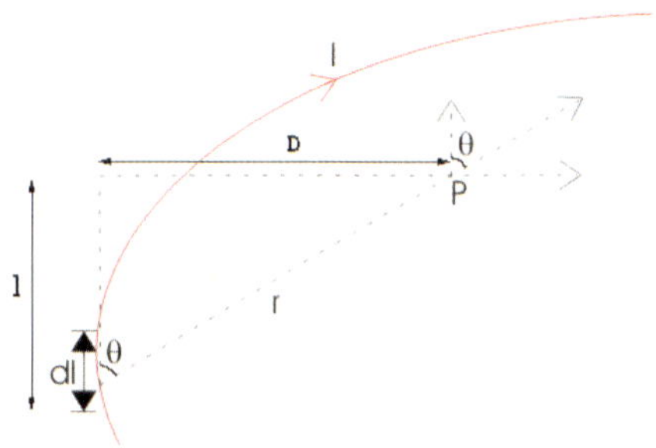

If D is the perpendicular distance of the point P form the wire, then

$$r\sin\theta = D \; or \; r = \frac{D}{\sin\theta}$$

Now, the expression of flux density B at point P can be rewritten as,

$$B = \frac{I\mu_0\mu_r}{4\pi}\int\frac{\sin\theta}{r^2}dl = \frac{I\mu_0\mu_r}{4\pi}\int\frac{\sin^3\theta}{D^2}dl$$

$$Again, \frac{1}{D} = \cot\theta \Rightarrow l\,D\cot\theta$$

As per the figure above,

$$Therefore, dl = -D\csc^2\theta d\theta$$

Finally the expression of B comes as,

$$B = \frac{I\mu_0\mu_r}{4\pi}\int\frac{\sin^3\theta}{D^2}\left[-D\csc^2\theta d\theta\right]$$
$$= -\frac{I\mu_0\mu_r}{4\pi D}\int\sin^3\theta\csc^2\theta d\theta$$
$$= -\frac{I\mu_0\mu_r}{4\pi D}\int\sin\theta d\theta$$

This angle θ depends upon the length of the wire and the position of the point P. Say for certain limited length of the wire, angle θ as indicated in the figure above varies from θ_1 to θ_2. Hence, magnetic flux density at point P due to total length of the conductor is,

$$B = \frac{I\mu_0\mu_r}{4\pi D}\int_{\theta_1}^{\theta_2}\sin\theta d\theta$$
$$= -\frac{I\mu_0\mu_r}{4\pi D}\left[-\cos\theta\right]_{\theta_1}^{\theta_2}$$
$$= \frac{I\mu_0\mu_r}{4\pi D}\left[\cos\theta_1 - \cos\theta_2\right]$$

Let's imagine the wire is infinitely long, then θ will vary from 0 to π that is $\theta_1 = 0$ to $\theta_2 = \pi$. Putting these two values in the above final expression of Biot Savart law, we get,

$$B = \frac{I\mu_0\mu_r}{4\pi D}\left[\cos 0 - \cos\pi\right] = \frac{I\mu_0\mu_r}{4\pi D}\left[1-(-1)\right] = \frac{\mu_0\mu_r}{2\pi D}I$$

Equation

Electric Currents (Along a Closed Curve/Wire)

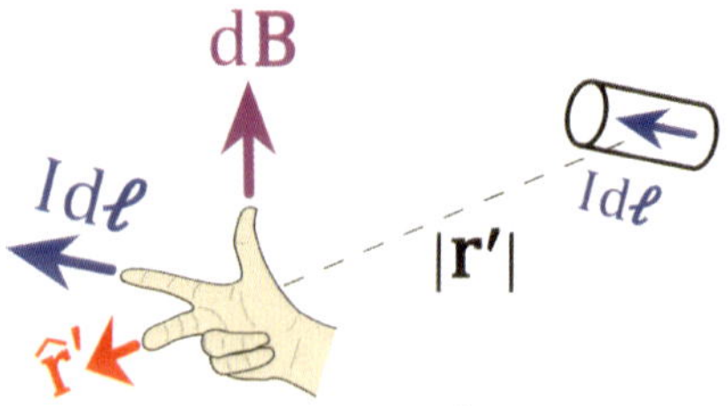

Shown are the directions of $Id\ell$, $\hat{\mathbf{r}}'$, and the value of $|\mathbf{r}'|$

The Biot–Savart law is used for computing the resultant magnetic field B at position r in 3D-space generated by a *steady* current *I* (for example due to a wire). A steady (or stationary) current is a continual flow of charges which does not change with time and the charge neither accumulates nor depletes at any point. The law is a physical example of a line integral, being evaluated over the path *C* in which the electric currents flow (e.g. the wire). The equation in SI units is

$$\mathbf{B}(\mathbf{r}) = \frac{\mu_0}{4\pi}\int_C \frac{Id\ell \times \mathbf{r}'}{|\mathbf{r}'|^3}$$

where $d\ell$ is a vector along the path C whose magnitude is the length of the differential element of the wire in the direction of *conventional current*. $\mathbf{r}' = \mathbf{r} - \ell$ is the full displacement vector from the wire element ($d\ell$) to the point at which the field is being computed ($\mathbf{r}$), and μ_0 is the magnetic constant. Alternatively:

$$\mathbf{B}(\mathbf{r}) = \frac{\mu_0}{4\pi}\int_C \frac{Id\ell \times \hat{\mathbf{r}}'}{|\mathbf{r}'|^2}$$

where $\hat{\mathbf{r}}'$ is the unit vector of $\mathbf{r}'$. The symbols in boldface denote vector quantities.

The integral is usually around a closed curve, since stationary electric currents can only flow around closed paths when they are bounded. However, the law also applies to infinitely long wires (as used in the definition of the SI unit of electric current – the Ampere).

To apply the equation, the point in space where the magnetic field is to be calculated is arbitrarily chosen ($\mathbf{r}$). Holding that point fixed, the line integral over the path of the electric current is calculated to find the total magnetic field at that point. The application of this law implicitly relies on the superposition principle for magnetic fields, i.e. the fact that the magnetic field is a vector sum of the field created by each infinitesimal section of the wire individually.

There is also a 2D version of the Biot-Savart equation, used when the sources are invariant in one direction. In general, the current need not flow only in a plane normal to the invariant direction and it is given by **J** (current density). The resulting formula is:

$$\mathbf{B}(\mathbf{r}) = \frac{\mu_0}{2\pi}\int_C \frac{(\mathbf{J}d\ell)\times\mathbf{r}'}{|\mathbf{r}'|} = \frac{\mu_0}{2\pi}\int_C (\mathbf{J}d\ell)\times\hat{\mathbf{r}}'$$

Electric Current Density (Throughout Conductor Volume)

The formulations given above work well when the current can be approximated as running through an infinitely-narrow wire. If the conductor has some thickness, the proper formulation of the Biot–Savart law (again in SI units) is:

$$\mathbf{B}(\mathbf{r}) = \frac{\mu_0}{4\pi}\iiint_V \frac{(\mathbf{J}dV)\times\mathbf{r}'}{|\mathbf{r}'|^3}$$

or, alternatively:

$$\mathbf{B}(\mathbf{r}) = \frac{\mu_0}{4\pi}\iiint_V \frac{(\mathbf{J}dV)\times\hat{\mathbf{r}}'}{|\mathbf{r}'|^2}$$

where dV is the volume element and **J** is the current density vector in that volume (in SI in units of A/m^2).

Constant Uniform Current

In the special case of a steady constant current I, the magnetic field **B** is

$$\mathbf{B}(\mathbf{r}) = \frac{\mu_0}{4\pi} I\int_C \frac{d\ell\times\mathbf{r}'}{|\mathbf{r}'|^3}$$

i.e. the current can be taken out of the integral.

Point Charge at Constant Velocity

In the case of a point charged particle q moving at a constant velocity v, Maxwell's equations give the following expression for the electric field and magnetic field:

$$\mathbf{E} = \frac{q}{4\pi\epsilon_0}\frac{1-v^2/c^2}{(1-v^2\sin^2\theta/c^2)^{3/2}}\frac{\hat{\mathbf{r}}'}{|\mathbf{r}'|^2}$$

$\mathbf{H} = \mathbf{v}\times\mathbf{D}$ or

$$\mathbf{B} = \frac{1}{c^2}\mathbf{v}\times\mathbf{E}$$

where $\hat{\mathbf{r}}'$ is the unit vector pointing from the current (non-retarded) position of the particle to the point at which the field is being measured, and θ is the angle between **v** and $\mathbf{r}'$.

When $v^2 \ll c^2$, the electric field and magnetic field can be approximated as

$$\mathbf{E} = \frac{q}{4\pi\epsilon_0} \frac{\hat{\mathbf{r}}'}{|\mathbf{r}'|^2}$$

$$\mathbf{B} = \frac{\mu_0 q}{4\pi} \mathbf{v} \times \frac{\hat{\mathbf{r}}'}{|\mathbf{r}'|^2}$$

These equations are called the "Biot–Savart law for a point charge" due to its closely analogous form to the "standard" Biot–Savart law given previously. These equations were first derived by Oliver Heaviside in 1888.

Magnetic Responses Applications

The Biot–Savart law can be used in the calculation of magnetic responses even at the atomic or molecular level, e.g. chemical shieldings or magnetic susceptibilities, provided that the current density can be obtained from a quantum mechanical calculation or theory.

Aerodynamics Applications

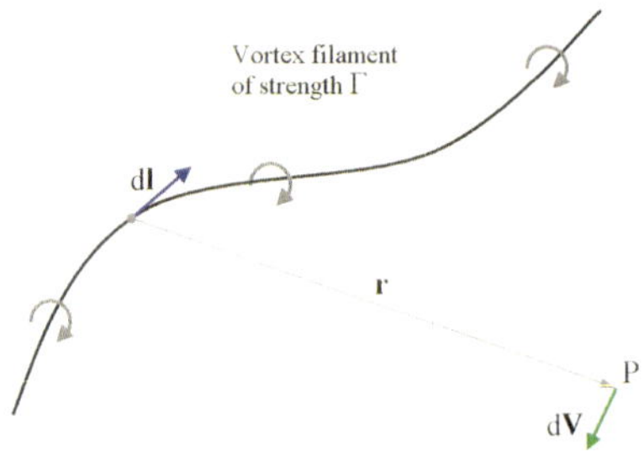

The figure shows the velocity (dV) induced at a point P by an element of vortex filament (dL) of strength Γ.

The Biot–Savart law is also used in aerodynamic theory to calculate the velocity induced by vortex lines.

In the aerodynamic application, the roles of vorticity and current are reversed in comparison to the magnetic application.

In Maxwell's 1861 paper 'On Physical Lines of Force', magnetic field strength H was directly equated with pure vorticity (spin), whereas B was a weighted vorticity that was weighted for the density of the vortex sea. Maxwell considered magnetic permeability μ to be a measure of the density of the vortex sea. Hence the relationship,

1. Magnetic induction current

   $$\mathbf{B} = \mu \mathbf{H}$$

   was essentially a rotational analogy to the linear electric current relationship,

2. Electric convection current

   $$\mathbf{J} = \rho \mathbf{v}$$

where ρ is electric charge density. B was seen as a kind of magnetic current of vortices aligned in their axial planes, with H being the circumferential velocity of the vortices.

The electric current equation can be viewed as a convective current of electric charge that involves linear motion. By analogy, the magnetic equation is an inductive current involving spin. There is no linear motion in the inductive current along the direction of the B vector. The magnetic inductive current represents lines of force. In particular, it represents lines of inverse square law force.

In aerodynamics the induced air currents are forming solenoidal rings around a vortex axis that is playing the role that electric current plays in magnetism. This puts the air currents of aerodynamics into the equivalent role of the magnetic induction vector B in electromagnetism.

In electromagnetism the B lines form solenoidal rings around the source electric current, whereas in aerodynamics, the air currents form solenoidal rings around the source vortex axis.

Hence in electromagnetism, the vortex plays the role of 'effect' whereas in aerodynamics, the vortex plays the role of 'cause'. Yet when we look at the B lines in isolation, we see exactly the aerodynamic scenario in so much as that B is the vortex axis and H is the circumferential velocity as in Maxwell's 1861 paper.

In two dimensions, for a vortex line of infinite length, the induced velocity at a point is given by ,

$$v = \frac{\Gamma}{2\pi r}$$

where Γ is the strength of the vortex and *r* is the perpendicular distance between the point and the vortex line.

This is a limiting case of the formula for vortex segments of finite length:

$$v = \frac{\Gamma}{4\pi r}\left[\cos A - \cos B\right]$$

where *A* and *B* are the (signed) angles between the line and the two ends of the segment.

Gauss's Law for Magnetism

Discoverers of the Law

Gauss's law for magnetism is a physical application of Gauss's theorem (also known as the divergence theorem) in calculus, which was independently discovered by Lagrange in 1762, Gauss in 1813, Ostrogradsky in 1826, and Green in 1828. Gauss's law for magnetism simply describes one physical phenomena that a magnetic monopole does not exist in reality. So this law is also called "absence of free magnetic poles".

People had long been noticing that when a bar magnet is divided into two pieces, two small magnets are created with their own south and north poles. This can be explained by Ampere's circuital law: the bar magnet is made of many circular currents rings, each of which is essentially a magnetic dipole; the macroscopic magnetism is from the alignment of the microscopic magnetic dipoles. Because a small current ring always generates an equivalent magnetic dipole, there is no way of generating a free magnetic charge. So far, no magnetic monopole has been found in experiments, despite that many theorists believe a magnetic monopole exists and are still searching for it.

However, as pointed out by Pierre Curie in 1894, magnetic monopoles can exist conceivably. Introducing fictitious magnetic charges to the Maxwell's equations can give Gauss's law for magnetism the same appearance as Gauss's law for electricity, and the mathematics can become symmetric.

Gauss's law for magnetism states that no magnetic monopoles exists and that the total flux through a closed surface must be zero. This page describes the time-domain integral and differential forms of Gauss's law for magnetism and how the law can be derived. The frequency-domain equation is also given. At the end of the page, a brief history of the Gauss's law for magnetism is provided.

Integral Equation

The Gauss's law for magnetic fields in integral form is given by:

$$\oint_S \mathrm{b} \cdot \mathrm{d}a = 0,$$

where:

- b is the magnetic flux

The equation states that there is no net magnetic flux b (which can be thought of as the number of magnetic field lines through an area) that passes through an arbitrary closed surface S. This means the number of magnetic field lines that enter and exit through this closed surface S is the same. This is explained by the concept of a magnet that has a north and a south pole, where the strength of the north pole is equal to the strength of the south pole. This is equivalent to saying that a magnetic monopole, meaning a solitary north or south pole, does not exist because for every positive magnetic pole, there must be an equal amount of negative magnetic poles.

Differential equation

Gauss's law for magnetic fields in the differential form can be derived using the divergence theorem. The divergence theorem states:

$$\int_V (\nabla \cdot \mathrm{f})\, dv = \oint_S \mathrm{f} \cdot \mathrm{d}a$$

where f is a vector. The right-hand side looks very similar to first equation . Using the divergence theorem, Equation first, is rewritten as follows:

$$0 = \oint_S \mathrm{b} \cdot \mathrm{d}a = \int_V (\nabla \cdot \mathrm{b})\, dv.$$

Because the expression is set to zero, the integrand (∇·b) must be zero also. Thus the differential form of Gauss's law becomes:

$$\nabla \cdot \mathrm{b} = 0.$$

Vector Potential

Due to the Helmholtz decomposition theorem, Gauss's law for magnetism is equivalent to the following statement:

There exists a vector field A *such that,*

$$\mathbf{B} = \nabla \times \mathbf{A}.$$

The vector field A is called the magnetic vector potential.

Note that there is more than one possible A which satisfies this equation for a given B field. In fact, there are infinitely many: any field of the form $\nabla\varphi$ can be added onto A to get an alternative choice for A, by the identity:

$$\nabla \times \mathbf{A} = \nabla \times (\mathbf{A} + \nabla\phi)$$

since the curl of a gradient is the zero vector field:

$$\nabla \times \nabla\phi = \mathbf{0}$$

This arbitrariness in A is called gauge freedom.

Field Lines

The magnetic field B, like any vector field, can be depicted via field lines (also called *flux lines*) – that is, a set of curves whose direction corresponds to the direction of B, and whose areal density is proportional to the magnitude of B. Gauss's law for magnetism is equivalent to the statement that the field lines have neither a beginning nor an end: Each one either forms a closed loop, winds around forever without ever quite joining back up to itself exactly, or extends to infinity.

Modification if Magnetic Monopoles Exist

If magnetic monopoles were discovered, then Gauss's law for magnetism would state the divergence of B would be proportional to the *magnetic charge density* ρ_m, analogous to Gauss's law for electric field. For zero net magnetic charge density ($\rho_m = 0$), the original form of Gauss's magnetism law is the result.

The modified formula in SI units is not standard; in one variation, magnetic charge has units of webers, in another it has units of ampere-meters.

Units	Equation
cgs units	$\nabla \cdot \mathbf{B} = 4\pi\rho_m$
SI units (weber convention)	$\nabla \cdot \mathbf{B} = \rho_m$
SI units (ampere-meter convention)	$\nabla \cdot \mathbf{B} = \mu_0\rho_m$

where μ_0 is the vacuum permeability.

So far no magnetic monopoles have been found, despite extensive search.

Derivation using Biot-Savart Law¶

Gauss's law can be derived using the Biot-Savart law, which is defined as:

$$b(r) = \frac{\mu_0}{4\pi}\int_V \frac{(j(r')dv)\times \hat{r}}{|r-r'|^2},$$

where:

- b(r) is the magnetic flux at the point r
- j(r ′) is the current density at the point r ′
- μ o is the magnetic permeability of free space.

Taking the divergence of both sides of above equation yields:

$$\nabla.b(r) = \frac{\mu_0}{4\pi}\int_V \nabla.\frac{(j(r')dv)\times \hat{r}}{|r-r'|^2}.$$

To carry through the divergence of the integrand in above equation , the following vector identity is used:

$$\nabla.(A\times B) = B.(\nabla\times A) - A.(\nabla\times B).$$

Thus, the integrand becomes:

$$\left[j(r').\left(\nabla\times\frac{\hat{r}}{|r-r'|^2}\right)\right]-\left[\frac{\hat{r}}{|r-r'|^2}.(\nabla\times j(r'))\right]$$

The above equation is zero as the curl of $\frac{\hat{r}}{|r-r'|^2}$ is zero. The equation below becomes zero because j depends on r' and ∇ depends only on r. Plugging this back into, $\nabla.b(r) = \frac{\mu_0}{4\pi}\int_V \nabla.\frac{(j(r')dv)\times \hat{r}}{|r-r'|^2}.$ the right-hand side of the expression becomes zero. Thus, we see that:

$$\nabla.b(r) = 0,$$

which is Gauss's law for magnetism in differential form.

Differential equation in the frequency-domain

The equation can also be written in the frequency-domain as:

$$\nabla.B = 0.$$

Ampère's Circuital Law

Ampère's law, one of the basic relations between electricity and magnetism, stating quantitatively the relation of a magnetic field to the electric current or changing electric field that produces it. The law is named in honour of André-Marie Ampère, who by 1825 had laid the foundation of electromagnetic theory. An alternative expression of the Biot-Savart law (*q.v.*), which also relates the magnetic field and the current that produces it, Ampère's law is generally stated formally in the language of calculus: the line integral of the magnetic field around an arbitrarily chosen path is proportional to the net electric current enclosed by the path. James Clerk Maxwell is responsible for this mathematical formulation and for the extension of the law to include magnetic fields that arise without electric current, as between the plates of a capacitor, or condenser, in which the electric field changes with the periodic charging and discharging of the plates but in which no passage of electric charge occurs. Maxwell also showed that even in empty space a varying electric field is accompanied by a changing magnetic field. In this more general form, the so-called Ampère-Maxwell law is one of the four Maxwell equations that define electromagnetism.

- Ampere's circuital law in magnetism is analogous to gauss's law in electrostatics
- This law is also used to calculate the magnetic field due to any given current distribution
- This law states that

  " The line integral of resultant magnetic field along a closed plane curve is equal to μ_0 time the total current crossing the area bounded by the closed curve provided the electric field inside the loop remains constant" Thus,

  $$\oint B.dl = \mu_0 I_{enc}$$

  where μ_0 is the permeability of free space and I_{enc} is the net current enclosed by the loop as shown below in the figure

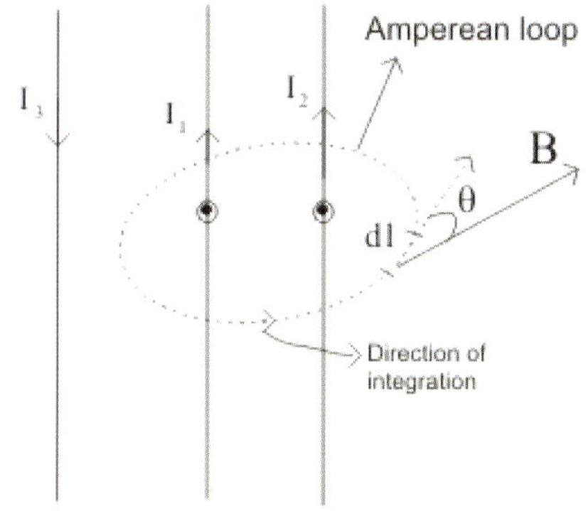

Ampere's law applied to a loop containing two long straight wires.

The circular sign in above equation means that scalar product B.dl is to be integrated around the closed loop known as Amperian loop whose beginning and end point are same

Anticlockwise direction of integration as chosen in figure is an arbitrary one we can also use clockwise direction of integration for our calculation depending on our convenience

To apply the ampere's law we divide the loop into infinitesimal segments dl and for each segment, we then calculate the scalar product of B and dl

B in general varies from point to point so we must use B at each location of dl

Amperion Loop is usually an imaginary loop or curve ,which is constructed to permit the application of ampere's law to a specific situation.

Proof of Ampere's Law

- Consider a long straight conductor carrying current I perpendicular to the page in upward direction as shown below in the figure

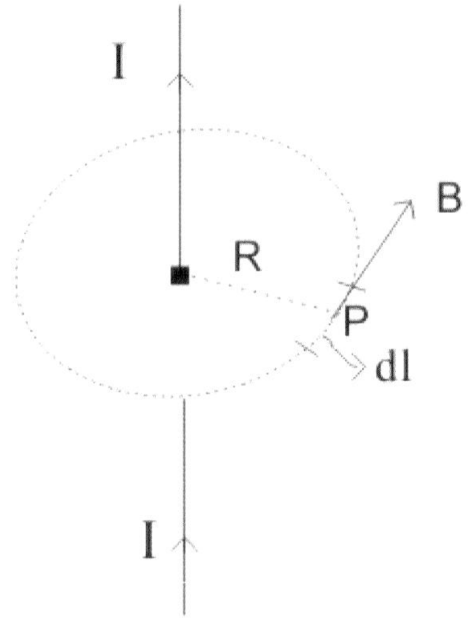

B is the magnetic field due to current carrying conductor at point P

- From Biot Savart law, the magnetic field at any point P which is at a distance R from the conductor is given by,

$$B = \frac{\mu_0 I}{2\pi R}$$

- Direction of magnetic Field at point P is along the tangent to the circle of radius R withTh conductor at the center of the circle .
- For every point on the circle magnetic field has same magnitude as given by ,

$$B = \frac{\mu_0 I}{2\pi R}$$

And field is tangent to the circle at each point.

- The line integral of B around the circle is,

$$\oint B.dl = \oint \frac{\mu_0 I}{2\pi R} dl = \frac{\mu_0 I}{2\pi R} \oint dl$$

since $\int dl = 2\pi R$ ie, circumference of the circle so,

$$\oint B.dl = \mu_0 I$$

This is the same result as stated by Ampere law.

- This ampere's law is true for any assembly of currents and for any closed curve though we have proved the result using a circular Amperean loop.
- If the wire lies outside the amperion loop, the line integral of the field of that wire will be zero

$$\oint B.dl = 0$$

but does not necessarily mean that B=0 everywhere along the path ,but only that no current is linked by the path.

- While choosing the path for integration ,we must keep in mind that point at which field is to be determined must lie on the path and the path must have enough symmetry so that the integral can be evaluated.

It alternatively says, the integral of magnetic field intensity (H) along an imaginary closed path is equal to the current enclosed by the path.

$$\oint \vec{B}.\vec{dl} = \mu_0 I$$

$$\Rightarrow \oint \frac{\vec{B}}{\mu_0}.\vec{dl} = I$$

$$\Rightarrow \oint \vec{H}.\vec{dl} = I$$

$$\left[\because \vec{H} = \frac{\vec{B}}{\mu_0}\right]$$

Let us take an electrical conductor, carrying a current of I ampere, downward as shown in the figure below.

Let us take an imaginary loop around the conductor. We also call this loop as amperian loop.

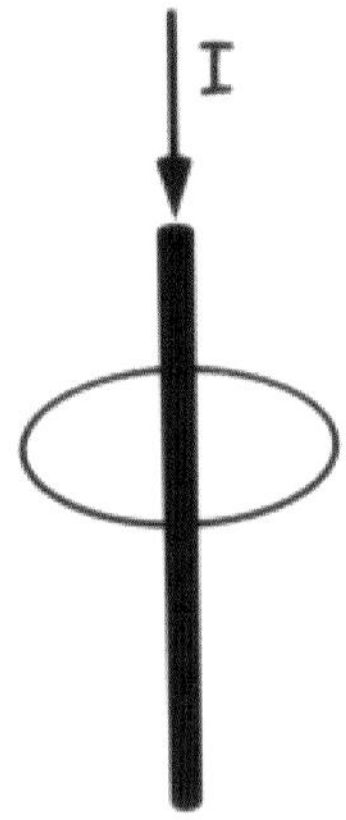

Let us also imagine the radius of the loop is r and the flux density created at any point on the loop due to current through the conductor is B.

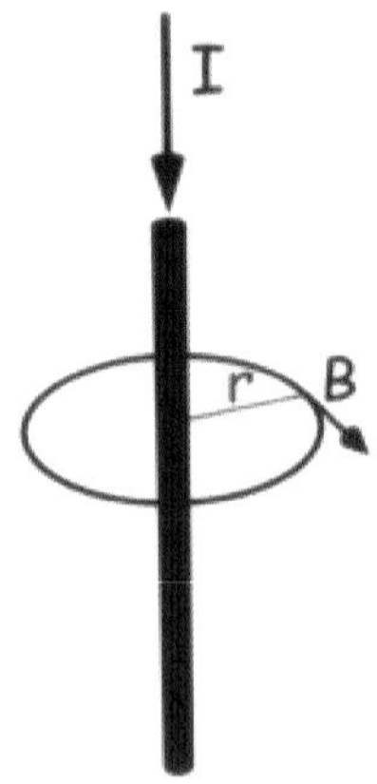

Let us consider an infinitesimal length dl of the amperian loop at the same point.

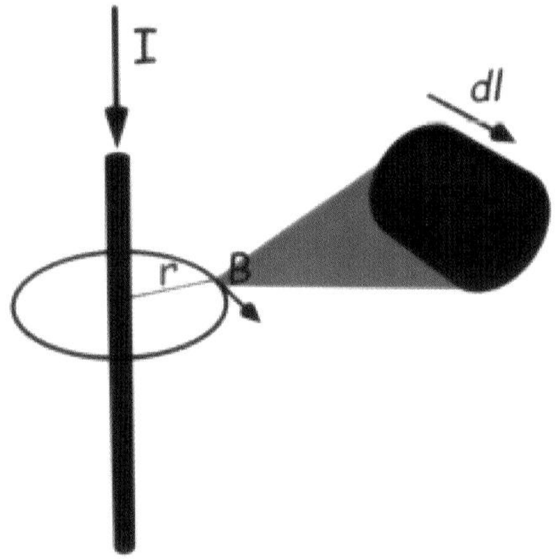

At each point on the amperian loop, the value of B is constant since the perpendicular distance of that point from the axis of conductor is fixed, but the direction will be along the tangent on the loop at that point.

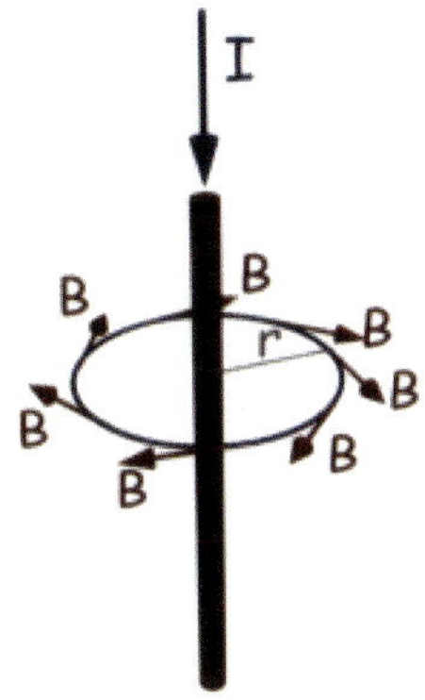

The close integral of the magnetic field density B along the amperian loop, will be,

$$\oint B\,.\,dl \quad [dot\ product]$$

$$\begin{bmatrix} \because Direction\ of\ B\ \&\ dl\ is\ same \\ at\ each\ point\ on\ the\ loop. \end{bmatrix}$$

$$= B\oint dl = B.(2\pi r)$$

Now, according to Ampere's Circuital Law

$$\oint B\,.\,dl = \mu_0 . I$$

Therefore,

$$2\pi r\ B = \mu_0\ I$$

$$\Rightarrow \frac{B}{\mu_0} = \frac{I}{2\pi r}$$

$$\Rightarrow H = \frac{I}{2\pi r}$$

Instead of one current carrying conductor, there are N number of conductors carrying same current I, enclosed by the path, then

$$H = \frac{NI}{2\pi r}$$

References

- Feynman, Richard P.; Leighton, Robert B.; Sands, Matthew (2006). The Feynman Lectures on Physics. 2. ISBN 0-8053-9045-6
- Magnetic-mom, iit-jee-magnetism: askiitians.com, Retrieved 19 March 2018
- Hiebert, W; Ballentine, G; Freeman, M (2002). "Comparison of experimental and numerical micromagnetic dynamics in coherent precessional switching and modal oscillations". Physical Review B. 65 (14). p. 140404. Bibcode:2002PhRvB..65n0404H. doi:10.1103/PhysRevB.65.140404

- What-is-magnetic-flux: circuitglobe.com, Retrieved 29 May 2018
- Yung, K. W.; Landecker, P. B.; Villani, D. D. (1998). "An Analytic Solution for the Force between Two Magnetic Dipoles" (PDF). Magn. Elec. Separation. 9: 39–52. doi:10.1155/1998/79537. Retrieved November 24, 2012
- Biot-Savart-law, science: britannica.com, Retrieved 09 July 2018
- Boyer, Timothy H. (1988). "The Force on a Magnetic Dipole". Am. J. Phys. 56 (8): 688–692. Bibcode:1988AmJPh..56..688B. doi:10.1119/1.15501
- Amperes-law, britannica.com: Retrieved 09 July 2018
- Purcell, Edward and Morin, David; Electricity and Magnetism 3rd Edition; Cambridge University Press, New York, 2013. ISBN 978-1-107-01402-2
- Ampere-circuital-law: electrical4u.com, Retrieved 19 June 2018

Permissions

All chapters in this book are published with permission under the Creative Commons Attribution Share Alike License or equivalent. Every chapter published in this book has been scrutinized by our experts. Their significance has been extensively debated. The topics covered herein carry significant information for a comprehensive understanding. They may even be implemented as practical applications or may be referred to as a beginning point for further studies.

We would like to thank the editorial team for lending their expertise to make the book truly unique. They have played a crucial role in the development of this book. Without their invaluable contributions this book wouldn't have been possible. They have made vital efforts to compile up to date information on the varied aspects of this subject to make this book a valuable addition to the collection of many professionals and students.

This book was conceptualized with the vision of imparting up-to-date and integrated information in this field. To ensure the same, a matchless editorial board was set up. Every individual on the board went through rigorous rounds of assessment to prove their worth. After which they invested a large part of their time researching and compiling the most relevant data for our readers.

The editorial board has been involved in producing this book since its inception. They have spent rigorous hours researching and exploring the diverse topics which have resulted in the successful publishing of this book. They have passed on their knowledge of decades through this book. To expedite this challenging task, the publisher supported the team at every step. A small team of assistant editors was also appointed to further simplify the editing procedure and attain best results for the readers.

Apart from the editorial board, the designing team has also invested a significant amount of their time in understanding the subject and creating the most relevant covers. They scrutinized every image to scout for the most suitable representation of the subject and create an appropriate cover for the book.

The publishing team has been an ardent support to the editorial, designing and production team. Their endless efforts to recruit the best for this project, has resulted in the accomplishment of this book. They are a veteran in the field of academics and their pool of knowledge is as vast as their experience in printing. Their expertise and guidance has proved useful at every step. Their uncompromising quality standards have made this book an exceptional effort. Their encouragement from time to time has been an inspiration for everyone.

The publisher and the editorial board hope that this book will prove to be a valuable piece of knowledge for students, practitioners and scholars across the globe.

Index